The Timeless Approach
Frontier Perspectives in 21st Century Physics

Series on the Foundations of Natural Science and Technology

Series Editors: C. Politis (*University of Patras, Greece*)
                   W. Schommers (*Forschungszentrum Karlsruhe, Germany*)

Published:

Vol. 1:   Space and Time, Matter and Mind: The Relationship between Reality and Space-Time
           by W. Schommers

Vol. 2:   Symbols, Pictures and Quantum Reality: On the Theoretical Foundations of the Physical Universe
           by W. Schommers

Vol. 3:   The Visible and the Invisible: Matter and Mind in Physics
           by W. Schommers

Vol. 4:   What is Life? Scientific Approaches and Philosophical Positions
           by H.-P. Dürr, F.-A. Popp and W. Schommers

Vol. 5:   Grasping Reality: An Interpretation-Realistic Epistemology
           by H. Lenk

Vol. 6:   Nano-Engineering in Science and Technology: An Introduction to the World of Nano-Design
           by M. Rieth

Vol. 9   The Timeless Approach
           Frontier Perspectives in 21st Century Physics
           by D. Fiscaletti

Forthcoming:

Vol. 7   Magneto Thermoelectric Power in Heavily Doped Quantized Structures
          by K. P. Ghatak

Series on the Foundations of Natural Science and Technology — Vol. 9

The Timeless Approach
Frontier Perspectives in 21st Century Physics

Davide Fiscaletti
SpaceLife Institute, Italy

NEW JERSEY · LONDON · SINGAPORE · BEIJING · SHANGHAI · HONG KONG · TAIPEI · CHENNAI · TOKYO

Published by

World Scientific Publishing Co. Pte. Ltd.
5 Toh Tuck Link, Singapore 596224
USA office: 27 Warren Street, Suite 401-402, Hackensack, NJ 07601
UK office: 57 Shelton Street, Covent Garden, London WC2H 9HE

Library of Congress Cataloging-in-Publication Data
Fiscaletti, Davide, author.
 The timeless approach : frontier perspectives in 21st century physics / Davide Fiscaletti, SpaceLife Institute, Italy.
  pages cm. -- (Series on the foundations of natural science and technology ; v. 9)
 Includes bibliographical references and index.
 ISBN 978-9814713153 (hardcover : alk. paper) -- ISBN 9814713155 (hardcover : alk. paper)
 1. Physics. 2. Space and time. I. Title. II. Title: Frontier perspectives in 21st century physics. III. Title: Frontier perspectives in twenty-first century physics. IV. Series: Series on the foundations of natural science and technology ; v. 9.
 QC21.3.F565 2015
 530--dc23
                                                            2015031766

British Library Cataloguing-in-Publication Data
A catalogue record for this book is available from the British Library.

Copyright © 2016 by World Scientific Publishing Co. Pte. Ltd.

All rights reserved. This book, or parts thereof, may not be reproduced in any form or by any means, electronic or mechanical, including photocopying, recording or any information storage and retrieval system now known or to be invented, without written permission from the publisher.

For photocopying of material in this volume, please pay a copying fee through the Copyright Clearance Center, Inc., 222 Rosewood Drive, Danvers, MA 01923, USA. In this case permission to photocopy is not required from the publisher.

In-house Editor: Christopher Teo

Typeset by Stallion Press
Email: enquiries@stallionpress.com

Printed in Singapore

To My Family,
For their patience and encouragement

Contents

Introduction	1
Chapter 1: About Time as the Numerical Order of Material Changes	17
Chapter 2: Three-Dimensional Euclid Space and Special Relativity	113
Chapter 3: Three-Dimensional non-Euclid Space as a Direct Information Medium and Quantum Phenomena	151
Chapter 4: About Quantum Cosmology in a Background Space as an Immediate Information Medium	230
Chapter 5: The Gravitational Space in an A-Temporal Quantum-Gravity Space Theory	253
Chapter 6: A Three-Dimensional Timeless Quantum Vacuum as the Fundamental Bridge Between Gravitation and the Quantum Behavior	304
Conclusions	396
References	413
Subject Index	443

Introduction

Time has always puzzled philosophers and scientists and, along with such concepts as space and matter, can be considered as a crucial concept of any physical theory, in particular of a fundamental one. Today, in the light of the most significant research on quantum gravity, we can say that the notion of time intended as a primary physical reality, as an independent physical dimension which has existence on its own, indeed has no fundamental citizenship in physics. This happened, above all, as a consequence of general relativity and its "deformable" time and of the fact, determined by quantum physics, that observed phenomena arise from a timeless quantum vacuum which acts as a fabric of reality, of the fact that in the quantum regime the ordinary forms of matter and energy are manifestations of this fundamental vacuum.

The beginning of the 21st century brought substantially a new vision into theoretical physics: particles move in a timeless space background and time as-measured with clocks exists only as an emergent mathematical entity indicating the numerical order of particles' motion. In this picture, the linear time — "past-present-future" — can be considered merely as a mind model inside which human beings experience the material world.

The whole development of theoretical physics can be seen as a continuous improvement of the models of space and time. No physical law can be formulated without being collocated in an opportune spatial-temporal arena. This fact stimulated many times in the history of physics the idea of a tight connection between the physical processes and the global arena in which they take place. This connection must be intended in a realistic sense, and implies a series of radical assumptions with regards to the fundamental substructure and the nature of space and time — which delineate a series of deep perspective changes about our way of working with the physical theories in the research of a unifying view of the world [1]. From Newton's absolute model of the space-time background to the non-Euclidean and conformal geometries of the different classical and quantum geometrodynamics, all the theories of physics are characterized by the notion of a "process" — intended as the evolution of a set of observables in a specific arena [2].

Newtonian absolute conception was the first attempt to ascribe a fundamental role to space and time in order to explain the properties of matter. In Newtonian physics space and time are absolute entities and define a stage over which the processes take place, the various dynamical entities being the actors. Isaac Newton founded classical mechanics on the basis of the idea that space is something distinct from body and that time is something that passes uniformly without regard to whatever happens in the world. According to Newton's view time passes in space and is not part of space. It was Newton's great achievement to create a conceptual clarification of the notion of time in his *Philosophiae naturalis principia mathematica* from 1687, where he introduced the concept of time as a physical quantity having a primary, independent existence:

> "An absolute, true, and mathematical time exists that, from its own nature, passes equally without relation to anything external (and thus without reference to any change of matter or way of measuring time) and by another name is called duration; relative, apparent, and common time, is some sensible and external (whether accurate or unequable) measure of duration by the means of motion, which is commonly used instead of true time; such as an hour, a day, a month, a year." [3].

Thus, Newtonian clocks are devices that provide some sensible and external (whether accurate or unequable) measure of duration by the means of motion, which is commonly used instead of true time. Moreover, Newton regarded space and time as real entities with their own manner of existence as needed by God's existence (more specifically, his omnipresence and eternality).

The concept of an absolute time t was introduced by Newton as a necessary element for his formulation of mechanics and indeed is still close to the way many people think about time. However, by definition one cannot directly access Newton's absolute time. Newton's absolute time is unobservable, it is an abstraction about which we have only an imprecise knowledge due to inadequate realizations by concrete physical processes. We never exactly know the "true" duration between two events. Any realistic physical clock emulates Newtonian time t merely down to a suitable scale; it approximates t within a certain accuracy. Below that scale higher-order physical effects, systematic or statistical errors influence the performance of any clock. Nonetheless, Newtonian mechanics is based on the assumption that such an unobservable, absolute time with a unique topological and metrical structure exists, in terms of which the dynamics is defined.

If Newton believed that absolute time exists on its own as a primary physical reality, independently of the existence of material clocks, a contemporary of Newton, Leibniz suggested a different view, known as the relational view of time, according to which events are more fundamental than moments, which do not exist in their own right [4]. According to Leibniz neither absolute time nor space exist: the primary role is to be ascribed to matter and its properties while space and time are to have only the merely relational function of a stage of coordinates. The world is to be understood in terms of more fundamental entities that fuse space and matter into relative configurations of simultaneously existing bodies. Bodies and events define points and instants by conferring their identity upon them, and thus enabling them to serve as the loci of other bodies and events. In Leibniz's view, the world does not contain things, it is things. Time is merely supposed to be the succession of such instantaneous configurations, and not something that flows quite independently of the bodies in the universe. The dynamics is exclusively based on observable elements. The disjoint viewpoints of Newton and Leibniz yield a historical debate whether time and space exist as real objects (absolute), or whether they are merely orderings upon actual objects (relational). In particular, it was held in a famous correspondence between Leibniz and Clark, a follower of Newton, known as the Leibniz–Clark correspondence.

150 years after, Mach developed Leibniz's ideas in a much more radical sense, by formulating the thesis that it is the whole matter of the universe that exerts a natural physical action, that the law of inertia is relative to the stars [5]. Mach can be considered the first scientist who claimed that Newtonian time cannot be determined directly by experiment, and for this reason cannot be of any practical and hence scientific use [5, 6]. According to Mach, time has to be regarded as a mere metaphysical concept. Mach was the first who suggested that one should only use those notions in physics which have a direct empirical meaning. In Mach's picture, all statements concerning the temporal behavior of an object necessarily refer to a clock time, namely to a specific physical process. The impression that absolute Newtonian time exists, arises from many apparently periodic processes in nature (like the motion of celestial bodies or the oscillations of a pendulum) which seem to march in step with Newtonian time: when they are used to measure time, Newton's laws are found to be valid. According to Mach these processes only feign the existence of an abstract time, for instance we cannot decide if the period of a pendulum varies in Newtonian time.

In the 20th century three fundamental theories have changed the understanding and explanation of the physical phenomena: special relativity, general relativity and quantum theory. The special theory of relativity — formulated by Lorentz, Poincarè, Einstein and Minkowski — has determined the first important changes in the Newtonian picture. In special relativity, space and time are intrinsically linked, united into a manifold called "space-time", a smooth four-dimensional (4D) continuum. In the special theory of relativity Einstein described electromagnetic phenomena with the formalism of 4D space created by German mathematician Hermann Minkowski. On the basis of the mathematical expression of the fourth coordinate $X_4 = ict$ (where i is the imaginary unit, c is light speed and t is time), the Minkowski arena can be considered as a 4D space [7]. In the context of special relativity, the Minkowskian 4D space is the stage which serves as the arena for all of physics.

The Minkowski metric dictates the field equations and restricts the forms of interaction terms in the action.

A fundamental result of special relativity lies in the fact that the absolute Newtonian time is replaced by a class of related times, the Lorentz times. There is one external time for each Lorentz observer. Special relativity has modified the notion of time in several surprising ways, regarding in particular the relativity of simultaneity. The unique temporal order of events is lost, determining new phenomena such as time dilatation or the twin paradox. Clocks are affected by motion, and since physical time is measured by clocks, one has to conclude that time is stretched, as well. However, from a fixed Lorentz observer's viewpoint time remains a distinguished, absolute, external, and global parameter. Therefore, despite time being no more absolute and the idea of absolute simultaneity being abandoned (which really shatters some Newtonian ideas), just as in Newtonian physics, also in special relativity, there is still a fixed space-time which serves as the arena for all of physics. The Minkowskian arena implies that in special relativity there is still a fixed space-time which serves as the arena of physical processes. It is the stage on which the drama of evolution unfolds. Actors are particles and fields. The stage constrains what the actors can do. But the actors cannot influence the stage: Minkowskian geometry characteristic of special relativity is immune from change. The Newtonian construct of a fixed background stage over which physics happens is not altered by special relativity.

A more profound change in the picture of the world happened with Einstein's general relativity. The central discovery of general relativity is that Newton's space-time and the gravitational field are the same thing and that the dynamics of the gravitational field and any other dynamical object is fully relational, in the Aristotelian–Cartesian sense. Matter curves space-time. The space-time metric is no longer fixed. This is a crucial paradigm shift: one has to abandon the idea that space-time is fixed, immune to change, one has to encode gravity into the very geometry of space-time, thereby making this geometry dynamical. With general relativity Einstein abandoned the tenet that geometry is inert and made it a physical entity that interacts with matter. As a consequence, general relativity treats time in a peculiar way, as compared to pre-general relativistic physics. The absolute time of Newton has disappeared. In its place, there are different possible internal times, related to specific physical clocks. Only the evolution with respect to these physical clocks makes sense. The physical content of Einstein's equations of general relativity is that in general relativistic phenomena there is no independent time variable. In general relativity, clocks do not measure the idealized coordinate time t that appears as the argument of the field variable, in particular of the metric $g_{\mu\nu}$ but measure the proper time s associated with a given world line $\gamma = \gamma^\mu(\tau)$, which is defined by $s = \int_\gamma d\tau \sqrt{g_{\mu\nu}(\gamma(\tau)) \frac{d\gamma^\mu}{d\tau} \frac{d\gamma^\nu}{d\tau}}$. Given a solution of the field equations, proper time provides the definition of a time quantity along every timelike worldline. It is a complicated nonlocal function of the gravitational field. Proper time does not capture all the properties of Newtonian time, it is neither spatially global nor external or unique, but dynamical and state-dependent. If one uses Poincare's principle of simplicity (according to which time has to be defined in such a way that the fundamental laws of physics become as simple as possible, namely the equations of motion in classical mechanics take on their simple Newtonian form) the proper time can be regarded as the preferred time in general relativistic systems (in general relativity the geodesic equations of motion — which replace inertial motion — take on their simplest form when the system is parametrized with respect to an affine parameter, which precisely recovers proper time), and thus one may assume that there exist ideal clocks, whose rates are unaffected by acceleration and which measure proper time along their worldlines (this is the content of the so-called clock hypothesis of general relativity). However, it must be emphasized that, in general relativity, a clock is a physical object and thus reacts on the metric and changes the geometry. In contrast to nongeneral relativistic physics, where one can choose a clock which is completely disentangled from the proper system and solve the dynamics for the clock independently, here one has to consider the full coupled gravity-clock system in order to predict the evolution in clock time. Therefore, none of the (weaker) notions of time available in general relativity incorporates all features of Newtonian time, and none of them can generally be used to evolve the system globally, namely leading to a fully deparametrized dynamical system. A general relativistic system cannot be formulated as a Hamiltonian system evolving in a physical time parameter (at least such a formulation is neither known nor expected). This leads to the statement that general relativity is fundamentally a timeless theory in the sense that its physical interpretation is not based in a direct way on the concept of time. The notion of time plays a secondary role. In synthesis, one can conclude that, in general relativity, change is not described as evolution of physical variables as a function of a preferred independent observable time variable. Instead, it is described in terms of a relation among equal footing variables. In general relativity, there is not a preferred and observable quantity that plays the role of independent parameter of the evolution, as there is in nonrelativistic mechanics. General relativity describes the relative evolution of observable quantities, not the evolution of quantities as functions of a preferred one.

On the other hand, another relevant 20th century scientific revolution happened with quantum theory. Quantum mechanics is perhaps the 20th century physical theory that have contributed most profoundly to change our understanding of the universe and at the same time have created the most complex interpretative problems as regards what it says about the world. Indeed quantum mechanics introduces much more spread scenarios and perspectives than those offered by every previous physical theory. From quantum mechanics we have learned that every dynamical object is made up of quanta and can be in probabilistic superposition states. Quantum mechanics has taught us that in microphysics the different physical quantities (for example energy, momentum) have not a continuous spectrum, but can assume only a discrete set of values, namely are quantized. The quantum theory determined many changes in the description of the world by showing in particular that subatomic particles are nonlocally connected and can be in entangled states. In the EPR-Bell phenomena, in fact, objects separated in space and time show space-like correlations that are inexplicable in classical terms, and that characterize the quantum statistics from which the most interesting properties of the structure of matter derive. The nonlocal correlations characterizing the behavior of subatomic particles do not transport energy, for this reason they do not violate the laws of relativity. In other words, quantum nonlocality turns out to be compatible with the existence of a limit speed: the quantum formalism cannot be used to send superluminal signals. This condition defines a sort of "pacific coexistence" between relativity and quantum physics [8]. Nonetheless, it becomes natural to put ourselves the following fundamental question: is the space-time model used in relativity appropriate to describe the peculiarities of the geometry of the quantum world? The problem of reconciling the classical image of Einstein's space-time and quantum nonlocality becomes therefore to construct a dynamic model of the space-time background in which the quantum processes also find a place and to characterize the geometrical properties of this background, in other words to develop a quantum geometrodynamics. This research line clearly takes its inspiration from the philosophical and geometrical approach of general relativity.

Moreover, just like in Newtonian physics, in standard quantum mechanics time is postulated as a special physical quantity and plays the role of the independent variable of physical evolution. The Schrödinger equation, just like Newton's equation, is introduced on the basis of the underlying assumption that an idealized, absolute time t in which the dynamics is defined exists. Both in the Schrödinger formulation and in the Heisenberg formulation of standard quantum mechanics, time is not represented as an operator acting on the relevant Hilbert space, namely it is not a physical observable. Rather, it is regarded as part of an *a priori* given classical background with a well-defined value. Time in quantum mechanics is a Newtonian time, i.e., an absolute global time. In fact, the two main methods of quantization, namely, Dirac's canonical quantization method and Feynman's path integral method are based on classical constraints which become operators annihilating the physical states, and on the sum over all possible classical trajectories, sum over histories, respectively. Both quantization methods rely on the Newtonian global and absolute time [9]. The absolute character of time in quantum mechanics turns out to be crucial for its interpretation, i.e., matrix elements are evaluated at fixed time and the internal product on the Hilbert space is required to be conserved in time. Newtonian time is further needed to make sense of a complete set of observables commuting at a fixed value of time, or of the canonical equal-time commutation relations. Analogously to classical mechanics, the system can be described by a phase space and a Hamilton operator. The Hamiltonian generates the evolution relative to the classical external Newtonian time.

As in the classical case, also in quantum mechanics one may ask if there are suitable observables, quantum clocks, which can be used as internal times to approximate the Newtonian background time. In this regard, it was shown by Unruh and Wald [10] that quantum observables, which serve as good quantum clocks, do not exist. No quantum clock provides an adequate measure of the Newtonian time. More precisely, these two authors excluded the existence of an observable, T, which has a nonvanishing probability of running forward in Newtonian time, and a vanishing probability of running backward, provided that the energy of the physical system is bounded from below. There is no good quantum clock in the sense that its observed values increase monotonically with Newtonian time.

The transition from quantum mechanics to (special) relativistic quantum field theories can be realized by replacing the unique absolute Newtonian external time by a set of time-like parameters associated with the naturally distinguished family of relativistic inertial frames. Relativistic quantum field theories are formulated on a fixed background, the Minkowski space and the peculiar time variable of a particular Lorentz frame is used to describe the evolution, which is generated by a Hamiltonian. They are consistent with the equivalence of Lorentz frames, in virtue of the Poincarè covariance due to which quantum field theories constructed in different Lorentz frames are unitarily equivalent. Altogether, the external Lorentzian time is also the well-established concept of time prevalent in quantum field theories. Therefore, in (special) relativistic quantum field theories time continues to be treated as a background external parameter.

So, as regards the features of time, the crucial point at which we arrive inside contemporary physics is the following: time is absolute in quantum theory (in quantum mechanics as well as in special relativistic quantum field theories), but dynamical in general relativity. What, then, happens if one seeks a unification of gravity with quantum theory or, more precisely, seeks an accommodation of gravity into the quantum framework? Quantization methods, when applied to general relativity, lead to the Wheeler–DeWitt (WDW) equation [11, 12], a second order functional differential equation, which leads to the so-called frozen formalism time-problem in quantum gravity, according to which, in apparent contradiction with everyday experience, nothing at all seems to happen in the universe. General relativity does not seem to possess a natural time variable, while quantum theory relies quite heavily on a preferred time. Just as a consequence of the fact that the nature of time in quantum gravity is not yet clear, the quantum constraints of general relativity do not contain any time parameter, and it emerges the time arbitrariness problem, the so-called "problem of time" of quantum gravity [13–15].

Now, at the beginning of the new millennium, certain conceptual confusions in different physical theories, models and ideas are evident. The most important problems lie in the fact that special relativity, general relativity and quantum mechanics — despite their relevant successes from the predictive point of view — created a picture of the physical world which is somewhat unclear, incomplete and fragmented. In particular, Einstein's general theory of relativity and quantum theory — which perhaps can be considered the two greatest achievements of 20th century physics — are indeed mutually incompatible. To sum, quantum theory turns out to be very accurate, efficient and successful as far as the quantum phenomena do not interact with space-time, which plays the role of a background entity. Nevertheless, it breaks down when space-time is dynamical and therefore interacts with the quantum phenomena. Thus our basic understanding of nature is not only incomplete and fragmented — but inconsistent. One clearly needs a new consistent theory describing the so-called "quantum gravity domain" which provides a unitary picture of the universe, which takes the principles of both quantum mechanics and general relativity into account, and reduces to them in appropriate limits. Finding a new synthesis and a unitary picture of the universe represents therefore the most important challenge in today's fundamental physics.

As regards the problem regarding the unification of general relativity and quantum mechanics into a coherent picture, Macìas and Camacho recently emphasized that it is not possible to reconciliate and integrate into a common scheme the absolute and nondynamical character of Newtonian time of canonical quantization and path integral approaches with the relativistic and dynamical character of time in general relativity [16]. What is needed is a radical change of perspective either in general relativity or in quantum mechanics. According to Macìas and Camacho, one needs either a theory of gravity with a nondynamical Newtonian time, or a quantum theory with a dynamical time in its construction. A possibility is the formulation of a quantization procedure in which time has a more flexible behavior.

Despite 20th century theoretical physics has opened important perspectives in the exploration of new territories (such as the meaning of matter at the Planck scale and the role of the quantum information), it is characterized by significant foundational problems which the inner unity of physics knowledge depend on. According to the author of this book, the most relevant foundational problems regard the arena of the universe and its geometry. In the light of the developments of theoretical physics in the fields of special relativity, general relativity and quantum physics, what must be considered the real, fundamental arena of physical processes? What is the real background of physics? What is space? What is the geometry that rules physical processes? What is matter? What is time? What is the meaning of "being somewhere"? What is the meaning of "now"? What is the meaning of "moving"? These are some of the foundational questions that one must face in order to find a novel picture of the world and to find a new unifying picture in physics.

We need to find a new answer to these questions — different from Newton's answer — which took into account what we have learned about the world with quantum mechanics and general relativity and, at the same time, involves also significant changes of perspectives.

Assuming nonlocality as the ultimate visiting card of quantum mechanics, one can think that there are two possible ways in order to build a fundamental theory which describes the geometrodynamics and the arena of physical processes:

(a) the quantum geometry is assumed as primary and nonlocal, and therefore it is necessary to introduce additional hypotheses about its deep structure, or

(b) the space-time manifold must be considered an emergence of the deepest processes situated at the level of quantum gravity. In this regard, one can mention here, at least, Sacharov's original proposal of deducing gravity as "metric elasticity" of quantum vacuum [17] and the more recent one by Consoli on ultra-weak excitations in a condensed manifold as a model for gravity and Higgs mechanism [18].

According to these two approaches, the entire connected and local structure of both space-time and the Shannon–Turing information we use to compute the events in it can be seen as the explicit order of a hidden, implicit order, which acts as a "fabric of reality" at a subquantum level, fundamentally discrete and noncommutative [19].

The idea of a structure of relations and thus of a fundamental stage, subtending the observable forms of matter, energy and space-time was defined by J. A. Wheeler as "quantum foam," with the precise intent of evoking the erosion of traditional notions toward the Planck scale typical of quantum gravity. Since the second half of the 20th century, different strategies have been adopted in order to provide a unifying picture of general relativity and quantum mechanics. In this regard, the various versions of string theory, despite a certain success in overcoming some of the impasses of particle physics, show an ongoing lack of a strong unifying principle. According to these approaches, it has been suggested that the space-time manifold is the result of the interaction between p-branes, and that the acquisition of the masses in Higg's Ocean finds its natural explanation in the mechanisms of uncurling and compactification (see for example [20–21]). Indeed, the majority of the versions of string theory, just like quantum field theory, which is its closest relative, works with a flat Minkowski space-time, while a correct, authentically relativistic theory (in the sense of general relativity) should be independent of the background, or not presuppose any metric signature. Various theories have these requirements. One is Penrose's twistor theory [22]. To use Penrose's own words, "a twistor is an object similar to the two-faced Janus, unitary but with one face turned towards quantum mechanics and the other towards general relativity." More precisely a twistor is an object without mass and charge and with spin, invariant for the conformal group, so as to find again the light-cone of Minkowski's space-time. The famous representation by Robinson is based on a stereographic projection of Clifford's algebra that defines the structure of twistors and allows the essential characteristics of the dynamic non-local "fragments" of space-time to be intuitively taken from it.

Another very elegant theory that has the right relativistic requirements is Rovelli's and Smolin's "loop quantum gravity" [23] which suggests that at the most fundamental level, physical space is composed by elementary grains, a net of intersecting loops, having the Planck size. The loops are closed field lines that do not depend on the coordinate system and therefore provide the basis for a relational description of space-time in the spirit of Leibniz. Loop quantum gravity presupposes a very particular space-time structure at the Planck scale: the operators associated with area and volume in fact have a discrete spectrum, giving birth to a complex and fascinating graph structure and thus furnishing a discrete combinatorial view of physics. Taking account of the results of loop quantum gravity, it turns out to be permissible to assume that space has a granular structure at Planck scale.

As regards interesting attempts to provide a unifying picture of general relativity and quantum mechanics at the Planck scale, one should also mention Kleinert's, Jizba's and Scardigli's recent model of a quantized space-time as a crystal lattice [24] and Preparata's

Planck lattice where the quantum foam structure itself acts as the Higgs mechanism and allows the emerging of a spectrum of masses selected by the lattice [25].

In this book we will show that, on the basis of some recent research, a novel picture of the physical world can be portrayed in which the assumption of a fundamental nonlocal quantum geometry and the view that space-time emerges from the deepest processes situated at the level of quantum gravity can be in a certain sense embedded and unified inside the same approach. This novel approach emerges by starting from the basic hypothesis that space-time is not the fundamental physical arena of the universe.

There is no experimental evidence whatsoever, to support the view that space-time exists as the basis of a fundamental physical reality. Space-time cannot be considered a primary physical structure, independent of matter. While on the existence of space — as stated by Newton's view — there is no doubt (space is a physical medium in which matter exists), instead for the view of a temporal dimension intended as a primary physical reality there is no experimental evidence. Time's running in space on its own was never experimentally detected.

Already Einstein in his book, *Relativity: The Special and General Theory*, suggested that it makes no sense to think about space-time as an independent physical entity. The basic medium into which material change occurs is the gravitational field, namely a "space-matter structure", in other words a gravitational space. Space-time can be thus seen only as a structural quality of a gravitational field: "Space-time does not claim existence on its own but only as a structural quality of the [gravitational] field" [26].

Our experience of space is based on observations of material objects. Our experience of time derives from the changes and movements of matter into the universe. Time cannot be considered as a primary physical reality, as a quantity that is independent from the rest, as a quantity which flows on its own in the universe. Time is a mathematical coordinate which describes the motion of material objects. Time is a quantity associated with the stream of material changes and motions occurring in a timeless gravitational space and enters into existence when we measure it through clocks. Clocks run in a timeless gravitational space. With clocks we measure the numerical order of material changes that run into the universe.

This book introduces, in the light of relevant current research, interesting and novel perspectives as regards time and thus the arena of physical processes and its geometry (both in special relativity and in quantum mechanics and in the quantum gravity domain and about the quantum vacuum). The view analyzed in this book is that an absolute, idealized temporal dimension that flows on its own, as an independent variable of evolution, does not exist in the universe. In the light of relevant current research, in this book the author suggests that, at a fundamental level, the arena of physical processes is a timeless background, that time is not a primary physical reality but is an emergent mathematical quantity which measures the numerical order of material changes. With clocks we measure numerical order of material change namely motion in a three-dimensional (3D) space. In the physical world time is exclusively a mathematical quantity. The universe is not changing in time, on the contrary time as a numerical order t of change runs in the universe. At a fundamental level, the universe is timeless. As it is shown clearly in the various chapters of this book, the view of time explored by the author opens interesting and suggestive perspectives as regards the treatment of relevant foundational problems of theoretical physics.

This book is structured in the following manner. In Chapter 1 we will analyze some important aspects of the problem of time and we will show how our view of time as a numerical order of material changes allows us, on the basis of some current research, to provide a unifying key of reading of two fundamental approaches to time developed respectively by Barbour and Rovelli. In Chapter 2, we will review our suggestive interpretation of special theory of relativity in a 3D Euclid space, which introduces the possibility to derive the Lorentz transformations of Einsteinian special relativity from more fundamental laws which imply that, at a fundamental level, the arena of special relativity is a 3D Euclid space where time, at a fundamental level, exists only as a numerical order and the duration of material changes is a scaling function of the numerical order. In Chapter 3 we will take under examination the quantum domain illustrating, in particular, our view of a 3D space which acts as a direct medium of quantum information transfer between subatomic particles, in the form of a symmetrized quantum potential. Moreover, we will show that quantum processes can be interpreted as a modification of the geometrical properties of the 3D space, with respect to the Euclidean character, expressed in terms of an opportune entity called "quantum entropy". In Chapter 4 we will extend the approach of the background space as immediate information medium to the quantum gravity domain (in the context of WDW equation). Finally, in Chapters 5 and 6 we will review two models proposed recently by the author as regards the interpretation of gravity in the context of a timeless background where time exists only as a numerical order of material changes: an a-temporal quantum-gravity space theory and a 3D quantum vacuum view (defined by a energy density and state-reduction processes of creation–annihilation of quantum particles) as the fundamental levels of physical reality. We will show that these two models allow us to provide, at two different deep levels of reality, a suggestive reading of quantum mechanics and to interpret gravity and the quantum behavior as two aspects of the same source. The treatment made in Chapters 5 and 6 regarding the quantum gravity domain and a timeless quantum vacuum may moreover be considered as relevant attempts to achieve J. A. Wheeler's program, It from Bit (or Qbit), the possibility to describe the emergent features of space, time and matter as constrained and vehicled expressions of an informational matrix "at the bottom of the world" [27]. Although some improvements are obviously needed in order to clarify some aspects, the perspectives opened by the a-temporal quantum-gravity space theory and the timeless 3D quantum vacuum model constitute relevant steps in the process of search of a unifying approach describing the different regimes and thus towards a more and more satisfactory understanding of nature.

Chapter 1

About Time as the Numerical Order of Material Changes

1.1 The Problem of Time

In Newtonian physics as well as in conventional quantum mechanics, time is postulated as a special physical quantity that plays the role of the independent variable of physical evolution. Newton or Hamilton equations, as well as the Schrödinger equation, are introduced on the basis of the underlying assumption that an idealized, absolute time, external to the system under consideration, exists in which the dynamics is defined. In Newtonian physics as well as in conventional quantum theory, time is treated as a background idealized parameter which is used to mark the evolution of the system. However, it is an elementary observation that we never really measure an absolute time, that an idealized, absolute time never appears in laboratory measurements: we rather measure the frequency, speed and numerical order of material changes. What we realize in every experiment is to compare the motion of the physical system under consideration with the motion of a peculiar clock described by a peculiar tick. In other words, according to experimental results, the duration of material motions does not have a primary physical existence, independent of the motion of matter.

The results of several authors suggest that the space background of physical processes is timeless, that time cannot be considered as a primary physical reality that flows on its own in the universe. These research seem to suggest that time is an emergent quantity, which measures the numerical order of physical events.

In 1883, Ernst Mach wrote: "It is utterly beyond our power to measure the changes of things by time. Quite the contrary, time is an abstraction, at which we arrive by means of the changes of things". In this beautiful quote from *The Mechanics* [5], Mach laid out what was called his second principle: that time must be considered as a measure of the changes of things [28]. At the beginning of 20th century, John McTaggart discussed in his famous paper *Unreality of Time* that time is not a physical reality in which things exist. According to McTaggart's view, events do not pass, they just are: there is no passage of time, there is no moving present, the mere idea of a flowing time simply does not make any sense [29]. The idea of a timeless universe was also discussed by Einstein and Gödel in the second half of the 20th century. In particular, in 1949, Gödel postulated a theorem that stated: "In any universe described by the theory of relativity, time cannot exist" [30].

Though the argument of time as a coordinate of motion seems simple, elegant and epistemologically sound, it took nearly 100 years since the original quote by Mach before Barbour and Bertotti were able to formulate this principle into a mathematically rigorous theory of time in 1982 [31]. The reason for this delay cannot be attributed to technical complications, since the mathematics have been well understood since Jacobi, but rather to conceptual confusion surrounding how Mach's principles are implemented in general relativity. Though it was clear that Einstein was heavily influenced by Mach's ideas, general covariance proved to be misleading as a way of implementing relationalism. However, through the papers of Barbour and collaborators [31–34], we have now a clear picture of how Mach's ideas can be used to derive general relativity. From these works, we know that, in regards to time, Mach's second principle can be implemented in general relativity by using Jacobi's principle to determine the classical dynamics of the system.

There are many different aspects associated to the problem of time and many different ways of stating each of them [15, 35]. Due to its key role, the implications of the problem of time have been studied in detail from philosophical, conceptual, structural and mathematical points of view.

A detailed review of the most significant aspects related to the problem of time can be found in Isham's classic paper [13] and in Kuchař's paper [15] and, more recently, in Anderson's fascinating papers: *The Problem of Time in Quantum Gravity* [36] and *Problem of Time: Facets and Machian Strategy* [37]. In [36], Anderson underlines that the lesson of background independence which can be taken by general relativity can be realized as a freedom from absolute structures, according to the following strategies:

— Temporal relationalism, which concerns time being not primary for the universe as a whole (Leibniz's view). This can be implemented by reparametrization-invariance in the absence of extraneous time-like variables. In this picture, the conundrum of our apparent local experience of time is resolved in the Machian way "time is abstracted from change";
— Configurational relationalism, which consists in regarding the configuration space of dynamical objects to possess a group of transformations that are physically irrelevant. A classic example are the translations and rotations with respect to absolute space.

As regards the features of the problem of time and its attempts of solution, even more fascinating is the paper [37]. In this paper, before all, Anderson illustrates the eight following fundamental facets of the problem of time in canonical approaches (already mentioned by Isham [13] and Kuchař [15]):

— the Frozen Formalism Problem, according to which the Wheeler–DeWitt (WDW) equation which emerges from the quadratic Hamiltonian constraint of general relativity at the quantum level is a stationary or frozen equation;
— the thin sandwich, which regards the mathematical problem of treating the finite region between the hypersurfaces defined by different induced metrics, in analogy with the quantum mechanical setup of transition amplitudes between states at two different times;

— the Functional Evolution Problem, which consists in the fact that in general relativity more constraint terms might be unveiled than those corresponding to WDW equation and the constraint equation associated to the thin sandwich problem;
— the Problem of Observables, which concerns with finding enough quantities to describe the physics, these observables being defined as commutants with all of a theory's first-class constraints;
— the Foliation Dependence Problem, which lies in the fact that, while at the classical level, in virtue of general relativity's coordinate independence, the observable inner product combinations of wavefunctions and operators maintain foliation independence, it is not clear how or whether this can be attained in general at the quantum level;
— the Spacetime Reconstruction Problem, which refers to recovering space-time from assumptions of just space and/or a discrete ontology and thus consists in finding functionals that do not depend on any background foliation on the space-time manifold or, if this is not possible, understanding how to handle the situation and what it means in space-time terms;
— the Global Problem of Time, regarding the difficulties in choosing an "every-where-valid" time-function;
— the Multiple Choice Problem, which regards the existence of inequivalent quantum theories derived from different choices of time-function as a consequence of the fact that the canonical equivalence of classical formulations of a theory does not imply unitary equivalence of the quantizations of each.

Then, in the first part of this paper, Anderson considers the three fundamental strategies for resolving the problem of time (in a similar way to the classification of the various approaches to the problem of time in quantum gravity made by Isham in [13] and Kuchař in [15]), namely

— *tempus ante* quantum models, in which time exists prior to quantization, for example, internally at a classical level, as a functional of the canonical variables in the gravitational theory

or is provided by coupling matter clocks and reference fluids, or as a unimodular momentum conjugate to the cosmological constant;
— *tempus post* quantum models, where time emerges at the quantum level in the Klein–Gordon interpretation or in the third quantization, where the solutions of the WDW equation are turned into operators, or in the semi-classical interpretation, where time is an emergent quantity, an approximate, semi-classical concept whose definition depends on the quantum state of the system;
— *tempus nihil est* models, in which time has no fundamental role, such as: the naïve Schrödinger interpretation, in which time enters as an internal coordinate function of the three-geometry, and is represented by an operator that is part of the quantization of the complete three-geometry; Page's and Wooters' conditional probabilities approach, which provides a way to grasp the idea that the passage of time should be identified with correlations inside the system rather than reflecting changes with respect to an external parameter; or the consistent histories approach (which implies the notion of 'history' in a way that avoids having to make any direct reference to the concept of time); or the so-called frozen time formalism models, based on the consideration that observables in quantum gravity are operators that commute with all the constraints, and are therefore constants of the motion, 'timeless' entities.

Even more relevant, in the author's opinion, is the second part of the paper [37], in which Anderson suggests new perspectives of re-reading of the fundamental features of the problem of time. In fact, he shows that the temporal relationalism, which consists of adopting Leibniz's 'there is no time for the universe as a whole' principle as a desirable tenet of background independence and of closed universes, underlies the Frozen Formalism Problem, that the Configurational Relationalism, via the notion of Best Matching, generalizes the Thin Sandwich Problem, that the Problem of Observables becomes the Problem of Beables, and that the Functional Evolution Problem becomes the Constraint Closure Problem. Anderson also outlines how each of the Global and Multiple Choice Problems of Time have their own plurality of facets. Finally, he provides a local resolution to the Problem of Time which can be considered Machian and has three levels: classical (which resolves configurational relationalism by implementing best matching), semiclassical (where time is abstracted by a suitably semiclassical quantum change) and a combined semiclassical-histories-timeless records scheme (in which one deals with a path-type approach of records, localized subconfigurations of a single instant that contain information/correlations, which can provide, in a purely timeless approach, an underlying dynamics or history).

As also the same papers [13, 15, 37] put in evidence, the problem of time occurs, above all, as a consequence of the fact that the "time" of general relativity and the "time" of quantum theory are mutually incompatible, thus determining several problems when one tries to unify these two theories in a single unifying framework. In this way, the problem of time has become one of the most extensively investigated topics in quantum gravity.

Despite the several properties often ascribed to time (as well as the fundamental facets and strategies of solution, such as those analyzed by Anderson in [37]), one can say that the two fundamental matters concerning the debate on the features of time are probably the following: whether or not time is a fundamental quantity of nature (namely, a physical primary parameter with respect to which change occurs), and how the clock time of metrology emerges in the experimental description of dynamics. Inside quantum gravity, the most common facet of the problem of time is perhaps the frozen formalism problem which shows up in attempting canonical quantization of general relativity (or of many other gravitational theories that are likewise background-independent), due to the general relativistic Hamiltonian constraint being quadratic but not linear in the momenta. In this regard, a simple way of stating the frozen formalism part of the problem of time (regarding the debate if it must be considered a primary physical reality or not) is to note that the Wheeler–DeWitt (WDW) equation regarding the wavefunctional Ψ of the universe:

$$\left[(8\pi G)G_{abcd}(\hat{g}^3)p^{ab}p^{cd} + \frac{1}{16\pi G}\sqrt{g}(2\Lambda - {}^{(3)}R(\hat{g}^3))\right]\Psi\left(g^3\right) = 0,$$
(1.1)

where $G_{abcd} = \frac{1}{2}\sqrt{g}\left(g_{ac}g_{bd} + g_{ad}g_{bc} - g_{ab}g_{cd}\right)$ is the supermetric, p^{ab} are the momentum operators related to the 3-metric g_{ab}, Λ is the cosmological constant, G is the gravitational constant, depends in a configuration basis only on the 3-metric g^3 and variations with respect to it. Since in the WDW equation (1.1) the time parameter which defines the foliation of the space-time does not appear, the equation introduces the so-called frozen formalism problem of time in quantum gravity, according to which, in apparent contradiction with everyday experience, nothing at all seems to happen in the universe. On the basis of equation (1.1), the formal wavefunctional $\Psi\left[g^3\right]$ depends only on the configuration space variables g^3 and is completely independent of any variable one could interpret as time. As a result, solutions to the WDW equation are stationary states. This fact leads to difficulties, for example, in forming an inner product under which evolves unitarily and with which one can define a clear notion of probability. As regards the timeless nature of the WDW equation, interesting considerations have been made recently by Gryb in the paper *Jacobi's principle and the disappearance of time*:

> "The timelessness of the Wheeler–DeWitt equation can be seen as resulting from using Jacobi's principle to define the dynamics of 3-geometries through superspace. [...] If one uses Jacobi's principle in non-relativistic particle mechanics, one finds that the analogue of the WDW equation is the time-independent Schrödinger equation. The result that one gets a quadratic scalar constraint on the wavefunction whose solutions are stationary states comes from using Jacobi's principle to implement Mach's second principle in order to define time in a relational way."

The most accepted interpretation of the problem of time of the WDW equation is that time must be reintroduced into the quantum theory by means of an auxiliary physical entity whose values can be correlated with the values of other physical entities. This correlation allows, in principle, to analyze the evolution of physical quantities with respect to the "auxiliary internal time". Since there is no clear definition of the auxiliary internal time, one can only use the imagination to choose a quantity as the time parameter. For instance, if we have a physical quantity which classically evolves linearly with time, then it could be a good candidate for an auxiliary internal time. Although the linearity seems to be a reasonable criterion, it is not a necessary condition. Examples of this type of auxiliary internal time are the very well analyzed minisuperspaces of quantum cosmology [38, 39]. In particular, one could select the auxiliary internal time as one of the scale factors of homogeneous cosmological models. The volume element which is a combination of scale factors would also be a good choice since in most cases it evolves linearly in cosmic time and reproduces the main aspects of cosmological evolution. The volume element has also been used recently in loop quantum cosmology [40, 41] as auxiliary internal time. Certain low energy limits in string theory contain a tachionic field which linearly evolves in time and, consequently, could be used as auxiliary internal time for quantization [42]. It is not clear at all if the procedure of fixing an auxiliary internal time can be performed in an exact manner and, if it can be done, whether the results of choosing different auxiliary internal times can be compared and are somehow related [15, 43, 44]. Moreover, a controversial point is whether such an auxiliary internal time can be used to relate the usual concepts of space-time.

In quantum gravity, the problem of time is perhaps as old as the subject itself. This was already remarked by Bryce DeWitt in his classic paper *The Quantum Theory Of Gravity* [45] in which he presented the constraint equations for canonical quantum general relativity, namely the constraints corresponding to the vanishing of momenta associated to the Lapse and the Shift functions

$$p\Psi = 0, \qquad (1.2)$$

$$p^i\Psi = 0, \qquad (1.3)$$

the Hamiltonian constraint, represented by the WDW equation

$$\left[(8\pi G)\, G_{abcd}\left(\hat{g}^3\right) p^{ab} p^{cd} + \frac{1}{16\pi G}\sqrt{g}\left(2\Lambda - {}^{(3)}R\left(\hat{g}^3\right)\right)\right]\Psi\left(g^3\right) = 0 \tag{1.1}$$

and the vector Diffeomorphism constraint

$$\chi^i \Psi = 0. \tag{1.4}$$

DeWitt noted that in the constraints equations for canonical quantum general relativity, if we were to consider the wave function Ψ as a function $\Psi\left(x, x^0\right)$ of the spatial coordinates and time, we see that the expectation value of the field operator, the momentum conjugate, undergoes the following peculiar relation: $\Psi^+ p^{ij}(x,0)\Psi = \Psi^+ p^{ij}(x, x^0)\Psi$. This leads him to conclude that:

> "... Since the statistical results of any set of observations are ultimately expressible in terms of expectation values, one therefore comes to the conclusion that nothing ever takes place in quantum gravity dynamics, that the quantum theory can never yield anything but a static picture of the world."

Thus the WDW equation (1.1), in the standard interpretation, leads to the so-called "frozen formalism" which apparently tells that nothing should happen in canonical quantum gravity. So how can one obtain at least a partial solution to this problem regarding time? How can one extract dynamics from a system that is totally constrained? One may preliminarily observe that the problem of the frozen formalism arises fundamentally due to the fact that the Hamiltonian is but a sum of constraints, and that they in themselves saturate the theory. Due to the absence of the unconstrained Hamiltonian, we are left with no operator which can generate infinitesimal time translations of the system. To face the problem of time, different strategies have been adopted. Among the several approaches that have come to the forefront of current research, two are markedly similar. One, initiated and developed by Barbour, is the so-called "No-time" interpretation: this view, does away with time from the very outset. The other, pioneered by Rovelli, seeks to "forget time" conceptually. In this chapter, we want to show that, on the basis of relevant current research, one can provide a unifying key of reading of Barbour's approach and of Rovelli's view. The starting point to obtain this result is represented by the view, developed by the author with the fellow researcher Sorli, according to which time does not exist as a primary physical reality but exists only as a measuring system of the numerical order of material changes.

1.2 Time in the Jacobi–Barbour–Bertotti Theory

Dirac's, Arnowitt's, Deser's and Misner's discovery to cast general relativity into Hamiltonian form which implies that it is not the four-metric $g_{\mu\nu}$ of space-time that evolves but the three-metric g_{ij} of space, as well as DeWitt's equally remarkable discovery that the canonical quantum theory of a closed universe is static, can be considered as the fundamental starting points which explain how Barbour's time research began. At the beginning of the 60's, Barbour became acquainted with Mach's writings and was studying the papers in which Einstein created general relativity in order to implement Mach's principle. Surprised by Einstein's failure to attack the Machian proposal directly, Barbour resolved to go back to 'Machian first principles' in order to obtain a fully relational theory *ab initio*. In such a theory, duration should be derived from change, position should be defined relative to the universe at large, and only relative sizes within the universe should have objective meaning (if all scales were doubled overnight, no change could be observed). Of these three principles, which represent the axioms of a truly relational theory, indeed Einstein had consciously worked to implement just the second, the relativity of position and only indirectly.

Barbour formulated a basic mathematical framework in which these axioms could be implemented in 1974 [46]. This led to an extended collaboration with Bruno Bertotti and to the publication of the paper *Mach's Principle and the Structure of Dynamical Theories* in 1982 [31], in which Barbour and Bertotti had important help from Karel Kuchař. This paper developed a universal framework in which one can automatically develop theoretical schemes to meet the three essential relational requirements listed above. In these schemes, it is important that the relativity of duration is achieved in a manner that is quite different from that used to implement relativity of position and size, which is done by a process which is called "best matching" (in this regard, the reader can find details also in [47]). In contrast, time is eliminated entirely from the basic kinematical structure of the theory, which is formulated as a geodesic theory on configuration space.

The paper [31] showed that in the case of a closed universe, general relativity fully meets the requirements of relativity of duration and position. It also showed that the discoveries of Dirac, Arnowitt, Deser and Misner, and DeWitt were simple direct consequences of these relational aspects of general relativity, which had long remained hidden in its original space-time formulation. In particular, the disappearance of time in canonical quantum gravity discovered by DeWitt turned out to be a direct consequence of the relativity of duration. Barbour also became convinced that the quantum implications of the relational structure of general relativity could perhaps explain the arrow of time and lead to a complete explanation of our sense of a passage of time in an ontologically timeless universe. Barbour eventually wrote about this in his famous book *The End of Time* [48].

However, at that time it seemed to Barbour that the quantum implications were still more significant, as a consequence of the relativity of size, which had not been further explored in the paper [31]. In 1999, Barbour showed how this could be done for particles in Euclidean space and the extension to dynamical geometry was then performed, initially by Niall O' Murchadha and Barbour [49, 50] and then in more detail by Barbour, O' Murchadha, Edward Anderson, Brendan Foster, and Bryan Kelleher [32, 34]. All these research demonstrated two fundamental things: first, the dynamical object in geometrodynamics is not the four-dimensional (4D) metric $g_{\mu\nu}$ but the three-dimensional (3D) metric g_{ij}; second, not all structure that appears in the space-time form of Einstein's theory is necessarily ontological, it is emergent. The basic assumption of Barbour's approach was therefore: geometrodynamics is to be formulated in terms of the 3D metric of space g_{ij}, but only its angle-determining part is to be regarded as physical. Its scale factor, which determines a local proper distance, emerges from the relational form of its dynamics just as local inertial frames of reference and local proper time arose in the earlier work from the relativity of position and duration. Moreover, while Einstein required 4D general covariance (involving four arbitrary functions of space-time in the transformation laws, or space-time diffeomorphism invariance) and Weyl required not only that but also 4D conformal invariance (a fifth arbitrary function), in contrast Barbour and his collaborators retained 3D diffeomorphism invariance, which implements relativity of position, and replaced Einstein's freedom in the definition of simultaneity (foliation invariance) by a 3D conformal invariance, which ensures (local) relativity of size. Thus, Barbour's approach required invariance with respect to four arbitrary functions but swapped one of them (foliation invariance) for 3D conformal invariance. In this way, Barbour's approach lead to the so-called "no time" interpretation of physical events.

In the "No Time" interpretation developed by Barbour and his collaborators Foster and O' Murchadha in [33], Leibnizian Temporal Relationalism can be considered the starting point and is implemented by reparametrization invariance alongside an absence of auxiliary time-like variables such as Newtonian absolute time or the general relativistic lapse. The subsequent product-type action, modelable in this sense by the Jacobi action of mechanics,

$$S = \sqrt{2} \int ds\sqrt{E-V}, \tag{1.5}$$

where ds is the configuration space line element, V the potential energy and E the total energy, implies a purely quadratic constraint as a primary constraint. As shown in [31, 48, 51–53], Barbour's Leibnizian Temporal Relationalism can be seen as setting up a classical problem of time, which is resolved (at least in principle) in a "time is to be abstracted from change" Machian manner by the Jacobi–Barbour–Bertotti (JBB) emergent time,

$$t^{\text{JBB}} = t^{\text{JBB}}(0) + \int ds/\sqrt{E-V}. \tag{1.6}$$

In Barbour's approach, a "time" intended as primary physical reality is eliminated from the action and is replaced by an unphysical evolution parameter, with respect to which one can differentiate coordinates to derive velocities, and the action is invariant under an arbitrary change of this parameter, namely is reparameterization invariant. Thus eliminating time really leads to getting rid of a preferred time scale, i.e. the system refers to no external clock. If one considers λ to be the evolution parameter, this condition implies that $\lambda \to f(\lambda)$ leaves the action invariant. Applying this condition and using the metric

$$g_{ab} = -2V(q)\delta_{ab}, \qquad (1.7)$$

on a configuration space coordinatized by the q's, where the conformal factor $V(q)$ corresponds to the potential, the JBB theory is constructed on the basis of the action

$$T(\lambda) = \sum_{j=1}^{M} \frac{m_j}{2}\left(\frac{dq_j^i(\lambda)}{d\lambda}\right)^2$$

$$S_{\text{JBB}} = \int_{\lambda_0}^{\lambda_f} d\lambda\, 2\sqrt{(T(\lambda))\left(E - V\left(q_j^i\right)\right)}, \qquad (1.8)$$

where $Tt(\lambda) = \sum_{j=1}^{M} \frac{m_j}{2}\left(\frac{dq_j^i(\lambda)}{d\lambda}\right)^2$ is the kinetic energy of an M particle system, $V(q_j^i(\lambda))$ is the potential energy (that does not depend explicitly on λ) and E is the constant part of V and can be understood as the total energy of the system. The index i ranges from 1 to d while j ranges from 1 to M. The action (1.8) can be called as the JBB action. The JBB action satisfies the Jacobi geodesic variational principle, which establishes that the geodesics on the configuration space extremize the action. It is invariant under reparameterizations of λ. As a consequence, its apparent dependence on λ is artificial, in other words the action S_{JBB} is independent of anything can be called a time parameter. It does, however, depend on a path γ in configuration space which is gauge invariant as it represents a collection of points in the configuration space and is independent of any parameter that one might use to parameterize it. Nevertheless, the introduction of the parameter λ, and of all the gauge redundancy that goes along with it, opens the possibility to access well-known techniques in order to determine the correct measure for the path integral in the canonical quantization.

Starting from the action (1.8), a variation with respect to q^i leads to the following Newton-type equations of motion:

$$m\frac{d^2 q^i}{d\tau_{\text{BB}}^2} = -\frac{\partial V}{\partial q^j}\eta^{ij}, \quad (1.9)$$

where η^{ij} is the flat metric with Euclidean signature and

$$\tau_{\text{BB}} = \int_{\lambda_0}^{\lambda_f} \frac{\sqrt{T}}{\sqrt{E-V}} d\lambda \quad (1.10)$$

is a reparamaterization invariant quantity first referred to as ephemeris time by Barbour and Bertotti in the papers [31, 51] in analogy to the operational definitions of time first adopted by astronomers [54]. On the basis of the ephemeris time defined by Eq. (1.10), one can say that the system under consideration becomes its own relational clock: the ephemeris time provides indeed a metric-trans-temporal notion of identity between two subsequent configurations, and time is here completely emergent from the dynamics. According to Eqs. (1.9) and (1.10), the following re-reading of the Newtonian time becomes permissible: the ephemeris time τ_{BB} is equivalent to the Newtonian time but is defined in terms of a length in configuration space equipped with a suitably defined metric. As a consequence, τ_{BB} provides a measure of duration corresponding to the relative change in the positions of the particles in the system and, thus, is a precise realization of Mach's second principle. This system of M particles defined by the action (1.8) and governed by the equations of motion (1.9) can be called a Barbour–Bertotti (BB) clock since it provides us with a way of measuring τ_{BB}. In other words, from the perspective of the JBB theory, the analysis of the processes is performed through the following procedure. One starts with an action that depends only on the gauge invariant path γ, which represents the relative positions of particles in the universe, and of an arbitrary potential $V(q)$ defined only on the configuration space. After writing down the classical equations of motion, one may define a gauge invariant quantity called time (which can be thought of as a length of γ) to describe how the q's coordinatizing the configuration space change relative to each other. In the end, one recovers equations of motion equivalent to those of Newton's theory, for a fixed energy E, in terms of this invariant quantity. However, in this theory it is not necessary to define an absolute Newtonian time: the time emerges as a convenient tool for keeping track of the relative positions of particles in a system.

Moreover, it is interesting to remark that Barbour's approach based on Eqs. (1.8) and (1.9) may be expressed also into a Hamiltonian formulation. By defining the canonical momenta

$$p_i = \frac{\partial L_{\text{JBB}}}{\partial \dot{q}_i} = \sqrt{\frac{E-V}{T}} m\dot{q}^j \eta_{ij}, \tag{1.11}$$

one obtains that the canonical Hamiltonian is identically zero:

$$H_c = p_i \dot{q}^i - L\left(q^i, p_i\right) = 0. \tag{1.12}$$

Here, the constraint takes the form

$$\mathrm{H} = \frac{p^2}{2m} + V(q) - E = 0, \tag{1.13}$$

which can be thought of as a kind of circle identity in phase space. The total Hamiltonian is then proportional to the constraint

$$H_T = H_c + N(q,p)\mathrm{H} = N(q,p)\mathrm{H}. \tag{1.14}$$

In this Hamiltonian formulation of JBB theory, Hamilton's equations of motion corresponding to the Newton-type equations of motion (1.9) are respectively

$$\dot{q}^i = \{q^i, H_T\} = N\frac{p_j}{m}\eta^{ij}, \tag{1.15}$$

$$\dot{p}^i = \{p^i, H_T\} = -N\frac{\partial V}{\partial q^i}. \tag{1.16}$$

On the other hand, in Barbour's approach, the emergence of time from timeless dynamics is linked crucially to the assumptions made about the structure of space. In this regard, it is important to emphasize that Euclidean, Riemannian and conformal geometry lead to distinct specific results.

As Barbour showed in the recent essay *The Nature of Time* [55], the concept of duration and the theory of clocks emerge from the timeless reparametrization-invariant Jacobi principle for the orbit of the Newtonian N-body problem in a Euclidean space expressed by equation:

$$\delta A_J = 0, \quad A_J = 2 \int d\lambda \sqrt{(E-V) \sum_i \frac{m_i}{2} \vec{x}_i^I \cdot \vec{x}_i^I}, \quad \vec{x}_i^I = \frac{d\vec{x}_i}{d\lambda}, \tag{1.17}$$

where λ is an arbitrary monotonic parameter that labels the points of the orbit, E is the constant total energy, and V is the potential energy of the system. The two key features of the timeless reparametrization-invariant Jacobi principle (1.17) are the following: (1) the particle positions are the vectors $\vec{x}_i$ in Euclidean space, which determine the nature of the emergent time; (2) the square root of the integrand, which makes the Lagrangian a metric on the configuration space, the geodesics of which are the Newtonian orbits in the configuration space with total energy E, ensures that there is no time in the kinematic foundations of the theory. Time only emerges as a parameter that simplifies the equation of the timeless geodesics that follows from (1.17) and is

$$\frac{d}{d\lambda}\left(\sqrt{\frac{E-V}{T}} m_i \frac{d\vec{x}_i}{d\lambda}\right) = -\sqrt{\frac{T}{E-V}} \frac{\partial V}{\partial \vec{x}_i}, \tag{1.18}$$

where T is the cofactor of $E - V$ in (1.17). As regards Eq. (1.18), a great simplification may be obtained by choosing the arbitrary label λ such that always $T = C(E-V)$, where the constant C sets the unit of the distinguished label λ. In this particular case, Eq. (1.18) becomes Newton's second law and thus the parameter λ emerges as the Newtonian time derived from change as Mach required. Moreover, the explicit expression for the increment δt of this emergent time is

$$\delta t = \sqrt{\frac{\sum_i m_i \delta \vec{x}_i \cdot \delta \vec{x}_i}{2(E-V)}}. \tag{1.19}$$

In [55], Barbour showed that mechanical clocks only march in step — and hence have any utility — if they are constructed in such a way that they measure this emergent time defined by (1.19). Jacobi's principle, properly interpreted for a closed system, provides the complete theory of Newtonian time. Moreover, the form of the expression (1.19) is linked crucially with the properties of Euclidean space, which lead to the appearance of the scalar product in the numerator and the distance dependence that appears in V. In this sense, the nature of time is determined in dynamics by the structure of space.

By continuing to follow Barbour in [55], as regards general relativity since space no longer has a Euclidean but rather the much richer Riemannian structure, the nature of the emergent time turns out to be correspondingly richer. Vacuum general relativity (without cosmological constant and for a spatially closed universe) can be derived from the Baierlein–Sharp–Wheeler (BSW) action [56]

$$A_{\text{BSW}} = \int d\lambda \int d^3x \sqrt{gRG^{ijkl}\left(g^I_{ij} - \zeta_{(i;j)}\right)\left(g^I_{kl} - \zeta_{(k;l)}\right)},$$

$$g^I_{ij} = \frac{dg_{ij}}{d\lambda}, \qquad (1.20)$$

where g is the determinant of the three-metric g_{ij}, R is the 3D scalar curvature, G^{ijkl} is the DeWitt supermetric, and the three-vector field ζ_i in the Killing form $\zeta_{(i;j)}$ is the generator of three-diffeomorphisms used in best matching to implement relativity of position in geometrodynamics. The BSW action (1.20) is defined on the space of Riemannian three-metrics on a closed three-manifold M. Variation with respect to ζ_i yields the Arnowitt–Deser–Misner (ADM) momentum constraint. The parameter λ is arbitrary and together with the square root ensures that the kinematic foundation of the theory is timeless.

It is important to remark that the square root is taken locally, namely a quadratic expression is formed at each point of space and then its square root is integrated over space. The local character of the square root implies that the equations of motion which are derived from (1.20) are simplified by analogy with what happens for Jacobi's principle if the value of λ is chosen locally such that

$$N = \sqrt{\frac{T}{4R}} = 1, \qquad (1.21)$$

where T is the cofactor of $\sqrt{gR}$ in (1.20). This then leads to an explicit expression for an emergent local proper time. The richer structure of Riemannian geometry as compared with Euclidean geometry therefore has the consequence that its dynamics leads to a more subtle nature of time. As O' Murchadha underlined, it is extremely difficult to construct consistent theories of the form (1.20) with local square root. In this picture, as shown in [33], the relativity of duration, implemented through the local square root, has numerous consequences and leads to the emergence of local proper time by very close analogy with the emergence of Newtonian time from the Jacobi action principle (1.17). The presence of the Killing term $\zeta_{(i;j)}$ in (1.20) implements relativity of position and leads to the ADM momentum constraint. The fundamental dynamical structure of general relativity is derived very directly, as explained in detail in [33]. In this picture, one has relativity of both duration and simultaneity.

The "no time" interpretation of JBB theory, in which time is replaced with a reparametrization invariant action that is integrated over an unphysical evolution parameter (and which is based on the JBB action (1.8) and the equations of motion (1.9)), leads therefore to the so-called Barbour–Foster–O' Murchadha (BFO) formalism of geometrodynamics where general relativity is treated as a theory of dynamical three spaces rather than a theory of three spaces embedded into an *a priori* assumed Lorentzian four manifold (spacetime) [33, 57]. In order to describe the BFO formalism of general relativity, one begins with a lagrangian for ADM approach to gravity (which splits space-time into spatial hypersurfaces associated with a preferred chosen time) given by the following relation

$$S = \int d\lambda d^3 x \left[\frac{1}{4N\sqrt{\gamma}} \left(k_{ab} k^{ab} - tr k^2 \right) - N\sqrt{\gamma} R \right], \qquad (1.22)$$

where N is the lapse function (which measures the rate of change of proper time with respect to the label time), γ_{ij} are the three-metrics, R is the 3D scalar curvature,

$$k_{ab} = \dot{\gamma}_{ab} - L_{\xi^a} \gamma_{ab}, \qquad (1.23)$$

where ξ is an arbitrary vector field with respect to which the Lie Derivative acting on the metric represents the action of the three-diffeomorphism group on the configuration space (that turns out to be equal to the shift of ADM gravity) and the over dot denotes differentiation with respect to an unphysical evolution parameter λ.

Varying with respect to N one obtains

$$N = \sqrt{\frac{k_{ab}k^{ab} - trk^2}{4R}}, \qquad (1.24)$$

which, by substitution in the action (1.22), leads to equation

$$S = \int d\lambda d^3x \sqrt{\gamma}\sqrt{R}\sqrt{T}, \qquad (1.25)$$

where

$$T = G^{qbcd}\left(\dot{\gamma}_{ab} - L_{\xi^a}\gamma_{ab}\right)\left(\dot{\gamma}_{cd} - L_{\xi^a}\gamma_{cd}\right), \qquad (1.26)$$

G being the DeWitt supermetric.

In the picture based on Eqs. (1.22)–(1.26), Barbour, Foster and O' Murchadha made general relativity a truly 3D theory by applying a consistency condition (constraint propagation) and the relaxation of the embeddability criterion (Dirac algebra). The Hamiltonian constraint of general relativity is

$$\text{H} = \int d^3x\left(NH + N^aH_a\right) = C\left(N\right) + C\left(N^a\right), \qquad (1.27)$$

where H is the Hamiltonian constraint, H_a is the momentum constraint, N^a is the shift function (which determines how the coordinates are laid down on the successive three-geometries),

$$C\left(N\right) = \int d^3x \left(\frac{1}{\sqrt{\gamma}}G^{abcd}p_{ab}p_{cd} - \sqrt{\gamma}R\right), \qquad (1.28)$$

and

$$C\left(N^a\right) = \int d^3x 2N^a\nabla_a p^{ab}, \qquad (1.29)$$

$$p_i = \frac{\partial L_{\text{JBB}}}{\partial \dot{q}_i} = \sqrt{\frac{E-V}{T}}m\dot{q}^j\eta_{ij} \qquad (1.30)$$

being the canonical momenta.

These constraints satisfy the following Dirac hypersurface deformation algebra

$$\{C(N^a), C(N^b)\} = C(L_N \neg N^b), \tag{1.31}$$

$$\{C(N^a), C(N)\} = C(L_{N^a} N), \tag{1.32}$$

$$\{C(N), C(N')\} = C\left(N' \gamma^{ab} \nabla_b N - N \gamma^{ab} \nabla_b N'\right), \tag{1.33}$$

which indicate that the constraints represent generators of tangential and normal deformations of the three spatial slice embedded into space-time. That is, the diffeomorphism constraint generates tangential deformations while the Hamiltonian constraint generates the normal deformations. The algebra of the constraints can be seen as the embeddability criterion for the space-like slices to be embedded into space-time. In the BFO approach the Hamiltonian constraint of general relativity emerges as from a square root identity of the local square root form of the action namely

$$\frac{1}{\sqrt{\gamma}}\left(p^{ab} p_{ab} - \frac{1}{2} tr p^2\right) - \sqrt{\gamma} R = 0, \tag{1.34}$$

and the diffeomorphism constraint can be derived from best matching where

$$\partial_\xi S_{\text{JBB}} = 0, \tag{1.35}$$

that leads to equation

$$-2 \nabla_a p^{ab} = 0. \tag{1.36}$$

Moreover, one has the consistency conditions

$$\dot{C}\left(\sqrt{\frac{T}{4R}}\right) = 0 \tag{1.37}$$

and

$$\dot{C}(\xi^a) = 0, \tag{1.38}$$

which mean that the Euler Lagrange equations obtained from varying the JBB action must propagate the constraints. Also, the role of the diffeomorphism constraint remains the same while the Hamiltonian constraint is now a generator of real physical evolution (in accordance with [58]). With this, Barbour's theory is fully consistent from the physical point of view.

Starting from Barbour's approach to general relativity, in the recent paper *Presympletic Geometry and the Problem of Time. Part 2* [59] Shyam and Ramachandra provided a detailded analysis and characterization of canonical gravity as a 3D constrained Hamiltonian theory in the picture of a presympletic dynamics inside a super phase space. The generic point of the super phase space is

$$z^I = \begin{pmatrix} p^{ab} \\ \gamma_{ab} \end{pmatrix} \qquad (1.39)$$

and

$$p_{ab} = -\sqrt{\gamma}\left(K_{ab} - \gamma_{ab} tr K\right) \qquad (1.40)$$

is the momentum conjugate to the metric that satisfies the Hamiltonian constraint, K's being the extrinsic curvature tensors. Shyam's and Ramachandra's model of presympletic dynamics is based on a presympletic equation of the form

$$(X_H)^b\big|_{\tilde{\Gamma}} = 0, \qquad (1.41)$$

where Γ is the total phase space and

$$X_H = \left[2\frac{N}{\sqrt{\gamma}}\left(p_{ab} - \frac{1}{2}\gamma_{ab}trp\right) + L_{\xi^a}\gamma_{ab}\right]\frac{\delta}{\delta\gamma^{ab}}$$

$$- \left[N\sqrt{\gamma}\left(R^{ab} - \frac{1}{2}\gamma^{ab}R\right) - \frac{N\gamma^{ab}}{2\sqrt{\gamma}}\left(p_{ab}p^{ab} - \frac{1}{2}trp^2\right)\right]\frac{\delta}{\delta p_{ab}}$$

$$+ \left[\frac{2N}{\sqrt{\gamma}}\left(p^{ac}p_c^b - \frac{1}{2}p^{ab}trp\right) + \sqrt{\gamma}\left(\nabla^a\nabla^b N - \gamma^{ab}\nabla^2 N\right)\right.$$

$$\left.+ L_{\xi^a}p^{ab}\right]\frac{\delta}{\delta p_{ab}} \qquad (1.42)$$

is the (locally) Hamiltonian vector field, R^{ab} is the Ricci tensor,

$$H = \int d^3x \left(\frac{N}{\sqrt{\gamma}}G^{abcd}p_{ab}p_{cd} - N\sqrt{\gamma}R - 2\xi^a\nabla_a p^{ab}\right) \qquad (1.43)$$

is the Hamiltonian defined on the phase space Γ. Moreover, in this picture, in virtue of relations

$$X_{\rm H}\left(H_a\right) = L_{\vec{\xi}} H_a + H\nabla_a N \approx 0, \tag{1.44}$$

$$X_{\rm H}\left(H_a\right) = L_N H + N\nabla_a H_a + 2\left(\nabla^a N\right) H_a \approx 0, \tag{1.45}$$

the constraint propagation condition corresponding to (1.37) and (1.38) is

$$f_{\rm H}^{\lambda}\left[\phi^J\left[z^I\right]\right] = \phi^J\left[z^I\right], \tag{1.46}$$

where

$$f_{\rm H}^{\lambda}\left[z^I\right] = e^{\lambda X_{\rm H}\left[z^I\right]} \tag{1.47}$$

is the canonical flow of the phase space variables, solution to the Cauchy problem

$$\begin{cases} f_{\rm H}^{0}\left[z^I\right] = z^I, \\ \dfrac{d}{d\lambda} f_{\rm H}^{\lambda}\left[z^I\right] = X_{\rm H}\left[z^I\right] \end{cases} \tag{1.48}$$

and

$$\phi^J\left[z^I\right] = \begin{pmatrix} \dfrac{1}{\sqrt{\gamma}} G^{abcd} p_{ab} p_{cd} - \sqrt{\gamma} R \\ -2\nabla_a p^{ab} \end{pmatrix} = 0 \tag{1.49}$$

are the constraints regarding the phase space corresponding to the super space.

Therefore, while in the original BFO approach the constraint propagation condition is propagated by the Euler Lagrange equations, Shyam's and Ramachandra's approach requires the constraints to be invariant under the flow of the Hamiltonian vector field on the constraint hypersurface. Moreover, by inserting the evolutionary part of the Hamiltonian vector field

$$\xi_{H(N)}\left[\gamma_{ab}\right]\Big|_{\tilde{\Gamma}} = 2\frac{N}{\sqrt{\gamma}}\left(p_{ab} - \frac{1}{2}\gamma_{ab}trp\right) \qquad (1.50)$$

into the equation

$$\phi^{J}\left[z^{I}\right] = \frac{1}{\sqrt{\gamma}}G^{abcd}p_{ab}p_{cd} - \sqrt{\gamma}R = 0, \qquad (1.51)$$

one can obtain the following expression for the Lapse function

$$N = \sqrt{\frac{G^{abcd}\xi_{H(N)}\left[\gamma_{ab}\right]\xi_{H(N)}\left[\gamma_{cd}\right]}{4R}}, \qquad (1.52)$$

which is exactly the same expression derived inside Barbour's approach. With the Lapse function (1.52), an ephemeris time can be defined by smearing this expression over the evolution parameter

$$\tau = N(\lambda, x) = \int\sqrt{\frac{G^{abcd}\xi_{H(N)}\left[\gamma_{ab}\right]\xi_{H(N)}\left[\gamma_{cd}\right]}{4R}}d\lambda. \qquad (1.53)$$

Now, after having analyzed how, in Barbour's approach, time emerges from a timeless dynamics in the context of Euclidean geometry of Newtonian physics and Riemannian geometry of general relativity, let us see what happens in a relational theory based on conformal geometry. In this regard, as shown clearly in the references [32, 34], one can obtain the most intriguing possibilities associated to a spatially closed universe by including in the BSW action (1.20) not only the Killing term $\zeta_{(i;j)}$, which implements relativity of position, but also a further term that implements relativity of size by best matching with respect to either full conformal transformations of the three-metric g_{ij} in accordance with

$$g_{ij} \to \varphi^4 g_{ij}, \quad \varphi = \varphi(x, \lambda) > 0, \qquad (1.54)$$

for such transformations that are untrestricted or preserve the total volume of the universe. In the case of the untrestricted transformations, one arrives to a theory that is very similar to general relativity but in which no expansion of the universe is possible. In the case of the transformations which preserve the total volume of the universe, it is possible to arrive to general relativity in the constant-mean-(extrinsic)- curvature (CMC) foliation that plays the key role in York's solution of the initial-value problem. The conformal best matching is implemented by varying φ by the free-end-point method and, in contrast to the diffeomorphism best matching, leads not only to a constraint but also to a nontrivial consistency condition that ensures propagation of the constraint. The constraint enforces the CMC condition on the initial hypersurface, while the consistency condition, which is a lapse-fixing condition, ensures that this condition propagates.

As shown by Barbour in [55], these results can be considered appealing for two main reasons. First, a local scale appears nowhere in the kinematic foundations of the theory in conformal superspace but is forced to emerge as a distinguished scale in the space of Riemannian three-metrics on a closed three-manifold by the best matching. This already happens on the initial hypersurface and is in strong contrast to the consequence of the diffeomorphism best matching, which does not lead to distinguished coordinates on the initial hypersurface but only in the propagation, for which Gaussian normal coordinates are distinguished. Second, the CMC condition, which introduces a unique definition of simultaneity, is produced dynamically by the variational principle. Despite this, the effective theory that emerges at the classical level is identical to general relativity in its predictions.

Thus, the diffeomorphism and conformal constraints lead to different behaviors. In [55], Barbour emphasizes that the reason for this difference, which could have considerable significance for quantum gravity and leads to the two striking features above, arises from a key property of the so-called 'bare' BSW action defined by relation:

$$A_{\text{bareBSW}} = \int d\lambda \int d^3 x \sqrt{g R G^{ijkl} \left(g_{ij}^I\right)\left(g_{kl}^I\right)}, \qquad (1.55)$$

which is simply the BSW action without the Killing term that implements diffeomorphism best matching. The action (1.55) is invariant under identical diffeomorphisms made at each λ: it is globally gauge invariant, because of the scalar nature of R under diffeomorphisms. However, since gR is not a conformal scalar, the action (1.55) is not globally conformally invariant and this fact has positive consequences in the picture of relational approaches in the sense that it allows a perfectly consistent theory with very interesting properties to be developed, namely just general relativity in the CMC foliation with the two angle-determining conformal degrees of freedom unambiguously identified as the true degrees of freedom of the theory.

In the conformal approach of relational geometrodynamics, as remarked by Barbour in [55], one can summarize the situation as follows. Here, one deals with far less kinematic structure than the standard approach, namely with an absolute minimum of assumptions about what exists. There is still a 'structured something' that evolves: all that exists in the classical theory is a sequence of conformal three-geometries. However, this sequence is determined by best matching from specification of a point and tangent vector in configuration space in such a manner that makes it possible to embed the sequence of conformal three-geometries into an Einsteinian spacetime in the CMC foliation. Local inertial frames of reference, local proper distance and local proper time are all emergent. In the spacetime representation, this 'additional structure' is assumed to have ontological status but is in fact nomological in nature, namely it is an expression of the fundamental dynamical law of the universe and not part of its 'substance' at all.

Finally, at the end of this paragraph, let us analyze the important development of Barbour's "no time" interpretation represented by the path integral quantization of JBB theory. The path integral of JBB theory has been originally explored by Hartle and Kuchař in [60]; connected with this treatment of the path integral is then the Becchi–Rouet–Stora–Tyupin (BRST) quantization which has been studied by Brown and York in [61] (as regards the BRST quantization of JBB theory one can find details also in the reference [62] by Teitelboim). Moreover, in [63], Henneaux and Teitelboim provided a BRST treatment of the relativistic particle (which is nearly mathematically identical to JBB treatment). As regards the path integral quantization of JBB theory, interesting results have been obtained recently by Gryb in the paper *Jacobi's principle and the disappearance of time* [64]. While the references [60–63] use ghost fields to do the gauge fixing, in his paper Gryb used directly the techniques developed by Faddeev and Popov [65] for gauge fixing the path integral so that one can be completely rigorous about boundary conditions. Moreover, Gryb compared the results regarding the path integral of JBB theory with those regarding a parametrized form of Newtonian mechanics (PNM), which, different from JBB theory, provides a time-dependent description to quantum processes. PNM is defined by a reparametrization invariant action of the form

$$S_{\text{PNM}}\left(q^{i}, q^{0}\right) = \int_{\lambda_0}^{\lambda_f} d\lambda \left[\frac{T\left(\dot{q}^{i}\left(\lambda\right)\right)}{\dot{q}^{0}\left(\lambda\right)} - \dot{q}^{0}\left(\lambda\right) V\left(q^{i}\left(\lambda\right)\right) \right], \quad (1.56)$$

inside an extended configuration space where q^0 is treated as an independent configuration space and satisfies the equation of motion

$$\dot{q}^0 = \sqrt{\frac{T}{E-V}}, \quad (1.57)$$

which is similar to JBB theory and, more precisely, is the definition of JBB's ephemeris time τ_{BB}. As regards the path integral of PNM theory (which is a function of two configuration space points q''^{δ} and q'^{δ}), Gryb found the following relation

$$k_{\text{PNM}}\left(q''^{\alpha}, q'^{\alpha}\right) = \int_{-\infty}^{+\infty} \frac{dp_0^0 \, d^3\vec{p}_0}{2\pi \, 2\pi} \frac{\Delta\lambda_0 dN_0}{2\pi} \prod_{K=1}^{N-1} \frac{dp_0^K \, d^3\vec{p}_K}{2\pi \, 2\pi}$$

$$\times \frac{\Delta\lambda_K dN_K}{2\pi} dq_K^0 d^3\vec{q}_K \frac{d\xi^K}{2\pi} [\text{FP}]_{\text{PNM}}$$

$$\times \exp\left\{ i \sum_{J=0}^{N-1} \Delta\lambda_J \left[\vec{p}_\alpha^J \cdot \dot{\vec{q}}_J^\alpha - N_J \left(\frac{\vec{p}_J^2}{2m} + p_0^J + V^J \right) \right. \right.$$

$$\left. \left. -\xi^J \left(f_J - \dot{q}_J^0 \right) \right] \right\}, \quad (1.58)$$

where

$$[\text{FP}]_{JBB} = \left| \left\{ f_M\left(q_K^i, p_i^K\right), \frac{\vec{p}_N^2}{2m} + V^N \right\} \right| \quad (1.59)$$

is the Faddeev–Popov determinant for PNM theory and the gauge fixing conditions are

$$G_K = f_K\left(q_K^i, p_i^K\right) - \dot{q}_0^K = 0. \tag{1.60}$$

Equation (1.58) can also be written as

$$k_{\text{PNM}}\left(q''^\alpha, q'^\alpha\right) = \int \frac{dE}{2\pi} e^{iE\tau} \tilde{k}_{\text{PNM}}\left(\vec{q}'', \vec{q}', \tau\right), \tag{1.61}$$

where

$$\tilde{k}_{\text{PNM}}\left(\vec{q}'', \vec{q}', E\right) = \int_{-\infty}^{+\infty} \frac{d^3\vec{p}_0}{2\pi} \frac{\Delta\lambda_0 dN_0}{2\pi} \prod_{K=1}^{N-1} \frac{d^3\vec{p}_K}{2\pi}$$

$$\times \frac{\Delta\lambda_K dN_K}{2\pi} d^3\vec{q}_K \frac{d\xi^K}{2\pi} [\text{FP}]_{\text{PNM}}$$

$$\times \exp\left\{i \sum_{J=0}^{N-1} \Delta\lambda_J \left[\vec{p}_J \cdot \dot{\vec{q}}_J - N_J \left(\frac{\vec{p}_J^2}{2m} - E + V^J\right)\right.\right.$$

$$\left.\left. - \xi^J \left(f_J - N_J\right)\right]\right\}. \tag{1.62}$$

In analogy, as regards JBB theory, in his paper [64] Gryb showed that, if one chooses the gauge fixing functions

$$G_k = f_k\left(q_K^i, p_K^i\right) - \frac{m\vec{p}_K \cdot \dot{\vec{q}}_K}{p_K^2} = 0 \tag{1.63}$$

inside a configuration space represented by a discrete lattice so that λ takes discrete values λ_K where capital roman indices range from 1 to some large number N, where $\vec{q}_K = \vec{q}(\lambda_K)$, the phase space path integral of JBB theory can be written as

$$k_{\text{JBB}}\left(\vec{q}'', \vec{q}', E\right) = \int_{-\infty}^{+\infty} \frac{d^3\vec{p}_0}{2\pi} \frac{\Delta\lambda_0 dN_0}{2\pi} \prod_{K=1}^{N-1} \frac{d^3\vec{p}_K}{2\pi}$$

$$\times \frac{\Delta\lambda_K dN_K}{2\pi} d^3\vec{q}_K \frac{d\xi^K}{2\pi} [\text{FP}]_{\text{JBB}}$$

$$\times \exp\left\{i\sum_{J=0}^{N-1}\Delta\lambda_J\left[\vec{p}_J\cdot\dot{\vec{q}}_J - N_J\left(\frac{\vec{p}_J^{\,2}}{2m} - E + V^J\right)\right.\right.$$
$$\left.\left. - \xi^J\left(f_J - \frac{m\vec{p}_J\cdot\dot{\vec{q}}_J}{p_J^2}\right)\right]\right\}, \qquad (1.64)$$

where the Faddeev–Popov determinant is

$$[\text{FP}]_{\text{JBB}} = \left|\left\{f_M - \frac{m\vec{p}_M\cdot\dot{\vec{q}}_M}{p_M^2}, \frac{\vec{p}_N^{\,2}}{2m} + V^N\right\}\right|. \qquad (1.65)$$

As regards the path integral of JBB theory, one of the three independent components of the vectors $\vec{q}_K$ is a pure gauge. Therefore, two of the boundary conditions on the components of the vectors $\vec{q}_K$ can be imposed in the usual way; instead, the third condition is imposed by letting the gauge degree of freedom vary freely and by choosing gauge fixing functions f_K that assure that the boundary conditions will be satisfied. By defining $x^{\vee} \equiv \frac{\vec{x}\cdot\vec{p}_0}{p_0}$ and $x^{\ni} \equiv \frac{|\vec{x}\times\vec{p}_0|}{p_0}$, for the gauge degrees of freedom the appropriate boundary conditions for q_0'' can be obtained choosing f_0 such that

$$\frac{m\left(q''^{\vee} - q'^{\vee}\right)}{p_0} + \sum_{J=0}^{N-1}\Delta\lambda_J\left[m\dot{\vec{q}}_J\cdot\left(\frac{\vec{p}_J}{p_J^2} - \frac{\vec{p}_0}{p_0^2}\right) - f_J\right] = 0. \qquad (1.66)$$

By applying the boundary conditions, the path integral of JBB theory (1.64) becomes

$$k_{JBB}\left(\vec{q}'', \vec{q}', E\right) = \int_{-\infty}^{+\infty}\frac{d\xi^0}{2\pi}$$
$$\times \left[\int_{-\infty}^{+\infty}\frac{d^3\vec{p}_0}{2\pi}\frac{\Delta\lambda_0 dN_0}{2\pi}d^3\vec{q}_0\delta\left(\phi_{q'} - \phi_0\right)\delta\left(\vec{q}'^{\ni} - q_0^{\ni}\right)\right]$$
$$\times \prod_{K=1}^{N}\frac{d^3\vec{p}_K}{2\pi}\frac{\Delta\lambda_K dN_K}{2\pi}d^3\vec{q}_K\frac{d\xi^K}{2\pi}[\text{FP}]_{JBB}$$

$$\times \exp\left\{i \sum_{J=0}^{N-1} \Delta\lambda_J \left[\vec{p}_J \cdot \dot{\vec{q}}_J - N_J \left(\frac{\vec{p}_J^2}{2m} - (E+\xi^0) + V^J\right)\right.\right.$$
$$\left.\left. - \xi^J \left(f_J - \frac{m\vec{p}_J \cdot \dot{\vec{q}}_J}{p_J^2}\right)\right]\right\} \exp\left(i\xi^0 \tau\right), \tag{1.67}$$

where

$$\tau = \frac{m(q''^\vee - q'^\vee)}{p_0} + \sum_{J=0}^{N-1} \left[m\dot{\vec{q}}_J \cdot \left(\frac{\vec{p}_J}{p_J^2} - \frac{\vec{p}_0}{p_0^2}\right)\right]. \tag{1.68}$$

Gryb's treatment shows that, in the path integral description, each path represents a different relational clock. JBB path integral based on Eqs. (1.64) and (1.67) shows that, in the full quantum theory, τ contributes to each term in the sum over all histories leading to a superposition of all possible clocks and thus one obtains a timeless theory. Therefore, by comparing the JBB time-independent theory to manifest time-dependent kernel of PNM theory, Gryb found that, in the quantum regime, the timelessness of JBB approach is a result of a superposition of all possible JBB-clocks which occurs when one sums over all possible histories. Nevertheless, in his treatment, Gryb found that a unique time can be recovered in the stationary phase approximation because, in this case, the path integral is dominated by contributions due to the unique classical history. In the stationary phase approximation, the path integral is dominated by the classical trajectory picking out a unique clock.

Taking into account that the bracketed expression after the $d\xi^0$ integral is nearly equal to $\tilde{k}_{\text{PNM}}(E+\xi^0)$, on the basis of Eq. (1.68), in the stationary phase approximation, by implementing the boundary conditions separately from putting constraints on the gauge fixing conditions and by pulling τ through the integral over all of phase space, a parameter time emerges in the quantum JBB theory. In the stationary phase approximation, one can approximate the kernel by a sum over the unique history that extremizes the action and thus one can approximate the kernel by a sum over the classical history. Because one no longer has an integral over all of phase space, τ can be moved through the bracketed expression. Furthermore, the boundary conditions are imposed by requiring the classical solution. Thus, one has succeeded in showing that the stationary phase approximation provides a theory with time. However, Gryb emphasized that this is not a theory with just any time. The emergent time must be given by Eq. (1.68) which is a specific function of the classical history. Using the boundary conditions and returning to the continuous limit, one finds that $\tau = \tau_{BB}$. That is, in Gryb's approach, the time that is emergent in the stationary phase approximation is exactly the ephemeris time of the classical theory. It is the time read off by a JBB-clock. Therefore, Gryb's treatment of JBB path integrals provides a consistent picture for how a time-dependent classical theory could be the limiting form of a time-independent quantum theory. Moreover, this picture also tells us how to obtain a unique notion of duration in the quantum theory. For quadratic potentials, the stationary phase approximation is exact and the Newtonian time along the classical trajectory can be considered as a unique notion of duration even in the full quantum theory. Taking a close look at the path integral thus provides us with a unique notion of time off shell (for certain potentials) and intuition as to why this notion of time breaks down in general. This intuition may be invaluable in suggesting new ways in which duration may be defined in a time-independent quantum theory.

In the stationary phase approximation, a unique energy can be computed for a unique time simply by inverting Eq. (1.10) and inserting the classical history. Specifically, Gryb showed that, in the stationary phase approximation, one obtains the equality

$$k_{\text{PNM}}\left(\vec{q}'',\vec{q}',t,E\left(t\right)\right) = e^{iE\tau}k_{\text{JBB}}\left(\vec{q}'',\vec{q}',t(E),E\right). \qquad (1.69)$$

On shell, the emergent time is determined through Eq. (1.10) uniquely by specifying the energy E and by imposing the boundary conditions and the classical equations of motion. Off shell however, Eq. (1.10) gives a different BB time for each history since an arbitrary history will lead, in general, to a very different value of τ_{BB} for a fixed energy. Because we sum over all histories, each contribution to the kernel will represent a different JBB clock leading to a kind of superposition of clocks. This superposition effectively integrates time out of the theory and leads to solutions of the quantum theory that are stationary states. From the path integral perspective, the mechanics responsible for this is very clear. The classical theory does have a unique notion of duration because, in the stationary phase approximation, there is only one time that gives an important contribution to the kernel: the JBB time. Decoherence suggests an alternative perspective for understanding this emergence. Gryb's treatment of JBB path integral suggests that, in the stationary phase approximation, the JBB time decoheres from the other components of the superposition making it a useful clock.

1.3 Time in Rovelli's Approach

Another very important and famous approach which faces the problem of time with a timeless formalism was developed by Rovelli in the context of the loop quantum gravity. In order to discuss the most relevant aspects of Rovelli's approach, we will follow the references [66–78].

Since the publication of the paper *Time in Quantum Gravity: An Hypothesis* in 1991 [66], Rovelli introduced the perspective that an idealized time is not defined at the fundamental level (at the Planck scale), namely that, in the quantum gravity regime, time should be simply forgotten. According to Rovelli's approach, the concept of an absolute time, as used in Hamiltonian mechanics as well as in Schrödinger quantum mechanics, is not relevant in a fundamental description of quantum gravity: it has to be replaced by arbitrary clock times in terms of which the dynamics may not be of the Schrödinger form. In the paper [66], Rovelli suggested this view of time in quantum gravity on the basis of two sets of considerations regarding, on one hand, general relativity and, on the other hand, quantum mechanics. In general relativity, the notion of an absolute time t is absent and is replaced by internal time variables T that represent physical clocks. Thus Rovelli emphasized that the choice of one of these clock times and its identification with the absolute time t of Hamiltonian mechanics is irrelevant and contrary to the principles of general relativity. On the other hand, Rovelli showed that quantum mechanics admits a natural extension, which can be defined as quantum mechanics without time, which can deal with systems in which an absolute Hamiltonian time t is not defined. On the basis of Rovelli's treatment in the paper [67], the time axiom can be eliminated from the Heisenberg picture without compromising the other axioms or the probabilistic interpretation of the theory. Rovelli's proposal implies then that the quantum physics of the presympletic systems that do not have a Hamiltonian version is governed by quantum mechanics without time. In the extension of quantum mechanics without time suggested by Rovelli, the Schrödinger equation, which in the Heisenberg picture is

$$\dot{\widehat{Q}} = i\hbar \left[\widehat{Q}, \widehat{H}\right], \qquad (1.70)$$

is replaced by the equations

$$\left[\widehat{Q}, \widehat{K}\right] = 0, \qquad (1.71)$$

where $\widehat{Q} = \widehat{Q}(T)$ are evolving constants that describe the evolution with respect to a physical clock variable and $\widehat{K}$ is the Hamiltonian constraint which satisfies the constraint equation

$$\widehat{K}\psi = 0, \qquad (1.72)$$

defining the physical state subspace. If the clock time happens to be a good Hamiltonian time, the basic Eq. (1.71) reduces to the Schrödinger equation for the evolving constants. Moreover, inside this picture Rovelli showed that unitary evolution and the Schrödinger equation may be obtained within an approximation.

Rovelli proposed therefore that quantum gravity is described by this quantum theory without time and that can be developed in this manner. The space H_{ph} of the solutions of the WDW equation is considered. A class of quantum operators that commute with the WDW constraints is constructed. Among these, there are observables that express the evolution with respect to a clock variable. The scalar product is then chosen in H_{ph} in such a way that the observables are self-adjoint. Rovelli emphasized also that this theory admits a choice of internal time T such that the evolution of the observables in T is given by a Schrödinger equation in T within a certain approximation.

On the basis of the treatment provided in the recent paper *Forget Time* [78], the main ideas and results of Rovelli's approach about time can be summarized as followed:

1. It is possible to formulate classical mechanics in a way in which the time variable is treated on equal footings with the other physical variables, and not singled out as the special-independent variable. This is the natural formalism for describing general relativistic systems.
2. It is possible to formulate quantum mechanics in the same manner. Moreover, Rovelli argues that this may be the effective formalism for quantum gravity.
3. The peculiar properties of the time variable are of thermodynamical origin, and can be captured by the thermal time hypothesis. Within quantum field theory, "time" is the Tomita flow of the statistical state ρ (expressing our incomplete knowledge of the physical system under consideration) in which the world happens to be, when described in terms of the macroscopic parameters we have chosen.
4. In order to build a quantum theory of gravity the most effective strategy is therefore to forget the notion of time all together, and to define a quantum theory capable of predicting the possible correlations between partial observables.

In [78], Rovelli's analysis puts in evidence that in a regime where quantum effects are negligible, all fundamental systems, including the general relativistic ones, can be described by making use of these concepts:

(i) The relativistic configuration space C, of the partial observables.
(ii) The relativistic phase space Γ of the relativistic states.
(iii) The evolution equation $f = 0$, where $f : \Gamma \times C \to V$ where V is a linear space.

The state in the phase space Γ is fixed until the system is disturbed. Each state in Γ determines (via $f = 0$) a motion of the system, namely a relation, or a set of relations, between the observables in C. Once the state is determined (or guessed), the evolution equation predicts all the possible events, namely all the allowed correlations between the observables, in any subsequent measurement. This language makes no reference to a special "time" variable. Here, as a consequence of what happens in general relativity, notions of evolution in time, observable at a fixed time, play no role.

As regards Hamiltonian mechanics, Rovelli showed that once the kinematics — that is, the space C of the partial observables q^a — is known, the dynamics — that is, Γ and f — is fully determined by giving a constraint surface Σ in the space Ω of the observables q^a and their momenta p^a. The constraint surface Σ is specified by $H = 0$ with a function $H : \Omega \to R^k$. By denoting $\tilde{\gamma}$ an unparametrized curve in Ω (observables and momenta) and γ its restriction to C (observables alone), the physical motion of the system under consideration is determined by the function H via the following:

Variational principle. A curve γ is a physical motion connecting the events q_1^a and q_2^a, if $\tilde{\gamma}$ extremizes the action

$$S[\tilde{\gamma}] = \int_{\tilde{\gamma}} p_a dq^a \qquad (1.73)$$

in the class of the curves $\tilde{\gamma}$ satisfying

$$H(q^a, p_a) = 0, \qquad (1.74)$$

whose restriction to C connects the events q_1^a and q_2^a.

No notion of time appears in this formulation suggested by Rovelli. The pair (C,H) where H is called the relativistic Hamiltonian, describes a relativistic dynamical system. All (relativistic and nonrelativistic) Hamiltonian systems can be formulated in this timeless formalism.

Moreover, taking account of Chiou's treatment in the paper *Timeless Path Integral for Relativistic Quantum Mechanics* [79], if one parametrizes the curve $\tilde{\gamma}$ with a parameter τ, the action (1.73) reads as

$$S[q^a, p_a, N_i] = \int d\tau \left(p_a(\tau) \frac{dq(\tau)}{d\tau} - N_i(\tau) H^i(q^a, p_a) \right), \qquad (1.75)$$

where the constraint (1.74) has been implemented with the Lagrange multipliers $N_i(\tau)$. Varying this action with respect to $N_i(\tau)$, $p_a(\tau)$

and $q^a(\tau)$ leads to the constraint equations

$$\frac{dq^a(\tau)}{d\tau} = N_j(\tau) \frac{\partial H^j(q^a, p_a)}{\partial p_a}, \quad (1.76)$$

$$\frac{dp_a(\tau)}{d\tau} = -N_j(\tau) \frac{\partial H^j(q^a, p_a)}{\partial q^a}. \quad (1.77)$$

For $k > 1$, a k-dimensional surface in C determines the motion of the system and different choices of the k arbitrary functions $N_j(\tau)$ determine different curves and parametrizations on the single surface that defines a motion. For $k = 1$, the motion of the system emerges as a 1D curve in C and different choices of $N_j(\tau)$ correspond to different parametrizations for the same curve. Different solutions of $q^a(\tau)$ and $p_a(\tau)$ for different choices of $N_j(\tau)$ provide gauge-equivalent representations of the same motion.

Along the solution curve, the change rate of H with respect to τ is given by

$$\frac{dH^i}{d\tau} = N_j \{H^i, H^j\}. \quad (1.78)$$

Here, the physical motion should remain on the constraint surface Σ, and thus must satisfy the condition

$$\{H^i, H^j\}|_\Sigma = 0 \quad (1.79)$$

for all i and all j, in order to be consistent.

Continuing the analysis of the paper [78], Rovelli then developed the formal structure of timeless quantum mechanics, a formulation of quantum mechanics (or a quantum version of relativistic classical mechanics) needed to describe systems where no preferred time variable is specified. This formulation of quantum theory is based on the following scheme:

(a) *Kinematical states.* Kinematical states are smooth functions $f(\alpha, t)$ on C and form a space S in a rigged Hilbert space $S \subset K \subset S'$ defined by the relativistic configuration space C, where $K = L^2[C, d\alpha dt]$ and S' is constituted by the tempered distributions over C.

(b) *Partial observables.* A partial observable is represented by a self-adjoint operator in K. The simultaneous eigenstates $|s\rangle$ of a complete set of commuting partial observables can be defined as quantum events.

(c) *Dynamics.* Dynamics is defined by a self-adjoint operator H in K, the (relativistic) Hamiltonian. The operator from S to S' defined as

$$P = \int d\tau\, e^{-i\tau H}, \tag{1.80}$$

can be called the projector. The transition amplitudes, which encode the entire physics of the dynamics, are represented by the matrix elements

$$W(s, s') = \langle s |\, P\, | s' \rangle. \tag{1.81}$$

(d) *Physical states.* A physical state is a solution of the quantum Hamiltonian constraint equation (namely the WDW equation)

$$\widehat{H}\psi = 0, \tag{1.82}$$

which is the quantum counterpart of Eq. (1.74).

(e) *Complete observables.* A complete observable A is represented by a self-adjoint operator on H. A self-adjoint operator A in K defines a complete observable if

$$[A, H] = 0. \tag{1.83}$$

(f) *Projection.* If the observable A takes value in the spectral interval I, the state ψ becomes then the state $P_I \psi$, where P_I is the spectral projector on the interval I. If an event corresponding to a sufficiently small region R is detected, the state becomes $|R\rangle$.

Some aspects of this scheme proposed by Rovelli for a timeless quantum theory were also analyzed in [79] by Chiou, who specified and clarified the following:

Measurements and collapse. If one performs a measurement of a partial observable $\hat{A}$, the outcome takes the value of one of the eigenvalues of $\hat{A}$ if the spectrum of $\hat{A}$ is discrete, or in a small spectral region (with uncertainty) if the spectrum is continuous. The measurement of a complete set of partial observables $\hat{A}_i$ simultaneously can be defined as a complete measurement at an "instance", and its outcome can be associated to a kinematical state $|\psi_\alpha\rangle)$ (which is a simultaneous eigenstate of $\hat{A}_i$ if the spectra of $\hat{A}_i$ are discrete). The physical state is said to be collapsed to $|\Psi_{\psi_\alpha}\rangle)$ by the complete measurement.

Prediction in terms of probability. If at one instance a complete measurement yields $|\psi_\alpha\rangle$, the probability that at another instance another complete measurement yields $|\psi_b\rangle$ is given by

$$P_{\beta\alpha} = \left| \frac{W[\psi_\beta, \psi_\alpha]}{\sqrt{W[\psi_\beta, \psi_\beta]}\sqrt{W[\psi_\alpha, \psi_\alpha]}} \right|^2, \qquad (1.84)$$

where

$$W[\psi_\beta, \psi_\alpha] = \int ds \int ds' \bar{\psi}_\beta(s) W(s, s') \psi_\alpha(s'). \qquad (1.85)$$

In particular, if the quantum events s form a discrete spectrum, the probability of the quantum event s given the quantum event s' is

$$P_{\beta\alpha} = \left| \frac{W(s, s')}{\sqrt{W(s, s)}\sqrt{W(s', s')}} \right|^2. \qquad (1.86)$$

If the spectrum is continuous, the probability of a quantum event in a small spectral region R given a quantum event in a small spectral region R' is

$$P_{RR'} = \left| \frac{W(R, R')}{\sqrt{W(R, R)}\sqrt{W(R', R')}} \right|^2, \qquad (1.87)$$

where

$$W(R, R') = \int_R ds \int_{R'} ds' W(s, s'). \qquad (1.88)$$

In order to develop a complete formulation of quantum gravity, as Rovelli emphasizes in [66], "the only way out that we see is to completely abandon the idea of absolute time", to assume that time is not a fundamental concept, which means that a theory of quantum gravity can be formulated and fully interpreted without the distinction of some flow of time. However, perception of a flow of time is an everyday-experience. Time plays an utmost crucial role in the way we experience and describe the world, and it appears in many well-established physical theories, wherein it seems to be an indispensable ingredient. So if time is accepted as a "real" component of the physical world, how can this be combined with the expected universal timelessness of Nature? How does time emerge to become such a characteristic part of our macroscopic world?

A solution to this problem was suggested by Rovelli in 1993 [76] for classical systems and then extended by Connes and Rovelli in 1994 [80] within the framework of quantum field theory (see also [23, 72, 81] for a review of this concept). It is constituted by the so-called thermal time hypothesis, according to which time is state-dependent and emerges on a thermodynamical scale. In Rovelli's and Connes' proposal, the familiar notion of time emerges at the thermodynamical level, when one gives an approximate statistical description of a system with a large number of degrees of freedom. In this picture, the physical time flow which seems to be perceived by human senses (and which seems to characterize the macroscopic world) is postulated to coincide with the equilibrium dynamics of the statistical state in which the system is found to be. In the thermal time hypothesis the key idea is to ascribe the time flow (the selection of a preferred flow) to statistical properties of the state. The relevant statistical distribution determines a flow which coincides with what we perceive as the physical flow of time. By measurements of macroscopic observables one can assign a density matrix to the system instead, which contains all the knowledge the observer gained about the system and permits a statistical treatment as the best description available. Following the reasoning of Connes and Rovelli it is this ignorance about the microscopic details of the system, expressed via the mixed state, which distinguishes a flow on the algebra of observables. This thermal time flow is selected in such a way that the state becomes a thermal equilibrium state relative to this flow. An equilibrium flow has a variety of physical properties, which is the main reason why this flow is postulated to be a physical time flow. The thermal time hyphotesis says that it is this flow which we perceive as the flow of time and which is best suitable parameter to evolve the system.

In order to provide a general description of the thermal time hypothesis, we base our treatment on Rovelli's recent paper [78]. Our incomplete knowledge of a system with a large number of degrees of freedom can be represented in terms of a statistical state ρ which is a positive function on the relativistic phase space Γ

$$\rho : \Gamma \to R^+, \tag{1.89}$$

which satisfies the normalization condition

$$\int_\Gamma ds \rho(s) = 1. \tag{1.90}$$

This function $\rho(s)$ represents the assumed probability density of the state s in Γ. The expectation value of any observable $A : \Gamma \to R$ in the state ρ is

$$\rho[A] = \int_\Gamma ds A(s) \rho(s). \tag{1.91}$$

According to the fundamental postulate of statistical mechanics, a system left free to thermalize reaches a time-independent equilibrium state that can be represented by means of the Gibbs statistical state

$$\rho_0(s) = N e^{-\beta H_0(s)}, \tag{1.92}$$

where $\beta = 1/T$ is a constant — the inverse temperature — and H_0 is the nonrelativistic Hamiltonian. Time evolution of the system is determined by the observable $A(t(s))$ where $s(t)$ is the Hamiltonian flow of H_0 on Γ. The correlation probability between $A(t(s))$ and B is given by

$$W_{AB}(t) = \int_\Sigma ds A(s(t)) B(s) e^{-\beta H_0(s)}. \tag{1.93}$$

While all mechanical predictions can be derived using the relativistic Hamiltonian H, which treats all variables on equal footing, Eqs. (1.92) and (1.93) of statistical mechanics single out the time as a special variable. While purely mechanical measurement cannot recognize the time variable, statistical or thermal measurements can.

According to Rovelli's approach, in a statistical context one has in principle an operational procedure for determining which one is the time variable. The operational procedure is the following: Measure ρ_0; compute H_0 from the following equation

$$H_0 = -\frac{1}{\beta}\ln\rho_0; \qquad (1.94)$$

compute the Hamiltonian flow $s(t)$ of H_0 on Σ: the time variable t is the parameter of this flow. The multiplicative constant in front of H_0 just provides the unit in which time is measured. On the basis of this unit, one can obtain which one is the time variable just by measuring ρ_0. This turns out to be in contrast with the purely mechanical context, which contains no operational procedure able to single out the time variable.

Then, considering a relativistic system where no partial observable is identified as the time variable, one finds that the statistical state describing the system is associated to a certain arbitrary state ρ. In this situation, one can define the quantity

$$H_p = -\ln\rho, \qquad (1.95)$$

which corresponds to a thermal time. The Hamiltonian defined by Eq. (1.95) may also be defined as thermal Hamiltonian. The thermal clock may be intended as any measuring device whose reading grows linearly with the Hamiltonian flow $s(t_p)$. Given an observable A, if one considers the one-parameter family of observables $A(t_p(s))$, it follows that the correlation probability between the observables $A(t_p(s))$ and B is

$$W_{AB}(t_p) = \int_\Sigma ds\, A(t_p(s))B(s)e^{-H_\rho(s)}. \qquad (1.96)$$

Here, one can remark that there is no difference between the physics described by Eqs. (1.92) and (1.93) and the one described by (1.95) and (1.96). Whatever the statistical state ρ is, there exists always a variable t_p, measured by the thermal clock, with respect to which the system is in equilibrium and physics is the same as in the conventional nonrelativistic statistical case.

This observation leads Rovelli to enunciate the following hypothesis:
The thermal time hypothesis. In nature, there is no preferred physical time variable t. There are no equilibrium states ρ_0 preferred *a priori*. Rather, all variables are equivalent; we can find the system in an arbitrary state ρ; if the system is in a state ρ, then a preferred variable is singled out by the state of the system. This variable is what we call time.

The main feature of the thermal time concept is therefore the following. According to Rovelli, the key observation is that given a (faithful normal) state

> "there exists always a [flow] [...], with respect to which the system is in equilibrium and whose physics is the same as in the conventional [...] statistical case" [23].

The idea is to select a preferred time flow in such a way that the resulting dynamical system is in thermal equilibrium. In that case the time evolution (expressed by the thermal time correlation functions), the equations of motion, etc. look the same as in well-known equilibrium situations described by ordinary time flows in nongenerally covariant quantum statistical mechanics. In other words, such a distinguished thermal time flow simply represents equilibrium dynamics. Consequently, the thermal time flow has all the properties that characterize an ordinary equilibrium time flow, it describes the same physics, and should thus be regarded as a physical time flow. Even more, because of its thermodynamical origin, thermal time is regarded as properly "flowing" and, thus, is supposed to explain and reproduce one of our most fundamental experiences with time.

The thermal time postulate proposes indeed a unifying perspective between the timelessness of a generally covariant quantum theory with the evidence of the flow of time. It acts on an intersection point which addresses various areas of physics. In particular, it must be emphasized that with the help of modular theory it is possible to define an intrinsic concept of time, which according to the cocycle Radon–Nikodým theorem is unique up to inner automorphisms. The canonical evolution associated with the algebra is the root of physical time. The deep mathematical results about the structure of von Neumann algebras were actually one main motivation for Connes and Rovelli to postulate the thermal time hypothesis, apart from the wish to develop a solution to the problem of time in quantum gravity. A state gives rise to completely different time flows when restricted to different members of the net of local algebras. The characteristics of the time evolution therefore may change substantially when probing the hypothesis on different regimes. Since the hidden symmetry characterizing the modular flow can generally be expected to open the door for highly unconventional flows (in comparison with an ordinary Hamiltonian flow in algebraic quantum field theory which is simply interpretable by its geometrical meaning), the thermal time hypothesis gives rise to an entirely new concept of time and, in the context of cosmological models, emerges as a promising candidate to explain the true nature of our time, which recovers familiar notions of time in important cases of physical interest. The fact that the time flow is chosen such that the system is always in equilibrium, according to Connes' and Rovelli's view, does not imply that evolution is frozen and one cannot detect any dynamical changes. What actually can be measured is the effect of fluctuations around the thermal state. Wightman and Streater showed that in conventional quantum field theories satisfying the Wightman axioms,[1] which have been proposed for charged fields of any spin, one can extract all physical

---

[1] In 1952, Garding and Wightman started to isolate those features of quantum field theory which could be stated in mathematically precise terms and to extract physically trustworthy postulates. This ended up in the Wightman axioms [82], which have been proposed for charged fields of any spin. They are the following:
(W1) The states of the theory are described by unit rays in a separable Hilbert space H, which carries a strongly continuous unitary representation $U(g)$ of the (universal covering of the) proper orthochronous Poincarè group satisfying the spectrum condition. Moreover, there is precisely one state Ω_0, the physical vacuum, which is invariant under $U(g)$ for all g belonging to the proper orthochronous Poincarè group.
(W2) The field ϕ is an operator valued tempered distribution over Minkowski space (the test function space is usually taken to be the Schwartz space). For all real-valued test functions f, the field operators $\phi[f]$ are essentially self-adjoint

information in terms of vacuum expectation values of products of field operators (n-point functions), namely by means of the single vacuum state Ω_0 [82]. In a (possibly generally covariant) algebraic quantum field theory in a state ω on the relevant von Neumann algebra R a nontrivial evolution in thermal time s whose flow is σ_s^ω can be expressed in terms of correlation functions of the form

$$F_{A,B}(s) = \omega(\sigma_s^\omega(B)A), \qquad (1.97)$$

for $A, B \in R$ and $(\sigma_s^\omega(B))|\Omega_\omega\rangle$ represents the amplitude of detecting if one prepares $A|\Omega_\omega\rangle$ (in the Gelfand–Naimark–Segal space). An admittedly speculative ansatz how a time asymmetry may be implemented in spite of the equilibrium dynamics generated by the thermal time can be found in [76].

Thus, according to Rovelli's thermal time approach, the ultimate entity that determines which variable is physical time is the statistical state, and not any *a priori* hypothetical "flow" that guides the system to a preferred statistical state. All variables are physically equivalent at the mechanical level. But if we restrict our observations to macroscopic parameters, and assume other dynamical variables are distributed according to a statistical state ρ, then a preferred variable emerges which has precisely the properties which characterize our macroscopic time parameter. When we say that a certain variable is "the time", we are not making a statement regarding the fundamental mechanical structure of reality. Rather, we are making a statement about the statistical distribution we use to describe the macroscopic properties of the system that we describe macroscopically. The "thermal time hypothesis" is the idea that what we call "time" is the thermal time of the statistical state in which the world happens to be, when described in terms of the macroscopic parameters we have chosen. To summarize, according to Rovelli's

---

operators f acting on a common dense domain D in H, which contains Ω_0 and is invariant under the action of $U(g)$ and $\phi[f]$.
(W3) Under the representation of the Poincarè group the field transforms as $U(g)\phi[f]U(g)^* = \phi[f_g]$; $f_g(x) = f(\Lambda^{-1}(x-a))$, for all $g(\Lambda, a)$ belonging to the proper orthochronous Poincarè group.
(W4) If the supports of f_1 and f_2 are space-like separated, $\phi[f_1]$ and $\phi[f_2]$ commute as self-adjoint operators.

view, time is the expression of our ignorance of the full microstate, a reflex of our incomplete knowledge of the state of the world.

Rovelli's thermal time hypothesis acts well in a number of cases. For example, if one considers the radiation field covariant cosmological model, with no preferred time variable and writes a statistical state representing the cosmological background radiation, then the thermal time of this state turns out to be precisely the Friedmann time [71]. Furthermore, as was underlined by Rovelli in [78] this hypothesis can be extended in a very direct way to the quantum context, and even more naturally to the quantum field theoretical context, where it leads also to a general abstract state-independent notion of time flow. In quantum mechanics, the time flow is given by

$$\alpha_t(A) = e^{itH_0} A e^{-itH_0}. \tag{1.98}$$

A statistical state is described by a density matrix ρ. The relation between a quantum Gibbs state ρ_0 and H_0 is same as in Eq. (1.85) namely

$$\rho_0(s) = Ne^{-\beta H_0}. \tag{1.99}$$

Correlation probabilities can be written as

$$W_{AB}(t) = Tr \lfloor e^{itH_0} A e^{-itH_0} B e^{-\beta H_0} \rfloor. \tag{1.100}$$

Given the generic state ρ the thermal Hamiltonian is

$$H_p = -\ln \rho, \tag{1.101}$$

and the thermal time flow is defined by relation

$$\alpha_{t_p}(A) = e^{it_p H_p} A e^{-it_p H_p}, \tag{1.102}$$

ρ being a KMS (Kubo–Martin–Schwinger) state, namely a state which satisfies

$$\rho \lfloor \alpha_t(A) B \rfloor = \rho \lfloor \alpha_{-it-i\beta}(B) A \rfloor \tag{1.103}$$

with respect to the thermal time flow it defines.

In the quantum field theory context, although Eq. (1.101) cannot be used and Eq. (1.99) makes no sense (because finite temperature states do not live in the same Hilbert space as the zero temperature states, and H_0 is a divergent operator on these states), in virtue of the Tomita theorem Gibbs states can still be characterized by Eq. (1.103): given any state ρ over a von Neumann algebra, there is always a flow α_t, called the Tomita flow of ρ such that Eq. (1.103) holds. This theorem allows us to extend Eq. (1.95), and thus the consequent thermal time hypothesis, also to quantum field theory. The thermal time flow α_{t_p} is defined in general as the Tomita flow of the statistical state ρ. Thus, in quantum field theory, what we call "the flow of time" corresponds simply to the Tomita flow of the statistical state ρ in which the world happens to be, when it is described in terms of the macroscopic parameters we have chosen. In the absence of a preferred time, any statistical state selects its own notion of statistical time. This determines a state-independent notion of time flow, which shows that a general covariant quantum field theory has an intrinsic dynamics, even in the absence of a Hamiltonian and of a time variable.

Therefore, according to Rovelli's approach, in a generally covariant quantum theory thermal time seems to be the fundamental notion of time available. However, if one takes into consideration, apart from the thermal time flow, another meaningful notion of time, such as a geometrical time flow with geometrical realization on the space-time manifold (as it is e.g. determined by a representation of a suitable subgroup of the Poincarè group), geometrical time and thermal time have to be related. The issue of comparing these two different flows, which was already addressed by Connes and Rovelli, was faced by Martinetti and Rovelli in 2003 [81]. In Martinetti's and Rovelli's proposal, when a physical time t is measured in a nongenerally covariant system being in an equilibrium state at inverse temperature β by a clock which measures just a rescaled physical time $\tilde{t} \propto t$, the state formally looks like an equilibrium state at the rescaled temperature. This observation was taken up by Martinetti and Rovelli and supplemented in the sense that also time-dependent rescalings should be permissible on the basis of the following "improved thermal time hypothesis":

Thermal Time Hypothesis — Addendum [81].

Consider a system in a state ω which admits a thermal time flow σ_s^ω as well as a geometrical time flow α_t. If the modular flow has a geometrical meaning and either flows are proportional to each other, where the proportionality constant may vary along the flow, a local notion of temperature is interpreted to be the ratio of the two flows, namely $\beta(s) = \frac{dt(s)}{ds} = \frac{||\partial_s||}{||\partial_t||}$, ∂_s and ∂_t being the corresponding tangent vector fields.

Provided that both times are just two different parametrizations of the same curve, when time is measured in geometrical time, the temperature is linked to its deviation from thermal time.

On the other hand, in the recent paper *Thermal time and Tolman–Ehrenfest effect: "temperature as the speed of time"*, Rovelli and Smerlak studied the notion of thermal time in the restricted context of stationary space-times, showing that the Tolman–Ehrenfest effect (namely the fact that, in the presence of gravity, temperature is not constant in space at thermal equilibrium) can be derived very simply by applying the equivalence principle to a key property of thermal time: at equilibrium, temperature is the rate of thermal time with respect to proper time — the 'speed of (thermal) time', in other words a sort of 'geometrical time' [83]. This entails in particular that the geometric properties of the thermal time flow of an equilibrium state remain valid in a curved space-time. In particular, according to Rovelli's and Smerlak's results, in a stationary space-time, the Tolman–Ehrenfest effect follows immediately from the following two properties of the thermal time: the fact that the thermal time flow generates a Killing symmetry of the metric, and the fact that the ratio between the flow of thermal time and the flow of proper time — the 'speed of (thermal) time' — is the temperature. The key feature of thermal time used in Rovelli's and Smerlak's derivation is the fact that the ratio between the flow of thermal time and the flow of proper time is the temperature. Although this fact appears somewhat tautological in flat space, it becomes consequential in a curved space-time, where the norm of the Killing field, and hence the thermal time flow, varies from point to point. While the global scale of thermal time remains arbitrary, the ratio between its flow in different space-time points (with respect to proper time) is physically meaningful. In the light of Rovelli's and Smerlak's results, the Tolman–Ehrenfest law can thus be stated very simply as: the stronger the gravitational potential, the faster the thermal time flow with respect to proper time, and hence the higher the temperature.

Finally, as regards the matter of the phenomenological arrows of time, it is important to mention that in the recent paper *Why do we remember the past and not the future? The 'time oriented coarse graining hypothesis'*, Rovelli suggested that the entropy of a system depends on the determination of a set of macroscopic observables, namely on a "coarse-graining" of the microscopic state space of the system, and thus past low-entropy depends on the coarse-graining implicit in our definition of entropy [84]. Rovelli's conjecture states that if the system is sufficiently complex and ergodic, for most paths $s(t)$ that satisfy the dynamics and for each choice of a direction in time, there is a family of observables A_n such that the entropy satisfies relation

$$\frac{dS_{s,A_n}}{dt} \geq 0, \tag{1.104}$$

namely any motion appears to have initial low entropy (and then nondecreasing entropy) under some coarse graining. If the conjecture holds, then we have a new way for facing the puzzle of the arrow of time: the universe was not in any special state in the past; rather, the universe is sufficiently rich to admit a description in terms of coarse grained observables with respect to which entropy increases. In the quantum regime, Rovelli's view of time oriented coarse graining implies that in any sufficiently rich quantum system (i.e. with a sufficient complex algebra of observables), given a generic state evolving in time as the wave function $\psi(t)$, there is a split of the system into subsystems such that the von Neumann entropy for the density operator ρ, measuring the average information the experimenter obtains in repeated observations of many copies of an identically prepared mixed state (given by

$$S_n(\rho) = \sum_{x=1}^{N} \lambda_x \ln \lambda_x, \tag{1.105}$$

where λ_x are the eigenvalues of the density matrix associated to the system), is low at initial time and increases in time. As a consequence, for any time evolution $\psi(t)$ there is a split of the system into subsystems such that the initial state has zero entropy. In this context, growing and decreasing of (entanglement) entropy is an issue about how the universe is split into subsystems, not a feature of the overall state of things. Entropic peculiarities of the past state of the universe should not be searched in the cosmos at large, but should be searched in the spilt, and this occurs also for the macroscopic observables, relevant for us. Time asymmetry, and therefore "time flow", might be a feature of a subsystem to which we belong, features needed for information gathering creatures like us to exist, and not features of the universe at large.

1.4 Some Relevant Recent Approaches of Timeless Description of Physical Events

Today, the idea that physical time cannot be considered as a primary physical reality is receiving more and more attention. For example, in the 1996's paper *Killing time* Woodward argues that Mach's principle leads to the conclusion that time, as we normally treat it in our common experience and physical theory, is not a part of fundamental reality [85]. As it was shown by Woodward, if inertia is relative and gravitationally induced, Mach's principle kills our conventional conception of time, makes difficult to avoid killing time. Also in the context of quantum mechanics, on the basis of the results of Woodward, the immediate interpretation of Mach's principle determines the weird consequences regarding timelessness: the price to pay in order to have clarity of understanding in quantum mechanics is to accept a radical timelessness.

In this chapter, we focus our attention on the following significant current approaches regarding a timeless description of physics: Chiou's timeless path integral approach for relativistic quantum mechanics; Palmer's view of a fundamental level of physical reality based on an Invariant Set Postulate; Elze's approach of time; Girelli's, Liberati's and Sindoni's toy model of a nondynamical timeless space as fundamental background of physical events; Caticha's approach of entropic time and of physical time; and, finally, Prati's model of physical clock time.

1.4.1 Chiou's timeless path integral approach for relativistic quantum mechanics

In the paper *Timeless path integral for relativistic quantum mechanics* [79], Chiou derives a timeless path integral approach for relativistic quantum mechanics by starting from the canonical formulation (Hilbert spaces and self-adjoint operators) described by Rovelli in [76]. Unlike the conventional path integral in which every path is parameterized by the time variable t, Chiou's path integral approach turns out to be completely independent of the parametrization for paths, manifesting the timeless feature. Furthermore, if the Hamiltonian constraint is a quadratic polynomial in the momenta p_a, the timeless path integral over the constraint surface Σ (specified by the relativistic Hamiltonian $H(q^a, p_a) = 0$ for the configurations variables q^a and their coniugate momenta p_a) reduces to the timeless Feynman's path integral over the (relativistic) configuration space.

By following Chiou's treatment in [79], all information of the quantum dynamics is encoded by the following transition amplitudes

$$W(q^a, q'^a) \approx \int_{-\infty}^{+\infty} d\tau \langle q^a | e^{-i\tau \hat{H}} | q'^a \rangle, \qquad (1.106)$$

where $\langle q^a | e^{-i\tau \hat{H}} | \rangle q'^a \rangle$ can be thought as the transition amplitude for a kinematical state $|q'^a\rangle$ to "evolve" to the state $|q^a\rangle$ by the "parameter time" τ (and $\sim$ is used to denote the equality up to an overall constant factor which has no physical significance). Equation (1.106) sums over $\langle q^a | e^{-i\tau \hat{H}} | q'^a \rangle$ for all possible values of τ, suggesting that $W(q^a, q'^a)$ is intrinsically timeless as the parameter time τ has no essential physical significance. For a fixed τ, if one introduces a parametrization sequence of the form $\tau_0 = 0, \tau_1, \ldots, \tau_{N-1}, \tau_N = \tau$ and identifying $q^a = q_N^a$ and $q'^a = q_0^a$, and uses $\sum_{n=1}^{N} \Delta \tau_n = \tau$ (where

$\Delta \tau_n = \tau_n - \tau_{n-1}$), one obtains

$$\langle q^a|e^{-i\tau \hat{H}}|q'^a\rangle = \left(\prod_{n=1}^{N-1}\int d^d q_n^a\right) \langle q_N^a|e^{-i\Delta\tau_N \hat{H}}|q_{N-1}^a\rangle$$
$$\times \langle q_{N-1}^a|e^{-i\Delta\tau_{N-1}\hat{H}}|q_{N-2}^a\rangle \cdots \langle q_1^a|e^{-i\Delta\tau_1 \hat{H}}|q_0^a\rangle. \tag{1.107}$$

For a given arbitrary small number ε, by increasing N, the parameter sequence can be always made fine enough such that $\{\tau_i\} \leq |\tau|/N \leq \varepsilon$. Consequently, one obtains

$$\langle q_{n+1}^a|e^{-i\Delta\tau_{n+1}\hat{H}}|q_n^a\rangle = \langle q_{n+1}^a|1 - i\Delta\tau_{n+1}\hat{H}(\hat{q}^a,\hat{p}_a)|q_n^a\rangle + O(\varepsilon^2). \tag{1.108}$$

For the generic case that the Hamiltonian operator $\hat{H}$ is a polynomial of $\hat{q}^a$ and $\hat{p}^a$ and is Weyl ordered, one has

$$\langle q^a|\hat{H}(\hat{q}^a,\hat{p}_a)|q'^a\rangle = \int \frac{d^d p_a}{(2\pi\hbar)^d}\exp\left[\frac{i}{\hbar}p_a(q^a-q'^a)\right]H\left(\frac{q^a+q'^a}{2},p_a\right). \tag{1.109}$$

Applying (1.109) to (1.108) one obtains

$$\langle q_{n+1}^a|e^{-i\Delta\tau_{n+1}\hat{H}}|q_n^a\rangle = \int \frac{d^d p_{na}}{(2\pi\hbar)^d} e^{ip_{na}\Delta q_n^a/\hbar} e^{-i\Delta\tau_{n+1}H(\bar{q}_n^a,p_{na})} + O(\varepsilon^2), \tag{1.110}$$

where $\bar{q}_n^a = \frac{q_{n+1}^a+q_n^a}{2}$ and $\Delta q_n^a = q_{n+1}^a - q_n^a$.

Making the parametrization sequence finer and finer (by decreasing ε or equivalently by increasing N) and at the end going to the limit $\varepsilon \to 0$ or $N \to \infty$, one can express (1.107) as

$$\langle q^a|\hat{H}(\hat{q}^a,\hat{p}_a)|q'^a\rangle = \lim_{N\to\infty}\left(\prod_{n=1}^{N-1}\int d^d q_n^a\right)\left(\prod_{n=1}^{N-1}\int \frac{d^d p_{na}}{(2\pi\hbar)^d}\right)$$
$$\times \exp\left(\frac{i}{\hbar}\sum_{n=0}^{N-1}p_{na}\Delta q_n^a\right)$$
$$\times \exp\left(-i\sum_{n=0}^{N-1}\Delta\tau_{n+1}H(\bar{q}_n^a,p_{na})\right). \tag{1.111}$$

In the limit $N \to \infty$ the points q_n and p_n can be considered as the sampled points of a continuous curve in $\Omega = T^*C$ given by $\tilde{\gamma}(\tau') = (q^a(\tau'), p^a(\tau'))$ which is parametrized by τ' and with the endpoints projected to C fixed by $q^a(\tau' = 0) = q'^a$ and $q^a(\tau' = \tau) = q^a$. Now, introducing the special notations for path integrals,

$$\prod_{n=1}^{N-1} \int d^d q_n^a \to \int Dq^a, \tag{1.112}$$

$$\prod_{n=1}^{N-1} \int \frac{d^d p_{na}}{(2\pi\hbar)^d} \to \int Dp^a, \tag{1.113}$$

and taking into account that for $N \to \infty$ the finite sums appearing in the exponents in (1.111) also converge to the integrals

$$\frac{i}{\hbar} \sum_{n=0}^{N-1} p_{na} \Delta q_n^a \to \frac{i}{\hbar} \int_{\tilde{\gamma}} p_a dq^a, \tag{1.114}$$

$$-i \sum_{n=0}^{N-1} \Delta \tau_{n+1} H(p_{na}, \bar{q}_n^a) \to -i \int_{\tilde{\gamma}} H(q^a(\tau'), p_a(\tau')) d\tau', \tag{1.115}$$

Eq. (1.111) can be expressed as

$$\langle q^a | \hat{H}(\hat{q}^a, \hat{p}_a) | q'^a \rangle = \int Dq^a \int Dp^a \exp\left(\frac{i}{\hbar} \int_{\tilde{\gamma}} p_a dq^a\right)$$
$$\times \exp\left(-i \int_{\tilde{\gamma}} H(q^a(\tau'), p_a(\tau')) d\tau'\right). \tag{1.116}$$

Here, one may notice that (up to the factor $i/\hbar$) the continuous limit in (1.114) is simply the line integral of the one-form $p_a dq^a$ over the curve $\tilde{\gamma}$, identical to Eq. (1.73) of Rovelli's approach, and is independent of the parametrization of τ. On the other hand, the integral in (1.115) depends on the parametrization of τ. Thus, in order to compute $W(q^a, q'^a)$ in Eq. (1.106), the integration over τ only hits the second exponential in (1.116) and the first exponential simply factors out. The integration of the second exponential over τ

leads to relation

$$\int_{-\infty}^{+\infty} d\tau \exp\left(-i \int_{\tilde{\gamma}} H(q^a(\tau'), p_a(\tau'))d\tau'\right)$$
$$= \delta\left(\int_{\tilde{\gamma}} H(q^a(\bar{\tau}), p_a(\bar{\tau}))d\bar{\tau}\right), \tag{1.117}$$

where $\bar{\tau} = \tau'/\tau$. The algebra associated with Eq. (1.117) implies that only the paths restricted to the constraint surface (namely, $\tilde{\gamma} \in \Sigma$, or equivalently $H(\tau') = 0$ for all τ' along the path) contribute to the path integral for $W(q^a, q'^a)$. With the position $\Delta\tau_n = \hbar^{-1}N_n\Delta\tau_n'$, by summing over different parametrizations $W(q^a, q'^a)$ can be expressed as

$$W(q^a, q'^a) \approx \int Dq^a \int Dp_a \int DN \exp\left(\frac{i}{\hbar}\int_{\tilde{\gamma}} p_a dq^a\right)$$
$$\times \exp\left(-\frac{i}{\hbar}\sum_{n=0}^{N-1}\Delta\tau'_{n+1}N_{n+1}H(\bar{q}^a_n, p_{na})\right), \tag{1.118}$$

where

$$\prod_{n=0}^{N-1}\int_{-\infty}^{+\infty}dN_{n+1} \to \int DN. \tag{1.119}$$

In the continuous limit, (1.118) can be written as the path integral

$$W(q^a, q'^a) \approx \int Dq^a \int Dp_a \int DN \exp$$
$$\times \left(\frac{i}{\hbar}\int_{\tilde{\gamma}}\left(p_a\frac{dq^a}{d\tau'} - N(\tau')H\right)d\tau'\right). \tag{1.120}$$

Here a procedure of integration over N can be carried out to obtain the delta functional:

$$\int DN \exp\left(\frac{i}{\hbar}\int_{\tilde{\gamma}} N(\tau')H d\tau'\right) \approx \delta[H] \equiv \lim_{N\to\infty}\prod_{n=0}^{N-1}\delta(H(\bar{q}^a_n, p_{na})), \tag{1.121}$$

and thus the path integral (1.120) can be written in an alternative form as

$$W(q^a, q'^a) \approx \int Dq^a \int Dp_a \delta[H] \exp\left[\frac{i}{\hbar} \int_{\tilde{\gamma}} p_a dq^a\right], \qquad (1.122)$$

where insertion of the delta functional $\delta[H]$ confines the path to be in the constraint surface (i.e. $\tilde{\gamma} \in \Sigma$). As regards the exponent in Eq. (1.122), it is interesting to remark that the phase appearing in it is identical to the classical action defined in (1.73) of Rovelli's approach (divided by $\hbar$); in analogous way, one may notice that the phase appearing in the exponent of Eq. (1.120) is identical to the classical action in (1.75) with $k = 1$. This means that each path in Σ contributes with a phase, which is the classical action divided by $\hbar$.

In the path integral formalism based on Eq. (1.120) (or the equivalent equation (1.122)) the transition amplitude $W(q^a, q'^a)$ is described as the sum, with the weight $\exp(iS/\hbar)$ (where S is the classical action of $\tilde{\gamma}$), over all arbitrary paths $\tilde{\gamma}$ which are restricted to Σ and whose projection γ to C connect q'^a and q^a. The fundamental feature of Chiou's formulation based on Eq. (1.120) (or the equivalent equation (1.122)) lies in its timeless picture. In fact, none of q^a turns out to be restricted to be monotonic along the paths and the parametrization for the paths has no physical significance as can be seen in the expression of Eq. (1.122), which is completely geometrical and independent of parametrizations.

In the special case that the classical Hamiltonian is quadratic polynomial in p_a namely

$$H(q^a, p_a) = \sum_a \alpha_a p_a^2 + \sum_a \beta_a p_a q^a + \sum_a \gamma_a p_a + V(q^a) \qquad (1.123)$$

(α_a, β_a and γ_a being constant coefficients and $V(q^a)$ being the potential which depends only on q^a) and the corresponding Hamiltonian operator is Weyl ordered, the transition amplitude admits a path integral formalism over the configuration space, whereby the functional integration over N is modified as $\prod_{n=0}^{N-1} \int_{-\infty}^{+\infty} d\sqrt{N_{n+1}} = \int D\sqrt{N}$. One obtains in this way the timeless Feynman's path integral over the paths in the configuration space C given by the following relation

$$W(q^a, q'^a) \approx \int Dq^a \int D\sqrt{N} \exp \frac{i}{\hbar} \int_{\tilde{\gamma}} d\tau'$$
$$\times \left(\sum_a \frac{N}{4\alpha_a} \left[\frac{\dot{q}^a}{N} - \beta_a q^a - \gamma_a \right]^2 - NV(q^a) \right), \quad (1.124)$$

where the velocity $\dot{q}^a \equiv \frac{dq^a}{d\tau'}$ is the continuous limit of $\frac{\Delta q_n^a}{\Delta \tau'_{n+1}}$. The configuration space path integral (1.124) sums over all arbitrary paths $\gamma \in C$ whose endpoints are fixed at q'^a and q^a, and each path contributes with a phase, which is identical to the classical Lagrangian function (divided by $\hbar$). The functional variations on (1.124) with respect to $\sqrt{N}$ and q^a lead to the classical Hamiltonian constraint and equation of motion.

The formulation of timeless path integral suggested by Chiou in the paper [79] is intuitively appealing and advantageous in many aspects as it generalizes the action principle of relativistic classical mechanics by replacing the classical notion of a single trajectory with a sum over all possible paths. It is easy to see that the classical solution contributes most to the transition amplitude. Various approximation methods developed in (conventional) path integral approaches can be readily adapted to the timeless description. Moreover, the timeless path integral approach introduces a new perspective to explain how the conventional quantum mechanics emerges from relativistic quantum mechanics within a certain approximation and may provide new insights into the problem of time. Specifically, for strictly deparametrizable systems, relativistic quantum mechanics and conventional quantum mechanics are different at the level of kinematics but identical at the level of dynamics; for nonstrictly deparametrizable systems, on the other hand, they are different for both kinematics and dynamics. The formulation of timeless path integral can be directly extended for the dynamical systems with multiple constraints as given in the following equation

$$W(q^a, q'^a) \approx \int Dq^a \int Dp_a \prod_{i=1}^{k} \delta[H^i] \exp \left[\frac{i}{\hbar} \int_{\tilde{\gamma}} p_a dq^a \right] \quad (1.125)$$

if the constraint operators $\hat{H}^i$ commute. Equation (1.125) constitutes the direct generalization of Eqs. (1.120) and (1.122). In the path integral, each path in Σ contributes with a phase, which is the classical action given in (1.75) divided by $\hbar$. Functional variation on (1.125) with respect to N_i, q^a and p_a yields the classical equations (1.74), (1.76) and (1.77).

For the case that $\hat{H}^i$ do not commute but form a closed Lie algebra, one has

$$W(q^a, q'^a) \approx \int d\mu(\vec{\theta}) \langle q^a | e^{-i\vec{\theta}\cdot\vec{\hat{H}}} | q'^a \rangle, \qquad (1.126)$$

where θ^i are coordinates of the Lie group G generated by $\hat{H}^i$. Starting from (1.126) and following the similar techniques as the ones used by Chiou in [79], one has the possibility to derive the timeless path integral, with the difference that here the nontrivial Haar measure $d\mu$ is involved and the nontrivial topology of G has to be taken into account. The timeless path integral may instead provide a better conceptual framework to start with for constructing the quantum theory. There is finally the problem to extend Chiou's approach of timeless path integral formalism for the quantum field theory described by Rovelli in the reference [75] in order to provide new insights into the issues of the connection between loop quantum gravity/loop quantum cosmology and "spin foam models".

1.4.2 *Palmer's view of the invariant set postulate*

Relevant current research are challenged with the view that space-time is the fundamental arena of the universe. They point out that the mathematical model of space-time does not correspond to a physical reality, and propose a "state space" or a "timeless space" as a fundamental arena.

For example, in the recent paper *A New Geometric Framework for the Foundations of Quantum Theory and the Role Played by Gravity*, Palmer underlines that, since quantum theory is inherently blind to the existence of state-space geometries, attempts to formulate unified theories of physics within a conventional quantum-theoretic framework are misguided, and that a successful quantum theory of gravity should unify the causal non-Euclidean geometry of spacetime with the a-temporal fractal geometry of state space [86]. In this paper, Palmer introduces a new geometric law of physics about the nature of physical reality based on an Invariant Set Postulate. The Invariant Set Postulate conjectures that states of physical reality are defined by a noncomputable fractal geometry I, embedded in state space and invariant under the action of some subordinate causal dynamics D_I. In other words, the Invariant Set Postulate subordinates the notion of the differential equation and elevates as primitive, a dynamically invariant fractal geometry in the state space of the universe. This geometry is used to define the notion of physical reality — states of physical reality are precisely those on the invariant set. Palmer showed that this set postulate has profound implications for our understanding of quantum physics and the corresponding role of gravity. The postulate is motivated by two concepts that would not have been known to the founding fathers of quantum theory: the generic existence of invariant fractal subsets of state space for certain nonlinear dynamical systems, and the notion that the irreversible laws of thermodynamics are fundamental rather than phenomenological in describing the physics of extreme gravitational systems. The Invariant Set Postulate posits the existence of a fractionally dimensioned subset I of the state space of the physical world (namely the universe as a whole). I is an invariant set for some presumed-causal (namely relativistic) deterministic dynamical system D_I; points on I, called also "world states", remain on I under the action of D_I. World states of physical reality are those, and only those, lying precisely on I. It is important to underline that in Palmer's theory, the subset I of the state space is more primitive than the deterministic dynamical system D_I. Given I, $D_I(t)$ maps some point $p \in I$ a parameter distance t along a trajectory of I. Crucially, D_I is undefined at points $\notin I$: if states of physical reality necessarily lie on I, then points $p \notin I$ in state space are to be considered literally "unreal". For practically relevant theories (such as quantum theory), the intricate structure of I is unknown and these points of unreality cannot be ignored. The features of the fractal state-space geometry I allows a causal realistic framework for quantum physics, providing a geometric basis for the essential contextuality and the role of the abstract Hilbert space, a possible realistic perspective on the essential role of the complex numbers, quaternions and a new re-reading of the standard mysteries of quantum theory. Moreover, in Palmer's approach gravity is seen as a manifestation of the heterogeneity in the geometry of the dynamically invariant set of state space. As regards the key question of how to represent quantum-theoretic states in a mathematically consistent way for the points of unreality $p \notin I$, the Invariant Set Postulate provides support to the search for a timeless description of physics: by treating the geometry of the invariant set as primitive introduces a fundamentally timeless perspective into the formulation of basic physics.

1.4.3 Elze's approach of time

H. T. Elze recently developed an approach based on the observation that "time passes" when there is an observable change, which is localized with the observer. On the basis of the "timeless" reparametrization invariant model of a relativistic particle with two compactified extradimensions, Elze studied classical Hamiltonian systems in which the intrinsic proper time evolution parameter is related through a probability distribution to the physical time, which is assumed to be discrete. In the paper *Quantum mechanics and discrete time from timeless classical dynamics* [87], he presented an attempt to show that owed to the inaccessability of globally complete information on trajectories of the system, the evolution of remaining degrees of freedom appears as in a quantum mechanical model when described in relation to the discrete physical time. In Elze's approach, the emergent discrete time naturally leads to a sort of "stroboscopic" quantization of the system [88–90]: when a continous physical time is not available but a discrete one is — like reading an analog clock under a stroboscopic light — then states of the system which fall in between subsequent clock "ticks" cannot be resolved. Such unresolved states form equivalence classes which can be identified with primordial Hilbert space states [88, 89, 91–93]. The residual dynamics then leads the evolution of these states through discrete steps. Under opportune circumstances, this results in unitary quantum mechanical evolution. Quantum theory thus appears to originate from "timeless" classical dynamics, due to the lack of globally complete information [90].

Essentially, Elze's view of time starts from the idea that some of the degrees of freedom of the system are employed to trigger a localized "detector", which can be defined invariantly. It amounts to attributing to an observer the capability to count discrete events. Then, the detector counts present an observable measure of discrete physical time. In the paper [90], Elze showed that the fundamental entities for the notion of a physical time are incidents, namely observable unit changes, which are recorded, and from which invariant quantities characterizing the change of the evolving system can be derived. This approach invokes compactified extradimensions in which a particle moves in addition to its relativistic motion in Minkowski space.

By following Elze's treatment in [90], one considers a (5 + 1)-dimensional model of a "timeless" relativistic particle of rest mass m with the action

$$S = \int ds L, \qquad (1.127)$$

where the lagrangian is defined by

$$L = -\frac{1}{2}(\lambda^{-1}\dot{x}_\mu \dot{x}^\mu + \lambda m^2), \qquad (1.128)$$

where λ stands for an arbitrary "lapse" function of the evolution parameter s, $\dot{x}^\mu = \frac{dx^\mu}{ds}$, $\mu = 0, 1, \ldots, 5$, the metric is $g_{\mu\nu} = \text{diag}(1, -1, -1, -1)$, and the two spatial coordinates $x^{4,5}$ are toroidally compactified. Setting the variations of the action to zero, one obtains the following equations of motion

$$\frac{\delta S}{\delta \lambda} = \frac{1}{2}(\lambda^{-2}\dot{x}_\mu \dot{x}^\mu - m^2) = 0, \qquad (1.129)$$

$$\frac{\delta S}{\delta x_\mu} = \frac{d}{ds}(\lambda^{-1}\dot{x}^\mu) = 0, \qquad (1.129a)$$

which are solved by

$$x^\mu(s) = x_i^\mu - p^\mu \int_0^s ds' \lambda(s') \equiv x_i^\mu + p^\mu \tau(s), \qquad (1.130)$$

where the conserved (initial) momentum p^μ is constrained to be on-shell and x_i denotes the initial position. Here also the fictitious proper time (function) τ appears and the lapse function introduces a gauge degree of freedom into the dynamics, which is related to the reparametrization of the evolution parameter s. In fact, the action (1.127) is invariant under the following set of gauge transformations:

$$s \equiv f(s'), \quad x^\mu(s) \equiv x'^\mu(s'), \quad \lambda(s)\frac{ds}{ds'} \equiv \lambda'(s'). \qquad (1.131)$$

It can be shown that the corresponding infinitesimal tranformations actually generate the evolution of the system. As a consequence, on the basis of these equations, there is no time in systems where dynamics is pure gauge, i.e. one has the "problem of time". In order to obtain and reproduce the evolution which obviously takes place in such systems, one needs a gauge invariant construction of a suitable time, which replaces the fictitious proper time τ. In Elze's picture, moreover, such construction is based on quasi-local measurements, since global information (such as invariant path length) is generally not accessible to an observer in more realistic, typically nonlinear or higher dimensional theories. In Elze's model, the physical time is constructed on the basis of the assumption that an observer in (3 + 1)-dimensional Minkowski space can perform measurements on the full (5 + 1)-dimensional trajectories, only within a quasi-local window to the two extradimensions $x^{4,5}$ (which are assumed to be toroidally compactified). In particular, the observer records the incidents ("units of change") when the full trajectory hits an idealized detector which covers a small convex area element on the torus described by the coordinates $x^{4,5}$. The detector can be defined invariantly and amounts to attributing to an observer the capability to count discrete events. Thus, the detector counts present an observable measuring system of a discrete physical time. In fact, in Elze's model physical time is an emergent discrete quantity related to the increasing number of incidents measured by the reparametrization invariant incident number:

$$I = \int_{S_i}^{S_f} ds^I \lambda(s^I) D(x^4(s^I), x^5(s^I)), \qquad (1.132)$$

where λ stands for an arbitrary "lapse" function of the evolution parameter s, $x^{4,5}$ describe the trajectory of the particle in the extradimensions, the integral is taken over the interval which corresponds to a given invariant path $x_i^\mu \to x_f^\mu$, and the function D represents the detector features. The physical time t has been therefore obtained by counting suitably defined incidents, namely coincidences of points of the trajectory of the system with appropriate detectors [88, 89]. This physical time determines stochastic features in the behavior of the external relativistic particle motion and is characterized by a discreteness in the sense that it is given by a nonnegative integer multiple of some unit time T, $t \equiv nT$. Moreover, by following Elze's treatment in the papers [87, 94], the discreteness of the physical time t is embedded through the following normalized probability distribution P to the proper time τ of the equations of motion:

$$P(\tau;t) = \exp(-S(\tau - t)), \int d\tau P(\tau;t) = 1. \qquad (1.133)$$

Thus, each physical system can be separated into degrees of freedom which are employed in the construction of a physical "clock", leading to the values of the physical time t, and remaining degrees of freedom evolving in the proper time τ. Neglecting the interaction between both components, and the details of the clock in particular, Elze's model describes the relation between physical and proper time by a probability distribution. In this situation, the Hamiltonian constraint only applies to the remaining degrees of freedom, while generally the system will be constrained as a whole.

Introducing a functional description of classical mechanics, the primordial state:

$$\langle \tau, \pi_a | t; t_0 \rangle = \int d\tau' \int_H D\Phi \exp\left[i \int_{\tau'}^{\tau+t} d\tau'' L_J \right.$$
$$\left. - S(\tau' - t_0) + i\pi_a \varphi^a(\tau + t) \right] \qquad (1.134)$$

About Time as the Numerical Order of Material Changes   77

and its (complex conjugated) adjoint state emerge as the fundamental elements of Elze's description, where the subscript "H" on the functional integral indicates the presence of the Hamiltonian constraint; $\varphi^a \equiv (q^1,\ldots,q^n;p^1,\ldots,p^n)$, $a = 1,\ldots,2n$ denote the classical phase space variables characterizing the system (q, p being the usual coordinates and conjugate momenta); π_a arise from exponentiating δ-functions involving φ^a at a fixed time; t_0 is the initial time; and $D\Phi = D\varphi D\lambda Dc D\bar{c}$ indicates the functional integration over all fields which enter the effective Lagrangian

$$L_J = \lambda_a(\partial_\tau \varphi^a - \omega^{ab}\partial_b H) + i\bar{c}_a(\delta_a^b \partial_\tau - \omega^{ac}\partial_c \partial_b H)c^b + J_a \varphi^a,$$
(1.135)

where J_a denotes an external source, c^a, $\bar{c}^a$ are anticommuting Grassmann variables, λ_a is an auxiliary variable. All physical (time-dependent) observables of the classical system can be calculated by starting from generic states of this form. Considering observables which are function(al)s of the phase space variables φ^a, one can define and compute:

$$\langle O[\varphi]|t\rangle = \int d\tau P(\tau;t) O\left[-i\frac{\delta}{\delta J(\tau)}\right] C\left[-i\frac{\delta}{\delta J(\tau)}\right] \log Z[J]_{J=0}$$
$$= Z^{-1} \int d\tau \int d\pi P(\tau)\langle t|\tau,\pi\rangle O[-i\partial_\pi] C[-i\partial_\pi]\langle \tau,\pi|t\rangle$$
$$= \langle \Psi(t)|\hat{O}(\varphi)\hat{C}(\varphi)|\Psi(t)\rangle \qquad (1.136)$$

where $Z = Z[0]$ is the generating functional and C the projector representing the Hamiltonian constraint, all states refer to $J = 0$ and

$$\hat{O}[\varphi] \equiv O[\hat{\varphi}], \quad \hat{C}[\varphi] \equiv C[\hat{\varphi}], \quad \hat{\varphi} \equiv -i\partial_\pi \qquad (1.137)$$

in the "τ,π" representation. Thus, a classical observable is represented by the corresponding function(al) of a suitably defined momentum operator. Furthermore, its expectation value at physical time t is represented by the effective quantum mechanical expectation value of the corresponding operator with respect to the physical time-dependent state under consideration, which incorporates the weighted average over the proper times, according to the distribution P. The evaluation of expectation values involves an integration over the whole τ-parametrized "history" of the states.

Finally, Elze showed that the generic states at physical times t and $t+T$ can be related to each other, making use of the classical Liouville operator propagating phase space variables in the parameter τ. In this regard, taking the conserved Hamiltonian constraint into account, one obtains a discrete physical time Schrödinger equation:

$$\langle \tau', \pi' | \Psi(t+T) \rangle = \int d\tau P(\tau) \exp\lfloor -i(\tau' + T - \tau)\hat{H}_Q(\pi', -i\partial_{\pi'}) \rfloor$$
$$\times \langle \tau, \pi' | \Psi(t) \rangle \qquad (1.138)$$

with the emergent Hamilton operator:

$$\hat{H}_Q(\pi, -i\partial_\pi) \equiv -\pi \cdot \omega \cdot \frac{\partial}{\partial \varphi} H(\varphi)|_{\varphi=-i\partial_\pi}, \qquad (1.139)$$

where H is a given classical Hamiltonian, ω is the symplectic matrix. Equipped with the Hamiltonian $\hat{H}_Q$ together with the operator $\hat{C}$ projecting out the constraint subspace, one has the possibility to analyze the stationary state eigenvalue problem related to Eq. (1.138). Here, the unusual structure of the Hamiltonian $\hat{H}_Q$ determines a spectrum which, in various examples, turns out to be too rich and, in particular, includes negative energy states and, as a consequence, does not possess a stable groundstate. However, it can be shown that a regularization (by discretization) of the operators involved overcomes this problem of the groundstate.

In Elze's approach, the states are collections of classical trajectories, as defined in Eq. (1.134) and there is not the usual close correspondence, for example, via coherent states. On the other hand, in this way a necessary element of quantum nonlocality enters the description of the underlying reparametrization invariant classical system. As a consequence, the classical limit of emergent quantum theories cannot give back the underlying model. Instead, the emergent quantum model must be seen as coarse-grained large-scale description of an underlying deterministic, dissipative classical dynamics. It is to have intrinsically a classical limit in accordance with the familiar correspondence principle.

1.4.4 Girelli's, Liberati's and Sindoni's model of a timeless nondynamical space as fundamental background

Girelli, Liberati and Sindoni recently developed a toy model where the Lorentzian signature and a dynamical space-time emerge from a flat nondynamical Euclidean space, with no diffeomorphisms invariance built in. In this sense, this toy-model provides an example where time (from the geometric perspective) is not fundamental, but simply an emerging feature. In order to analyze Girelli's, Liberati's and Sindoni's, we follow the paper [95].

Girelli's, Liberati's and Sindoni's approach explores the possibility that Lorentzian dynamics can be obtained as a property of the equations associated with the perturbations around some solutions of the equations of motion. In more detail, Girelli's, Liberati's and Sindoni's model suggests that at the basis of the arena of the universe there is some type of "condensation", so that the condensate is described by a manifold R^4 equipped with the Euclidean metric $\delta^{\mu\nu}$. Both the condensate and the fundamental theory are timeless. The condensate is characterized by a set of scalar fields $\Psi_i(x_\mu)$, $i = 1, 2, 3$. Their emerging Lagrangian is invariant under the Euclidean Poincarè group ISO(4) and has thus the general shape

$$L = F(X_1, X_2, X_3) = f(X_1) + f(X_2) + f(X_3),$$
$$X_i = \delta^{\mu\nu}\partial_\mu\Psi_i\partial_\nu\Psi_i. \qquad (1.140)$$

The equations of motion for the fields $\Psi_i(x_\mu)$ are

$$\partial_\mu\left(\frac{\partial F}{\partial X_i}\partial^\mu\Psi_i\right) = 0 = \sum_j\left(\frac{\partial^2 F}{\partial X_i \partial X_j}(\partial^\mu X_j) + \frac{\partial F}{\partial X_i}\partial_\mu\partial^\mu\Psi_i\right). \qquad (1.141)$$

The fields $\Psi_i(x_\mu)$ can be expressed as $\Psi_i = \psi_i + \varphi_i$ where φ_i are the perturbations which contain both the gravitational and matter degrees of freedom and the functions ψ_i are classical solutions of

Eq. (1.141). The lagrangian for ψ_i is given by

$$F(\bar{X}_1, \bar{X}_2, \bar{X}_3) + \sum_j \frac{\partial F}{\partial X_j}(\bar{X})\delta X_j + \frac{1}{2}\sum_{jk} \frac{\partial^2 F}{\partial X_j \partial X_k}(\bar{X})\delta X_j \delta X_k$$

$$+ \frac{1}{6}\sum_{jkl} \frac{\partial^3 F}{\partial X_j \partial X_k \partial X_l}(\bar{X})\delta X_j \delta X_k \delta X_l, \tag{1.142}$$

where $\bar{X}_i = \delta^{\mu\nu}\partial_\mu \psi_i \partial_\nu \psi_i$ and $\delta X_i = 2\delta_\mu \psi_i \partial^\mu \psi_i \partial_\mu \varphi_i \partial^\mu \varphi_i$.

Different choices of the solutions ψ_i lead to different metrics

$$g_k^{\mu\nu} = \frac{df}{dX_k}(\bar{X}_k)\delta^{\mu\nu} + \frac{1}{2}\frac{d^2 f}{(dX_k)^2}(X_k)\partial^\mu \psi_k \partial^\nu \psi_k. \tag{1.143}$$

This approach can lead naturally to a multi-metric structure. If one considers the specific class of equations of motion for which $\psi_i = \alpha^\mu x_\mu + \beta$, the SO(4) symmetry leads to $\bar{\psi} = \alpha x_0 + \beta$ which shows that the choice of the coordinate is completely arbitrary. Hence the Lorentzian signature can be obtained for the condition $\frac{df}{dX}(\bar{X}) + \frac{\alpha^2}{2}\frac{d^2 f}{(dX)^2}(\bar{X}) < 0$, $\frac{df}{dX}(\bar{X}) > 0$ and in this case the lagrangian becomes

$$L_{\text{eff}} = \sum_i \eta^{\mu\nu}\partial_\mu \varphi_i \partial_\nu \varphi_i, \tag{1.144}$$

where $\eta^{\mu\nu}$ is the Minkowski metric. Moreover, Girelli, Liberati and Sindoni showed that by means of the change of variables

$$\begin{pmatrix} \varphi_1 \\ \varphi_2 \\ \varphi_3 \end{pmatrix} = \Phi \begin{pmatrix} \phi_1 \\ \phi_2 \\ \phi_3 \end{pmatrix}, \tag{1.145}$$

with $\Phi^2 = \sum_i \phi_i^2 = l^2$ where l is a length scale related to Planck scale, the lagrangian (1.144) becomes

$$\int dx^4 L_{\text{eff}}(\varphi_i)$$

$$\to \int dx^4 \left(l^2 \eta^{\mu\nu}\partial_\mu \Phi \partial_\nu \Phi + \sum_i \Phi^2 \eta^{\mu\nu}\partial_\mu \phi_i \partial_\nu \phi_i + \lambda(|\phi|^2 - l^2) \right), \tag{1.146}$$

where λ is a Lagrange multiplier encoding the constraint $|\phi|^2 = l^2$, and a dynamical space-time emerges from L_{eff}, which is characterized by the Einstein–Fokker equations

$$R = 2\pi G_N T, \quad (1.147)$$
$$C_{\alpha\beta\gamma\delta} = 0, \quad (1.148)$$

where

$$R = \frac{6}{l^2} T, \quad (1.149)$$
$$T(\phi_i) = g^{\mu\nu} T_{\mu\nu}(\phi_i) = -\Phi^2 \sum_i \eta^{\mu\nu} \partial_\mu \phi_i \partial_\nu \phi_i, \quad (1.150)$$
$$g_{\mu\nu} = \Phi^2(x)\eta_{\mu\nu} \quad (1.151)$$

(which show that the gravitational degree of freedom is encoded in the scalar field Φ whereas matter is encoded in the fields ϕ_i) and where G_N is proportional to l^{-2}. In this picture, the Riemannian nature of the underlying fundamental signature allows the fundamental signature to be introduced without affecting the relativity principle at high energy. In synthesis, in Girelli's, Liberati's and Sindoni's model, through the change of variables (1.145), the free Lagrangian for scalar fields become a nonlinear sigma model coupled to Nordström gravity. Here, the Lorentz and the diffeomorphisms symmetries are emerging symmetries and thus are only approximate symmetries. These symmetries are broken when one takes into consideration the third-order terms in Eq. (1.142).

Despite the difficulty of an experimental test in this sense, this approach introduces the possibility that observers living in the emergent space-time could foresee that time is actually not fundamental. For example, a first obvious way to check that time is not actually fundamental is to be able to measure a Lorentz symmetry breaking contribution in high-energy (in the form of a nondynamical ether field). Another possibility could be of analyzing the strong gravitating regime. Indeed Girelli's, Liberati's and Sindoni's derivation obviously is valid for small perturbations φ_i, and hence small Φ, implying that in this framework one would predict strong deviations from the weak field limit of the theory whenever the gravitational field becomes very large. It would be interesting to see how these deviations actually do appear. This toy-model describes Nordström gravity, which is clearly a "nonphysical" approach since it implies there is no bending of light. It would be extremely interesting to be able to derive in a similar way general relativity, even though this seems to be a very difficult task (for example, it would probably require to have a mechanism that selects only a background signature and not the whole Minkowski metric as in Girelli's, Liberati's and Sindoni's original model, given that only the former is allowed to be nondynamical in general relativity). In this case, the purpose would be to obtain the emergence of a theory characterized by spin-2 gravitons (while in Nordström theory the graviton is just a scalar), thus opening a door to a possible conflict with the so-called Weinberg–Witten theorem [96]. However, there are many ways in which such a theorem can be evaded (in this regard, the reader can find details, for example, in the reference [97]). In particular, one may guess that similar models based on mechanisms like the one considered by Girelli, Liberati and Sindoni in the paper [95] could lead to a Lagrangian which shows Lorentz and diffeomorphisms invariance only as approximate symmetries for the lowest order in the perturbative expansion (while the Weinberg–Witten theorem assumes exact Lorentz invariance).

The emergence of time in Girelli's, Liberati's and Sindoni's toy-model also depends very much on the choice of the solution ψ. In fact the specific choice of ψ might even seem a bit contrived. Here one cannot choose freely ψ and this implies that other ingredients are needed to fully understand how a time-like direction emerges in a natural way out of an Euclidean manifold. The apparition of time could be predicted in a clear way if one would be able to show that the solution ψ is preferred for some reason.

Despite the two weak points here mentioned (namely the fact that Nordström's gravity is not equal to general relativity and the problem linked with the choice of ψ), Girelli's, Liberati's and Sindoni's toy-model introduces the crucial perspective of showing that time and gravity are not fundamentyal but only emergent and of identifying at least one solution ψ such that a dynamical space-time together with matter do emerge from a timeless nondynamical space. To summarize, the toy model developed by Girelli, Liberati and Sindoni has the merit to show that at a fundamental level space is a timeless condensate, that time as humans perceive it is only an emerging feature and that different solutions of the equations of motion of the fields characterizing this condensate determine different metrics of the space-time background. If in a timeless background different metrics are possible and time represents only an emerging feature, this means that, at a fundamental level, time cannot be considered a primary physical reality, and thus that the duration of material change has no existence of its own.

1.4.5 Caticha's view of entropic time and physical time

In the paper *Entropic dynamics, time and quantum theory* [98], Caticha proposed an "entropic dynamics" in which quantum mechanics is derived by applying a method of the maximum entropy. In Caticha's model, the basic assumption is that in addition to the particles under study having position x, the world includes extra things described by variables y that can be influenced by the particles and whose entropy $S(x)$ depends on x. The uncertainty in the values of y is described by distributions $p(y|x)$ in a statistical manifold M. The Schrödinger equation follows from the coupled dynamics of the particles and the variables y: the entropy $S(x)$ determines the dynamics of the particles x and, at the same time, the particles x determine the evolution of $S(x)$. One important aspect of this approach lies in the privileged role ascribed to position over and above all other observables, in a similar way to what happens in Nelson's stochastic mechanics. Both Caticha's entropic dynamics and Nelson's stochastic mechanics allow quantum theory to be derived as a kind of nondissipative Brownian motion. However, differently from stochastic mechanics, which operates at the ontological level (its goal is to develop a realistic interpretation of quantum theory as arising from a deeper, possibly nonlocal, but essentially classical reality), Caticha's entropic dynamics operates almost completely at the epistemological level: this approach makes predictions on the basis of limited information. Another important difference is that while in stochastic mechanics the existence of a universal Brownian motion and the idea that the current velocity is the gradient of some scalar function are assumed as starting-points, in entropic dynamics they are derived and not merely postulated. Finally, while stochastic mechanics is somewhat closer in spirit to Smoluchowski's approach to the theory of Brownian motion which involves keeping track of the microscopic details of molecular collisions through a stochastic Langevin equation and then taking appropriate averages, Caticha's entropic dynamics is closer to the Einstein approach and focuses on those elements of information that turn out to be directly relevant for the prediction of macroscopic effects. Caticha's approach does not assume any underlying dynamics whether classical, deterministic, or stochastic: both quantum theory and its classical limit are derived as examples of entropic inference.

A fundamental result of Caticha's "entropic dynamics" is that time is seen as a device to keep track of change, to keep track of the accumulation of small changes. Caticha's model shows how a dynamics governed by entropy naturally leads to an "entropic" notion of time. This model allows to derive something one might identify as an "instant", to provide a sense in which these instants can be "ordered" and to introduce a convenient concept of "duration" measuring the separation between instants. Moreover, in this model, an intrinsic directionality — an evolution from past instants towards future instants — and thus an arrow of time are derived directly and automatically and do not have to be externally imposed. This set of concepts developed by Caticha constitutes what is called "entropic time", which is modeled as an ordered sequence of instants with the natural measure of duration chosen to simplify the description of motion. In his paper [98], Caticha shows that whether the entropic order, the inferred sequence of states of the particle-clock composite agrees with the order in "physical" time or does not turn out to be quite irrelevant. It is only the correlations among the particles and the clock that are observable and not their "absolute" order.

By following Caticha's treatment in the paper [98], an entropic time as a sequence of instants is constructed in the following way. In entropic dynamics change is given, at least for infinitesimally short steps, by the transition probability $P(x'|x)$ in equation

$$P(x'|x) \approx \frac{1}{Z(x)} \exp\left[-\frac{\alpha(x)}{2\sigma^2}\delta_{ab}(\Delta x^a - \Delta \bar{x}^a)(\Delta x^b - \Delta \bar{x}^b)\right], \tag{1.152}$$

where $Z(x)$ is a function which contains factors independent of x', $x'^a = x^a + \Delta x^a$, α is a Lagrange multiplier which satisfies equation

$$\frac{\partial}{\partial \alpha}\log\zeta(x,\alpha) = -\frac{1}{2}\Delta\bar{l}^2, \tag{1.153}$$

where

$$\Delta\bar{l}^2 \equiv \langle\gamma_{ab}\Delta x^a \Delta x^b\rangle, \tag{1.154}$$

with $\gamma_{ab} = \frac{\delta_{ab}}{\sigma^2}$ being the metric of the Euclidean configuration space for a single particle,

$$\zeta(x,\alpha) = \int dx' e^{S(x') - \frac{1}{2}\alpha(x)\Delta\bar{l}^2}, \tag{1.155}$$

$$S(x') = -\int dy\, p(y|x') \log \frac{p(y|x')}{q(y)}, \tag{1.156}$$

being the entropy of the probability distribution $p(y|x')$ relative to an underlying measure $q(y)$ of the extra variable y associated with the particle under consideration. For finite steps the relevant piece of information is that large changes take place only as the result of a continuous succession of many very small changes.

By considering the nth step, in general we will be uncertain about both its initial and final positions, x and x' and we have

$$P(x') = \int dx\, P(x'|x)P(x). \tag{1.157}$$

If $P(x)$ happens to be the probability of different values of x at a given instant of entropic time t, then $P(x')$ can be seen as the probability of values of x' at a "later" instant of entropic time $t' = t + \Delta t$. Accordingly, if one writes $P(x) = \rho(x,t)$ and $P(x') = \rho(x',t')$, Eq. (1.157) can be rewritten as

$$\rho(x',t') = \int dx\, P(x'|x)\rho(x,t). \tag{1.158}$$

In Caticha's approach, Eq. (1.158) is used to define what one means by an instant: if the distribution $\rho(x,t)$ refers to an "initial" instant, then the distribution $\rho(x',t')$ defines what one means by the "next" instant. Thus, on the basis of Eq. (1.158), entropic time can be constructed, step by step, as a succession of instants. Successive instants are connected through the transition probability $P(x|x')$. Specifying the interval of time Δt between successive instants implies of tuning the steps or, equivalently, the multiplier $\alpha(x,t)$. In order to have a model of time that, like Newtonian time, flows "equably" everywhere, that is, at the same rate at all places and times, Δt must be independent of x and this can be obtained if $\alpha(x,t)$ is chosen so that

$$\alpha(x,t) = \frac{\tau}{\Delta t} = \text{constant}, \qquad (1.159)$$

where τ is a constant introduced so that Δt has units of time. In Caticha's view, Eq. (1.159) provides a justification for any definition of duration as a mathematical quantity for the description of motion. In this picture, the transition probability in Eq. (1.152) becomes

$$P(x'|x) \approx \frac{1}{Z(x)} \exp\left[-\frac{\tau}{2\sigma^2 \Delta t} \delta_{ab} (\Delta x^a - \Delta \bar{x}^a)(\Delta x^b - \Delta \bar{x}^b)\right], \qquad (1.160)$$

which can be recognized as a standard Wiener process. Here, one can also write

$$\Delta x^a = b^a(x) \Delta t + \Delta \omega^a, \qquad (1.161)$$

where the drift velocity $b^a(x)$ and the fluctuation $\Delta \omega^a$ are

$$\langle \Delta x^a \rangle = b^a \Delta t \quad \text{with} \quad b^a(x) = \frac{\sigma^2}{\tau} \delta^{ab} \partial_b S(x), \qquad (1.162)$$

and

$$\langle \Delta \omega^a \rangle = 0 \quad \text{and} \quad \langle \Delta \omega^a \Delta \omega^b \rangle = \frac{\sigma^2}{\tau} \Delta t \delta^{ab}. \qquad (1.163)$$

The constant $\frac{\sigma^2}{2\tau}$ plays the role of the diffusion constant in Brownian motion. The picture based on Eqs. (1.160)–(1.163) shows a formal similarity to Nelson's stochastic mechanics.

In the picture proposed by Caticha in the paper [98], the choice of an interval between instants is a matter of convenience — one chooses a notion of duration that reflects the translational symmetry of the configuration space. The result of the view of time as mathematical entity introduced to keep track of how small changes accumulate is that the underlying dynamics is standard diffusion and thus that the distribution ρ evolves according to Fokker–Planck equation

$$\frac{\partial \rho}{\partial t} = -\frac{\partial}{\partial x^a}(b^a \rho) + \frac{\sigma^2}{2\tau}\nabla^2 \rho, \quad (1.164)$$

where $\nabla^2 = \delta^{ab}\frac{\partial^2}{\partial x^a \partial x^b}$. The Fokker–Planck equation (1.164) can be rewritten as a continuity equation

$$\frac{\partial \rho}{\partial t} = -\frac{\partial}{\partial x^a}(v^a \rho), \quad (1.165)$$

where the velocity of probability flow — called also current velocity — is

$$v^a = b^a - \frac{\sigma^2}{2\tau}\delta^{ab}\frac{\partial_a \rho}{\rho}. \quad (1.166)$$

One can also introduce the osmotic velocity given by relation

$$u^a \equiv -\frac{\sigma^2}{2\tau}\partial^a \log \rho. \quad (1.167)$$

Its interpretation follows from $v^a = b^a + u^a$. The drift b^a given by Eq. (1.162), represents the tendency of the probability ρ to flow up the entropy gradient while u^a represents the tendency to flow down the density gradient. The situation is analogous to Brownian motion where the drift velocity is the consequence of the gradient of an external potential, while u^a emerges as a response to the gradient of concentration or chemical potential, the so-called osmotic force. The osmotic contribution to the probability flow is the actual diffusion current given by

$$\rho u^a = -\frac{\sigma^2}{2\tau}\partial^a \rho, \quad (1.168)$$

which can be recognized as Fick's law, with a diffusion coefficient given by $\sigma^2/2\tau$. Moreover, from Eqs. (1.162) and (1.167) one obtains

$$v^a = \frac{\sigma^2}{\tau}\partial^a \phi, \quad (1.169)$$

where $\phi(x,t) = S(x) - \log \rho^{1/2}(x,t)$.

To sum up, in Caticha's view entropic time has the merit to provide an order of the inferential sequence of small changes. The following question becomes so natural: what has this order with the order regarding a presumably more fundamental "physical" time? If so, why does 'entropic time' deserve to be called 'time' at all? The answer is that the systems include, in addition to the particles of interest, also another system that might be called the "clock". The goal is to make inferences about correlations among the particles themselves and with the various states of the clock. In this regard, Caticha points out that it is only the correlations among the particles and the clock that are observable and not their "absolute" order. By considering for example the case of a single particle, from the probability of a single step, given by Eq. (1.152), one can calculate the probability of any given sequence of (short) steps. Since the path is an ordered sequence of events in an entropic sense, when two events lie on the same path one can assert that one is earlier than the other in an entropic sense. The actual path, however, is uncertain: how can one compare possible events along different paths? The clock provides a criterion that allows us to decide whether an event x' reached along one path is earlier or later than another event x'' reached along a different path. This is where the clock comes in. The role of the clock can be played, for example, by a sufficiently massive particle. This assures that the clock follows a deterministic classical trajectory $x_C = \bar{x}_C(t)$ satisfying the classical Hamilton–Jacobi equation and that it remains largely unaffected by the motion of the particle. This implies that, in this picture, by "the time is t" one will just mean that "the clock is in its state $x_C = \bar{x}_C(t)$". One can say that the possible event that the particle reached x' along one path is simultaneous with another possible event x'' reached along a different path when both are simultaneous with the same state $\bar{x}_C(t)$ of the clock: then we say that x' and x'' happen "at the same time t". This justifies the use of the distribution $\rho(x, t)$ as the definition of an instant of time. Finally, one has to remark that important justifications for the assumptions underlying entropic dynamics lie in experiment. The ordering scheme provided by the entropic time allows one to predict correlations. Since these predictions, which are given by the Schrödinger equation, turn out to be empirically successful one can conclude that nothing deeper or more "physical" than entropic time is required. A similar claim has been made by Barbour in the book [48] in his relational approach to time in the context of classical dynamics.

1.4.6 Prati's model of physical clock time : towards the view of time as numerical order of dynamics

A fascinating recent paper on the problem of time is Prati's paper *The nature of time: from a timeless Hamiltonian framework to clock time metrology* [99]. In this paper, Hamiltonian mechanics, both in the classical domain and in quantum field theory, is rigorously well defined without the concept of an absolute, idealized time. Prati found that, by restricting the attention to closed systems, the Hamiltonian is time independent and the action principle can be expressed in terms of only the conjugate variables without the concept of time. Prati's approach provides thus a timeless formalism for Hamiltonian dynamics able to yield a universal definition of parameter time on the basis of the only generalized coordinates change in phase space, together with the Hamiltonian invariance on trajectories, and a variational principle.

Prati's treatment in [99] starts with the derivation of a parameter time from the Maupertuis principle in a picture in which Hamilton equations of motion are expressed in a timeless framework by using a variational principle on asynchronous varied trajectories. The time-independent Hamiltonian $H(\vec{q},\vec{p})$ is a function of the generalized 3D coordinates $\vec{p}$ and $\vec{q}$. The independence of H from t reduces the degrees of freedom to $2n - 1$. Here one must assume that a set of trajectories in the coordinates space μ for which H is constant exists. One starts from a generic and arbitrary parametrization λ of the points of the trajectories which is defined so that $q_i = q_i(\lambda)$ and $p_i = p_i(\lambda)$ where all such functions belong to C^2 on the interval $[\lambda_A, \lambda_B] \in R$. The Hamiltonian $H(p,q)$ does not depend explicitly on λ. A variational principle can be imposed on the trajectory by considering a variation which normally allows Hamilton equation from the Maupertuis principle to be derived in the context of asynchronous varied trajectories in canonical formalism. A new parametrization σ of the generalized coordinates and of λ is defined, under the condition that $\frac{d\lambda}{d\sigma} \neq 0$ on $[\sigma_A, \sigma_B]$, which is strictly related to the stationarity of the Hamiltonian.

Now, the stationarity of the action

$$A = \int p_i dq_i, \qquad (1.170)$$

where the Einstein summation on the repeated indexes is adopted and $i = 1, 2, 3$, implies that the Maupertuis variational principle reads as

$$\delta A = \delta \int p_i dq_i = 0. \qquad (1.171)$$

Moreover, the variation of the trajectories associated with the stationarity of the action (1.170) leads to

$$d\sigma = \left(\frac{\partial H}{\partial p_i}\right)^{-1} dq_i = -\left(\frac{\partial H}{\partial q_i}\right)^{-1} dp_i, \qquad (1.172)$$

under the hypothesis that $\left(\frac{\partial H}{\partial p_i}\right) \neq 0$ and $\left(\frac{\partial H}{\partial q_i}\right) \neq 0$. Equation (1.172) is different from the Hamilton equations since σ does not represent the macroscopic metric time. On the contrary, it only provides the natural parameterization of the system corresponding to the energy conservation. Since the trajectory and the Hamiltonian are assumed to be known, in Eq. (1.172) the only free parameter is the parametrization σ (associated with the properties of the Hamiltonian along the trajectory) which can be thus identified with the local parameter time of dynamics. This parameter is not the observable quantity measured by clocks; it provides the measurement of the change of the system along the trajectory. On the basis of Eq. (1.172), the quantity $d\sigma$ measures the amount of change along the two generalized coordinates q_i and p_i when energy conservation holds.

Then, Prati goes on by extending the action principle to quantum field theory in the context of the presympletic formalism. In the extended presymplectic formalism, the variational principle reads:

$$\delta A[\gamma] = \delta \int \theta = 0. \quad (1.173)$$

Here, a Hamiltonian operator $H = \int d^3x \mathrm{H}$, where H is the Hamiltonian density, acts as a constraint for quantum field dynamics. The Hamiltonian operator H acts as a constraint for quantum field dynamics. In quantum field theory, such a constraint corresponds to being on the mass shell. The action, in terms of the quantum fields $\psi_i(x)$ and the conjugate coordinates $\pi_i(x)$, can be rewritten as:

$$A = \int d^3x \int d\psi_i \pi_i, \quad (1.174)$$

where the Einstein summation on the repeated indexes is adopted. The roman index spans on the space dimensions 1, 2 and 3. Here, if one replaces the points of the trajectories $f(q_i, p_i) = 0$ in quantum field theory by space configurations of the generalized field $Q = (\psi_i(\vec{x}), \pi_i(\vec{x}))$ in the generalized coordinate space μ_Q, time appears as the natural parameterization of change in μ_Q. In the classical case neighboring position and momentum states are associated with the parameter σ, while in quantum field theory σ labels the generalized field with support in R^3. Two arrays of fields variate the quantum fields and their conjugate fields, respectively. As in the classical case, the extremality of the action is obtained under the following conditions:

$$d\sigma = \left(\frac{\delta \mathrm{H}}{\delta \pi_i}\right)^{-1}_{\psi_i} d\psi_i(\vec{x}) = \left(\frac{\delta \mathrm{H}}{\delta \psi_i}\right)^{-1}_{\pi_i} d\pi_i(\vec{x}). \quad (1.175)$$

Parameter time can be defined as the rate of change of the fields ψ_i and their conjugate variables π_i via the factors $\left(\frac{\delta \mathrm{H}}{\delta \pi_i}\right)^{-1}_{\psi_i}$ and $\left(\frac{\delta \mathrm{H}}{\delta \psi_i}\right)^{-1}_{\pi_i}$, respectively. The parameter σ is a real number by construction. The parameterization of the field distribution is locally achieved by associating neighboring configurations with the parameter σ. The parameter time is therefore constructed in analogy with the classical case even in the microscopic limit when small variations of the fields are taken into consideration. Such a construction is compatible with all the canonical Hamiltonian theories and naturally introduces a parameter time in timeless models. Moreover, the parameter time is by construction defined for those particles which are on the mass shell.

In synthesis, in Prati's approach, by restricting the attention to closed systems the Hamiltonian is time independent and the action principle can be expressed in terms of only the conjugate variables (Maupertuis action principle) without the concept of time. The assumption of being on the mass shell, or equivalently that the stationarity of energy holds, along the trajectory in the phase space, provides a parameterization which gives the ratio of change of conjugate variables (generalized coordinates q_i and p_i or generalized quantum fields ψ_i and π_i). Since all the observables are expressed in terms of such variables, σ parameterizes the whole algebra of observables.

The parameter time σ has the property of providing a privileged parameterization suitable for describing dynamics, but it is not an observable quantity. In order to explain the macroscopic experience of time in complex systems, an observable quantity T is built by Prati which realizes an experimentally measurable discrete approximation of σ. As regards the clock time T, Prati showed that in a timeless Hamiltonian framework a physical system S, if complex enough, can be separated in a subsystem S2 whose dynamics is described, and another cyclic subsystem S1 which behaves as a clock. In Prati's model clock time (which can be also called physical time) measured by macroscopic clocks is a coarse grained discrete quantity which can be defined in a system S complex enough to contain a subsystem S1 cyclic in the phase space. The cyclic subsystem acts as a clock reference used for the operative definition of time. Defining the clock time T, measured for example by atomic clocks, corresponds to label simultaneous occurences in the phase space of two or more subsystems where one is identified as the clock. The clock corresponds to the cyclic subsystem and the macroscopic time measured by a subsystem is a function of the subsystem itself. It is a matter of the experimentalist to choose suitable cyclic subsystems (macroscopic clocks) in order to provide a good approximation of the parameter time σ.

An important result of Prati's research is that, as a consequence of the gauge invariance (which transforms one parametric time into another in a way that they are all equivalent) the complex system S can be separated in many ways in a part which constitutes the clock and the rest. But what does it mean, in physical terms, that a complex physical system S can be separated in many ways in a subsystem whose dynamics is described and another subsystem which behaves as a clock? According to the author of this book, this means clearly that the time provided by each subsystem which acts as a clock cannot be considered as an absolute quantity, that the ticking of each subsystem which acts as a clock provides a different reference system in order to describe the dynamics and therefore that time as an idealized quantity that flows on its own does not exist: only the ticking of each subsystem acting as a clock exists as physical reality. This implies, in other words, that each subsystem which acts as a clock provides only a description of the dynamics of the other subsystem and that this description is tightly linked to ticking of the clock-subsystem. By changing the separation inside S we obtain a different measuring system of the dynamics taking place in S. In synthesis, one can say that each clock-subsystem provides only a measuring reference system for the dynamics of the other subsystem and that this reference system is not absolute; one can say that each subsystem which acts as a clock provides only the numerical order of the dynamics of the other subsystem.

Taking account of Prati's model, one can assert that clocks provide only the numerical order of material changes in a timeless space as a consequence of the gauge invariance, and thus in the sense that they are associated with subsystems that cannot be defined in an absolute way. It is as a consequence of the gauge invariance that a clock can be obtained in many ways by constructing a cyclic subsystem starting from a complex system and that the measuring references given by these different cyclic subsystems are all permissible from the physical point of view. Moreover, according to Prati's results, the states of the cyclic subsystem S1, which describe the dynamics of the subsystem S2, provide a discrete approximation of the parameter time, and this means that the numerical order of material changes has a discrete spectrum.

More precisely, according to Prati's model one has a parameter time σ which has the property of providing a privileged parameterization suitable for describing dynamics (but is not an observable quantity), and an observable quantity T which realizes an experimentally measurable discrete approximation of σ. It is this clock time T which provides the numerical order of material changes of the subsystem S2. Defining the clock time T, measured for example by atomic clocks, corresponds to label simultaneous occurrences in the phase space of two or more subsystems where one is identified as the clock. It is a matter of the experimentalist to choose suitable cyclic subsystems (macroscopic clocks) in order to provide a good approximation of the parameter time σ. As demonstrated by Oskay and other authors in [100] at the present time the most advanced available clock technology is given by single ion atomic clocks based on Al+/Hg+ with a fractional uncertainty of about $1 - 2 \times 10^{17}$. Adopting Prati's model for the construction of clocks, it derives for example that Planck time scale is an extrapolation, an extension of the concept of clock time beyond its field of definition. Following the terminology of Kofler and Brukner [101], macrorealism (i.e. the property of a system of being in one or more macroscopically distinct states) and classical (or semiclassical) laws emerge from quantum physics under the restriction of coarse-grained measurements. The description of time evolution of a system is necessarily semiclassic because the observer is tracking time with a macroscopic system whose fluctuations dominate on the short time scale. Indeed, T is expected to fail as a good approximation of σ in the fast decoherence process which occurs during a measurement.

On the basis of Prati's model, the numerical order of material changes can be defined mathematically in the following way. For a given $\bar{\sigma}$, a state ψ of the system S consists of the tensor product of the state $\psi_1(\bar{\sigma}) \in H_1$ and the state $\psi_2(\bar{\sigma}) \in H_2$ where H_1 and H_2 are the Hilbert spaces of the subsystems S1 and S2, respectively. Given the interval (σ_A, σ_B), by introducing the set:

$$\Omega(\sigma_A, \sigma_B) = \{\psi_2(\sigma) \in H_2 / \sigma \in (\sigma_A, \sigma_B)\}, \tag{1.176}$$

which is a particular set of states associates with the interval (σ_A, σ_B), $\Omega(\sigma_A, \sigma_B) \subset H_2$, physical time as numerical order of the material motion of the system S2 can be defined mathematically as a counter function k_{AB} that provides the number of states $\psi_2(\sigma) \in \Omega$ of the subsystem S2 whose dynamics is studied that satisfies an appropriate initial condition (namely the origin of measurement) $\psi_1(\sigma) = \bar{\psi}_1$ of the subsystem acting as a clock. The origin σ_0 of the parameter time σ is associated with the arbitrary initial states $\bar{\psi}_1 = \psi_1(\sigma_0)$ and $\bar{\psi}_2 = \psi_2(\sigma_0)$. Macroscopic time duration T^{S_1} of the interval (σ_A, σ_B) measured by the cyclic subsystem S1 (and therefore the numerical order of material change to describe the dynamics of S2) is given by the counter function k_{AB} that provides the number of states $\psi_2(\sigma) \in \Omega$ so that $\psi_1(\sigma) = \bar{\psi}_1$:

$$T^{S_1} \equiv k_{AB}. \qquad (1.177)$$

1.5 Time as a Measuring System of the Numerical Order of Material Changes and its Perspectives of Re-reading of Barbour's Theory and of Rovelli's Approach

On the basis of the recent research analyzed in Sec. 1.4, one can say that at a fundamental level the background space of physics is timeless, that the duration of physical events measured by clocks has not a primary existence but is an emergent quantity. Palmer's view of a fundamental level of physical reality based on an Invariant Set Postulate, Elze's approach of time, Girelli's, Liberati's and Sindoni's toy model of a nondynamical timeless space as fundamental background, Caticha's approach of entropic time and Prati's model of physical clock time seem all to point towards the same view: that the duration of physical events is not a fundamental reality, that physical clock/time exists only as a parameter measuring the order of the dynamics of processes inside a timeless background.

Palmer's approach of the Invariant Set Postulate which is based on the idea that states of physical reality emerge from a primitive noncomputable fractal geometry, and, in analogous way, Girelli's, Liberati's and Sindoni's model based on a flat nondynamical

Euclidean space introduce a fundamentally timeless perspective into the characterization of the background of the universe. Palmer's model and Girelli's, Liberati's and Sindoni's approach indicate clearly that time cannot be considered as a primary, fundamental physical reality.

On the other hand, Elze's research, Caticha's view and Prati's approach seem to explain in a clear way in what sense the ultimate level of physical processes is a timeless background: time is an emerging quantity, a mathematical quantity which derives from more fundamental physical realities. Essentially, Elze's view of time starts from the idea that some of the degrees of freedom of the system are employed to trigger a localized "detector", which can be defined invariantly. The detector counts provide an observable measure of discrete physical time. As we have seen in Sec. 1.4.3, in Elze's model physical time is an emergent discrete quantity related to the increasing number of incidents, i.e. observable unit changes, which are measured by the reparametrization invariant incident number given by Eq. (1.132). The physical time can be therefore obtained by counting suitably defined incidents, i.e. coincidences of points of the trajectory of the system with appropriate detectors. Thus, in Elze's model, the fundamental entities which lead to the notion of a physical time are observable unit changes, which are recorded, and the reparametrization incident number provides a mathematical measure of these observable unit changes. If the reparametrization incident number provides a mathematical measure of the observable unit changes, this means, from the physical point of view, that time as an idealized quantity that flows on its own does not exist: only the coincidences of points of the trajectory of the system with appropriate detectors exist as physical reality. On the basis of Elze's approach, it is therefore legitimate to conclude that each clock provides only the numerical order of the dynamics of the system of interest.

In Caticha's approach, time is seen as a device to keep track of change, to keep track of the accumulation of small changes and emerges from an entropic dynamics based on the distribution of transition probability (1.160) which evolves according to

Fokker–Planck equation (1.164). More precisely, Caticha's entropic time can be constructed, step by step, as a succession of instants, which are connected through the transition probability (1.160), on the basis of equation (1.158), where the distribution $\rho(x,t)$ refers to an "initial" instant and the distribution $\rho(x',t')$ defines what one means by the "next" instant. As a consequence, on the basis of Caticha's results regarding the entropic time defined by Eq. (1.158), the description of the dynamics of the particles of interest can be obtained studying the correlations among the particles themselves and with the various states of another physical system, namely the clock. As regards the dynamics of the coupled system constituted by the particles under consideration and the clock, it is only the correlations among these particles and the clock that are observable and not their "absolute" order. This means, from the physical point of view that the time provided by each system which acts as a clock (which thus indicates the correlations among the particles of interest and the clock) cannot be considered as an absolute quantity, that the ticking of a clock provides only a reference system in order to describe the dynamics of the particles of interest and therefore that time as an idealized quantity that flows on its own does not exist: only the ticking of each subsystem acting as a clock exists as physical reality. In particular, in Caticha's approach, the time provided by the clocks — which provides a mathematical measure of the dynamics of the system under consideration — has entropic features and origins. In synthesis, one can say that, on the basis of Caticha's model, each clock provides only a measuring reference system for the dynamics of the system of interest and that this reference system is not absolute; one can say that each clock provides only the numerical order of the accumulation of small changes of the system of interest (which emerge from an entropic dynamics).

Finally, in Prati's model of physical clock/time inside a timeless Hamiltonian framework, a physical system S, if complex enough, can be separated in a subsystem S2 whose dynamics is described, and another cyclic subsystem S1 which behaves as a clock; moreover, as a consequence of the gauge invariance the complex system S can be separated in many ways in a part which constitutes the clock and the rest. This means, from the physical point of view, that the time provided by each subsystem which acts as a clock cannot be considered as an absolute quantity, that the ticking of each subsystem which acts as a clock provides a different reference system in order to describe the dynamics and therefore that time as an idealized quantity that flows on its own does not exist. As a consequence of the gauge invariance, by changing the separation inside S we obtain different measuring systems of the dynamics taking place in S. On the basis of Prati's model, one can say that each subsystem which acts as a clock provides only the numerical order of material changes and the numerical order of material changes is defined in a precise mathematical way through Eqs. (1.176) and (1.177).

In synthesis, according to Palmer's view of a fundamental level of physical reality based on an Invariant Set Postulate, Elze's approach of time, Girelli's, Liberati's and Sindoni's toy model of a nondynamical timeless space as fundamental background, Caticha's approach of entropic time and Prati's model of physical clock time, one can conclude that material changes do not run in a time intended as a primary, fundamental physical reality, but instead time is a mathematical quantity which emerges from a more fundamental timeless background. The view according to which clocks represent measuring systems of the numerical order of material changes can be considered the most direct and natural perspective and development of all these current research: it is an a-temporal description of motion in physics. Moreover, the numerical order of material changes can be defined mathematically by three equivalent equations: the reparametrization invariant incident number (1.132), the entropic succession of instants (1.158) and the number k_{AB} of states of the Hilbert space of the system of interest whose dynamics is described — and which belong to the interval (1.176) that satisfies an appropriate initial condition (namely the origin of measurement) $\psi_1(\sigma) = \bar{\psi}_1$ of the subsystem acting as a clock.

Now, as the author of this book and Sorli have shown in the recent paper *Perspectives of the numerical order of material changes in timeless approaches in physics*, the view of time as an emergent quantity measuring the numerical order of material changes defined by Eqs. (1.132), (1.158) and (1.176) introduces the possibility to provide a unifying re-reading of the results obtained in the context of the two fundamental timeless approaches existing in the literature, namely the JBB theory and Rovelli's theory of time [102]. In Sec. 1.5.1, we will analyze how the view according to which time is a mathematical quantity measuring the numerical order of material changes corresponding to Elze's reparametrization invariant incident number (1.132) allows us to provide a suggestive unifying reading of the JBB theory and of Rovelli's approach. Then, in Sec. 1.5.2, by following the treatment made in [102], we will see in what way the idea of time as a mathematical quantity measuring the numerical order of material changes corresponding to Prati's number k_{AB} of states of the Hilbert space of the system of interest whose dynamics is described — and which belong to the interval (1.176) that satisfies an appropriate initial condition $\psi_1(\sigma) = \bar{\psi}_1$ of the subsystem acting as a clock — allows us to provide a suggestive unifying re-reading of the JBB theory and of Rovelli's approach.

1.5.1 About the JBB theory and Rovelli's approach in the view of time as numerical order of material changes corresponding to Elze's reparametrization invariant incident number

As we have seen in Sec. 1.2, the JBB theory is defined by the action (1.8) which leads to the Newton-type equations of motion (1.9) linked with the ephemeris time (1.10) providing a measure of duration corresponding to the relative change in the positions of the particles in the system, therefore being a precise realization of Mach's second principle. By introducing Elze's reparametrization incident number (1.132) in Eq. (1.10), one obtains:

$$\int_{\lambda_0}^{\lambda_f} \frac{\sqrt{T}}{\sqrt{E-V}} d\lambda = \int_{S_i}^{S_f} ds^I \lambda(s^I) D(x^4(s^I), x^5(s^I)). \quad (1.178)$$

According to Eq. (1.178), one can say that Barbour's ephemeris time provides a measure of duration — corresponding to the relative change in the positions of the particles — which is emergent from the dynamics in the sense that it derives from the numerical order associated with the reparametrization incident number measuring the coincidences of points of the trajectory of the system with appropriate detectors. As a consequence, the ephemeris time appearing in the Newton-type equations of motion (1.9) of the JBB theory reads merely as numerical order of material change determined by the reparametrization incident number measuring the coincidences of points of the trajectory of the system with appropriate detectors. In this picture, the JBB theory which is based on the JBB action (1.8) and the equations of motion (1.9) implies that time provides a measure of duration corresponding with the relative change in the positions of the particles in the system, thus being a precise realization of Mach's second principle because exists merely as numerical order of material change determined by the reparametrization incident number measuring the coincidences of points of the trajectory of the system with appropriate detectors. According to Eq. (1.178), Barbour's "no time" interpretation can be replaced with the view of a time as an emergent mathematical quantity which measures the numerical order of material change (corresponding with the reparametrization incident number measuring the coincidences of points of the trajectory of the system with appropriate detectors).

On the basis of the interpretation of the ephemeris time as mathematical quantity measuring the numerical order of material changes derived from Elze's reparametrization incident number, the BFO formalism of geometrodynamics where general relativity is treated as a theory of dynamical three spaces and which is based on Eqs. (1.22)–(1.38) may be itself seen as a consequence of the view of time as numerical order. In this picture, one can say that, in the BFO formalism, general relativity provides a description in terms of dynamical three spaces because the ephemeris time exists only as a mathematical quantity measuring the numerical order of material changes characterizing the system under consideration, corresponding to the reparametrization incident number measuring the coincidences of points of the trajectory of the system with appropriate detectors.

Moreover, Shyam's and Ramachandra's model of canonical gravity as a 3D constrained Hamiltonian theory in the picture of a presympletic dynamics inside a super phase space allows us to obtain an expression for the Lapse function, given by Eq. (1.52), which is exactly the same expression derived inside Barbour's approach, in the sense that this Lapse function is associated with an ephemeris time which corresponds to the reparametrization incident number measuring the coincidences of points of the trajectory of the system with appropriate detectors (and which thus measures only the numerical order of material changes of the system under consideration) according to equation

$$N(\lambda, x) = \int \sqrt{\frac{G^{abcd}\xi_{H(N)}[\gamma_{ab}]\xi_{H(N)}[\gamma_{cd}]}{4R}} d\lambda$$

$$= \int_{S_i}^{S_f} ds^I \lambda(s^I) D(x^4(s^I), x^5(s^I)). \qquad (1.179)$$

On the basis of Eq. (1.179), one can say that in Shyam's and Ramachandra's model a Lapse function is obtained which has the same expression derived inside Barbour's approach because it is linked with an ephemeris time which exists merely as a mathematical quantity derived from the reparametrization incident number measuring the coincidences of points of the trajectory of the system with appropriate detectors (and thus as a numerical order of material changes).

Finally, also Gryb's results regarding the path integral quantization of the JBB theory can be seen as a consequence of the view of time as a numerical order of material changes corresponding to the reparametrization incident number measuring the coincidences of points of the trajectory of the system with appropriate detectors. By inserting Eq. (1.132) into Eq. (1.66), this latest equations reads as

$$\tau = \frac{m(q''^{\vee} - q'^{\vee})}{p_0} + \sum_{J=0}^{N-1}\left[m\dot{\vec{q}}_J \cdot \left(\frac{\vec{p}_J}{p_J^2} - \frac{\vec{p}_0}{p_0^2}\right)\right]$$

$$= \int_{S_i}^{S_f} ds^I \lambda(s^I) D(x^4(s^I), x^5(s^I)). \qquad (1.180)$$

On the basis of Eq. (1.180), one can say that Gryb's treatment of JBB path integral based on Eqs. (1.62) and (1.65) shows that, in quantum theory, a parameter time intended as numerical order of material changes corresponding to Elze's reparametrization incident number (1.132) contributes to each term in the sum over all histories leading to a superposition of all possible clocks (namely to all the different measuring systems of the numerical order of material changes) and thus this is the fundamental reason which determines a timeless theory. Thus, in the quantum regime, the timelessness of JBB approach can be seen as a result of a superposition of all possible clocks providing all possible measuring systems of the numerical order of material changes corresponding to Elze's reparametrization incident number, which occurs when one sums over all possible histories. Nevertheless, in this treatment, Gryb's result that in the stationary phase approximation a unique time can be recovered, namely the path integral is dominated by the classical trajectory picking out a unique clock means that in the stationary phase approximation a time (1.180) emerges which exists only as a reference system measuring the numerical order material changes corresponding to Elze's reparametrization incident number measuring the coincidences of points of the trajectory of the system with appropriate detectors. According to equation (1.180), one can therefore say that, in Gryb's approach of JBB path integrals, a notion of duration in the quantum theory can be obtained which provides only a measure of the numerical order of material changes corresponding to the reparametrization incident number (1.132) measuring the coincidences of points of the trajectory of the system under consideration with appropriate detectors.

Now, let us take under consideration Rovelli's approach. Here, if one introduces Elze's view of time, the "thermal time hypothesis" can be seen as a consequence of the fact that clocks measure merely the numerical order of material changes corresponding to the reparametrization incident number (1.132). In this picture, in nature, there is no preferred physical time variable t, there are no equilibrium states ρ_0 preferred *a priori* just because time exists

About Time as the Numerical Order of Material Changes   103

only as a measuring system of the numerical of material changes corresponding to the reparametrization incident number (1.132). In our re-reading of Rovelli's view of time in terms of Elze's reparametrization incident number providing the numerical order of material changes characterizing the system under consideration, the thermal time t_p (provided by the thermal clock whose ticking grows linearly with the Hamiltonian flow $s(t_p)$) appearing in the fundamental equations (1.96) and (1.102) reads as

$$t_p = \int_{S_i}^{S_f} ds^I \lambda(s^I) D(x^4(s^I), x^5(s^I)), \quad (1.181)$$

namely derives just from the numerical order of material changes corresponding to the reparametrization incident number (1.132) measuring the coincidences of points of the trajectory of the system under consideration with appropriate detectors. According to the re-reading of Rovelli's approach based on Eq. (1.181), one can say that time is the expression of our ignorance of the full microstate, a reflex of our incomplete knowledge of the state of the world in the sense that it is associated with the numerical order of material changes corresponding to the reparametrization incident number (1.132).

In synthesis, by replacing time with the concept of numerical order of material changes associated with Elze's reparametrization incident number (1.132) one can provide a unifying reading both of the JBB theory of time based on Eqs. (1.9)–(1.10), (1.22)–(1.38), and of Shyam's and Ramachandra's results of canonical quantum gravity based on Eq. (1.52), and of Gryb's approach of JBB path integrals based on Eq. (1.66), and finally of Rovelli's results regarding the thermal time hyphotesis. In particular, by comparing Eqs. (1.178) and (1.181) one obtains

$$\int_{\lambda_0}^{\lambda_f} \frac{\sqrt{T}}{\sqrt{E-V}} d\lambda = \int_{S_i}^{S_f} ds^I \lambda(s^I) D(x^4(s^I), x^5(s^I)) = t_p, \quad (1.182)$$

which can be considered the fundamental law which provides a unifying re-reading of JBB theory and of Rovelli's model of time (as well as of some important developments of them). According to Eq. (1.182), one can conclude that Elze's reparametrization incident number measuring the coincidences of points of the trajectory of the system under consideration with appropriate detectors (and thus measuring the numerical order of material changes of the system under consideration) can be considered as the ultimate element which leads to Barbour's ephemeris time and to Rovelli's thermal time hypothesis.

1.5.2 About the JBB theory and Rovelli's approach in the view of time as numerical order of material changes corresponding to Prati's model of the number of states of the Hilbert space of the system of interest whose dynamics is described

In this paragraph, by following the treatment made in [102], we want to analyze the reading of the JBB theory and of Rovelli's approach provided by Prati's replacement of the concept of time with the number k_{AB} of states of the Hilbert space of the system of interest whose dynamics is described (and which belong to the interval (1.176) that satisfies an appropriate initial condition $\psi_1(\sigma) = \bar{\psi}_1$ of the subsystem acting as a clock).

By introducing Eq. (1.177) into Eq. (1.10), we obtain the fundamental starting equation of the JBB theory inside the picture of time as numerical order of material changes based on Prati's approach:

$$\int_{\lambda_0}^{\lambda_f} \frac{\sqrt{T}}{\sqrt{E-V}} d\lambda = k_{AB}. \qquad (1.183)$$

According to Eq. (1.183), one can say that Barbour's ephemeris time provides a measure of duration — corresponding to the relative change in the positions of the particles — which is emergent from the dynamics in the sense that it derives from the numerical order associated with the number k_{AB} of states of the Hilbert space of the system of interest whose dynamics is described (and which belong to the interval (1.176) that satisfies an appropriate initial condition $\psi_1(\sigma) = \bar{\psi}_1$ of the subsystem acting as a clock). On the basis of Eq. (1.183), the ephemeris time appearing in the Newton-type equations of motion (1.9) of the JBB theory reads merely as numerical order of material change determined by the number k_{AB} of states of the Hilbert space of the system of interest whose dynamics is described. In this picture, the JBB theory which is based on the JBB action (1.8) and the equations of motion (1.9) implies that time provides a measure of duration corresponding to the relative change in the positions of the particles in the system, thus being a precise realization of Mach's second principle because exists merely as numerical order of material change determined by the number k_{AB} of states of the Hilbert space of the system of interest whose dynamics is described. According to Eq. (1.183), Barbour's "no time" interpretation can be replaced with the view of a time as an emergent mathematical quantity which measures the numerical order of material change (corresponding to the number k_{AB} of states of the Hilbert space of the system of interest whose dynamics is described — and which belong to the interval (1.176) that satisfies an appropriate initial condition $\psi_1(\sigma) = \bar{\psi}_1$ of the subsystem acting as a clock).

On the basis of the interpretation of the ephemeris time as mathematical quantity measuring the numerical order of material changes derived from Prati's number k_{AB} of states of the Hilbert space of the system of interest whose dynamics is described, the BFO Murchadha formalism of geometrodynamics where general relativity is treated as a theory of dynamical three spaces and which is based on Eqs. (1.22)–(1.38) may be itself seen as a consequence of the view of time as numerical order in Prati's picture. In this picture, one can say that, in the BFO formalism, general relativity provides a description in terms of dynamical three spaces because the ephemeris time exists only as a mathematical quantity measuring the numerical order of material changes characterizing the system under consideration, corresponding to the number k_{AB} of states of the Hilbert space of the system of interest whose dynamics is described (and which belong to the interval (1.176) that satisfies an appropriate initial condition $\psi_1(\sigma) = \bar{\psi}_1$ of the subsystem acting as a clock).

Moreover, Shyam's and Ramachandra's model of canonical gravity as a three-dimensional constrained Hamiltonian theory in the picture of a presympletic dynamics inside a super phase space allows us to obtain an expression for the Lapse function, given by Eq. (1.52), which is exactly the same expression derived inside Barbour's approach, in the sense that this Lapse function is associated with an ephemeris time which corresponds to the number k_{AB} of states of the Hilbert space of the system of interest whose dynamics is described, and which thus measures only the numerical order of material changes of the system under consideration, according to equation

$$N(\lambda, x) = \int \sqrt{\frac{G^{abcd}\xi_{H(N)}[\gamma_{ab}]\xi_{H(N)}[\gamma_{cd}]}{4R}} d\lambda = k_{AB}. \quad (1.184)$$

On the basis of Eq. (1.184), one can say that in Shyam's and Ramachandra's model a Lapse function is obtained which has the same expression derived inside Barbour's approach because it is linked with an ephemeris time which exists merely as a mathematical quantity deriving from the number k_{AB} of states of the Hilbert space of the system of interest whose dynamics is described (and thus as a numerical order of material changes).

Finally, also Gryb's results regarding the path integral quantization of the JBB theory can be seen as a consequence of the view of time as a numerical order of material changes corresponding to the number k_{AB} of states of the Hilbert space of the system of interest whose dynamics is described (and which belong to the interval (1.176) that satisfies an appropriate initial condition $\psi_1(\sigma) = \bar{\psi}_1$ of the subsystem acting as a clock). By inserting Eq. (1.177) into Eq. (1.66), this latest equations reads as

$$\tau = \frac{m(q''^{\vee} - q'^{\vee})}{p_0} + \sum_{J=0}^{N-1} \left[m\dot{\vec{q}}_J \cdot \left(\frac{\vec{p}_J}{p_J^2} - \frac{\vec{p}_0}{p_0^2} \right) \right] = k_{AB}. \quad (1.185)$$

On the basis of Eq. (1.185), one can say that Gryb's treatment of JBB path integral based on Eqs. (1.62) and (1.65) shows that, in quantum theory, a parameter time intended as numerical order of material changes corresponding to Prati's number k_{AB} of states of the Hilbert space of the system of interest whose dynamics is described contributes to each term in the sum over all histories leading to a superposition of all possible clocks (namely to all the different measuring systems of the numerical order of material changes) and thus this is the fundamental reason which determines a timeless theory. Thus, in the quantum regime, the timelessness of JBB approach can be seen as a result of a superposition of all possible clocks providing all possible measuring systems of the numerical order of material changes corresponding to Prati's number k_{AB} of states of the Hilbert space of the system of interest whose dynamics is described. Nevertheless, in this treatment, Gryb's result that in the stationary phase approximation a unique time can be recovered, namely the path integral is dominated by the classical trajectory picking out a unique clock means that in the stationary phase approximation a time (1.185) emerges which exists only as a reference system measuring the numerical order material changes corresponding to Prati's number k_{AB} of states of the Hilbert space of the system of interest whose dynamics is described. According to Eq. (1.185), one can therefore say that, in Gryb's approach of JBB path integrals, a notion of duration in the quantum theory can be obtained which provides only a measure of the numerical order of material changes corresponding to Prati's number k_{AB} of states of the Hilbert space of the system of interest whose dynamics is described (and which belong to the interval (1.176) that satisfies an appropriate initial condition $\psi_1(\sigma) = \bar{\psi}_1$ of the subsystem acting as a clock).

Now, let us take under consideration Rovelli's approach inside Prati's picture based on Eqs. (1.176) and (1.177). Here, the "thermal time hypothesis" can be seen as a consequence of the fact that clocks measure merely the numerical order of material changes corresponding to Prati's number k_{AB} of states of the Hilbert space of the system of interest whose dynamics is described. In this picture, one can say that in nature there is no preferred physical time variable t, there are no equilibrium states ρ_0 preferred *a priori* just because time exists only as a measuring system of the numerical of material changes corresponding to Prati's number k_{AB} of states of the Hilbert space of the system of interest whose dynamics is described. In our re-reading of Rovelli's view of time in terms of Prati's number k_{AB} of states of the Hilbert space of the system of interest whose dynamics is described thus providing the numerical order of material changes characterizing the system under consideration, the thermal time t_p appearing in the fundamental equations (1.96) and (1.102) reads as

$$t_p = k_{AB}, \quad (1.186)$$

namely derives just from the numerical order of material changes corresponding to Prati's number k_{AB} of states of the Hilbert space of the system of interest whose dynamics is described (and which belong to the interval (1.176) that satisfies an appropriate initial condition $\psi_1(\sigma) = \bar{\psi}_1$ of the subsystem acting as a clock). According to the re-reading of Rovelli's approach based on Eq. (1.186), one can say that time is the expression of our ignorance of the full microstate, a reflex of our incomplete knowledge of the state of the world in the sense that it is associated with the numerical order of material changes corresponding to Prati's number k_{AB} of states of the Hilbert space of the system of interest whose dynamics is described.

In synthesis, by replacing time with the concept of numerical order of material changes associated with Prati's number k_{AB} of states of the Hilbert space of the system of interest whose dynamics is described one can provide a unifying reading both of the JBB theory of time based on Eqs. (1.9) and (1.10), (1.22)–(1.38), and of Shyam's and Ramachandra's results of canonical quantum gravity based on Eq. (1.52), and of Gryb's approach of JBB path integrals based on Eq. (1.66), and finally of Rovelli's results regarding the thermal time hypothesis. In particular, by comparing Eqs. (1.183) and (1.186) one obtains

$$\int_{\lambda_0}^{\lambda_f} \frac{\sqrt{T}}{\sqrt{E-V}} d\lambda = k_{AB} = t_p, \quad (1.187)$$

which can be considered the fundamental law which provides a unifying re-reading of JBB theory and of Rovelli's model of time (as well as of some important developments of them). According to Eq. (1.187), one can conclude that Prati's number k_{AB} of states of the Hilbert space of the system of interest whose dynamics is described (and thus which measures the numerical order of material changes of the system under consideration) can be considered as the fundamental element which yields Barbour's ephemeris time and Rovelli's thermal time hypothesis.

1.6 About Time as Numerical Order of Material Changes in Special Relativity and General Relativity

Isaac Newton founded classical mechanics on the view that space is something distinct from body and that time is something that passes uniformly without regard to whatever happens in the world. According to Newton's physics time passes in space and is not part of the space. Absolute, true and mathematical time, from its own nature, passes equably without relation to anything external, and thus without reference to any change or way of measuring of time.

On the existence of space as stated by Newton's view there is no doubt. Space is a physical medium in which matter exists. Instead for Newton's view on time there is no experimental evidence. Time's running in space on its own as a primary physical reality was never experimentally detected. Although in Newtonian physics as well as in standard quantum mechanics, time is postulated as a special physical quantity and plays the role of the independent variable of physical evolution, however, it is an elementary observation that we never really measure an idealized time, that an idealized, absolute time does not ever appear in laboratory measurements: we rather measure the frequency, speed and numerical order of material changes. What experimentally exists is only the motion of a system and the tick of a clock. What we realize in every experiment is to compare the motion of the physical system under consideration with the motion of a peculiar clock described by a peculiar tick.

According to the view proposed in Sec. 1.5, with clocks we measure numerical order of material change i.e. motion in space. In physical world time is exclusively a mathematical quantity. Universe is not changing in a time intended as primary physical reality, on the contrary time as a numerical order of change runs in the universe. At a fundamental level, universe is timeless. With clocks we measure velocity of material changes. In Newton's physics these velocities are absolute in the sense that clocks run with equal velocity in the entire universal space.

In the special theory of relativity, Einstein described electromagnetic phenomena with the formalism of 4D space created by German mathematician Hermann Minkowski. As in Newton physics also in special relativity clocks measure the numerical order of material changes. Experiments confirm that material changes (clocks' mechanisms included) have different velocity in different inertial systems. In special relativity velocity of material change is not absolute, it is relative. For more than 100 years misunderstanding of special relativity is that material changes run into a space-time intended as a fundamental physical reality and thus in a time intended as a part of space, and so that time is a dimension of a medium in which electromagnetic waves, particles and massive bodies move. Minkowski's formalism of the fourth coordinate $x_4 = ict$ (where i is the imaginary unit, c the speed of light and t time) shows clearly that space-time is a 4D timeless space. Definition of time "time is numerical order of change which runs in a 3D timeless space" resolves this century long misinterpretation: electromagnetic waves, particles and massive objects move exclusively in space, time is numerical order of their motion.

In general theory of relativity there is not a preferred and observable quantity that plays the role of independent parameter of the evolution, and thus Newton's absolute time is replaced by different possible internal times, related to specific physical clocks, just because clocks provide only a mathematical measure of the numerical order of physical events. Definition of time as "numerical order of change that runs in a 3D timeless space" resolves in a clear way the problem concerning the interpretation of the physical clocks inside general relativity. Since clocks can be defined as those instruments which measure the speed of the material changes and movements, the internal clocks/times of general relativity are only measuring systems of the numerical order of material change. Our definition of time as a mathematical coordinate that indicates the numerical order of material motion in space provides thus a suggestive, clear and simple re-reading of the internal clocks/times of general relativity.

On the other hand, as regards general relativity, recently Pavsic developed a Kaluza–Klein-type model in which the ordinary spacetime of general relativity is replaced with a configuration space C, a multidimensional manifold equipped with metric, connection and curvature [103]. Inside this model Pavsic showed that the ordinary general relativistic theory for a many particle system is only a special case that derives from a more general action in the configuration space C for a particular block diagonal metric. The most general action has the form

$$I[X^M] = M \int d\tau \left[\dot{X}^M \dot{X}^N G_{MN}(X^M) \right]^{1/2}, \qquad (1.188)$$

where $X^M \equiv X_i^\mu$ (with $\mu = 0, 1, 2, 3$ and $i = 1, 2, \ldots, N$, N being the number of the particles in the configuration) are the coordinates of the point particles of the system, M has the role of mass in C, τ is an arbitrary monotonically increasing parameter and G_{MN} is the metric. The action is proportional to the length of a worldline in C. The ordinary general relativistic theory for a many particle system derives from the action (1.188) in the special case

$$G_{MN} = \begin{pmatrix} g_{\mu\nu}(x_1) & 0 & 0 & \cdots \\ 0 & g_{\mu\nu}(x_2) & 0 & \cdots \\ 0 & 0 & g_{\mu\nu}(x_3) & \cdots \\ \cdot & \cdot & \cdot & \cdots \\ \cdot & \cdot & \cdot & \cdots \end{pmatrix}. \qquad (1.189)$$

Inside Pavsic's model, the metric in the configurations space C is not fixed but is dynamical, so that the total action contains a kinetic term for G_{MN}:

$$I[X^M, G_{MN}] = I_m + I_g, \qquad (1.190)$$

where

$$I[X^M] = \int d\tau M[\dot{X}^M \dot{X}^N G_{MN}(X^M)]^{1/2}$$
$$= \int d\tau M[\dot{X}^M \dot{X}^N G_{MN}(X^M)]^{1/2} \delta^D(x - X(\tau))d^D x,$$
(1.191)

and

$$I_g = \frac{1}{16\pi G_D} \int d^D x \sqrt{|G|} R, \qquad (1.192)$$

where R is the curvature scalar in C.

In Pavsic's model, the fact that, on the basis of Eqs. (1.188)–(1.192), the action of a system of N particles in the general relativistic domain does not depend explicitly on an idealized time but only on an arbitrary monotonically increasing parameter τ means that in general relativity time does not exist as a primary physical reality but is only a mathematical device and that the parameter τ represents just a measuring device of the mathematical numerical order of material changes characterizing the system of N particles under consideration.

In synthesis, on the basis of the relevant current research analyzed in this chapter, according to the author it is legitimate to assume that the fundamental arena in which material changes take place is a background space and time is not a primary physical reality but exists only as a numerical order of material changes. In the universe, material changes are running in space only while time is indeed a static concept. Past instants $t_{-n} \ldots t_{-2}, t_{-1}$, present moment t_0 and future instants $t_1, t_2 \ldots t_n$ exist only as a numerical order of material change in a background that, at a fundamental level, is timeless.

Chapter 2

Three-Dimensional Euclid Space and Special Relativity

On the basis of the treatment made in the previous chapter, the time appearing in equations of physical theories cannot be considered as a primary physical reality that acts as an independent variable of evolution but represents an abstract, mathematical parameter which is introduced as a measuring system of the numerical order of material motions. In this chapter, our aim is to make some considerations about a new interpretation of special theory of relativity in a picture where time exists only as a mathematical parameter measuring the numerical order of events.

As photon moves in space only and not in time (intended as a primary physical reality) in special relativity the fourth coordinate of Minkowski space-time is imaginary: $x_4 = ict$, given by the product of imaginary unit i, light speed c and numerical order t of an event. The fourth coordinate x_4 of Minkowski's space-time of special relativity is a mathematical coordinate which is introduced to describe the numerical order of a material motion into space. In special theory of relativity the fourth coordinate is different from the first three coordinates. In fact, the first three coordinates constitute physical realities in the sense that they correspond to the position of a material object in space; instead, the fourth coordinate is only a mathematical coordinate, is an abstract mathematical, imaginary entity that is used to describe the numerical order of events. As

113

a consequence, the space-time manifold of special relativity can be considered as a "physical-mathematical reality".

In virtue of the features of the fourth coordinate $x_4 = ict$, it is more correct to consider the Minkowskian background of special relativity as a four-dimensional (4D) physical-mathematical reality where the fourth dimension is a mathematical coordinate which describes the numerical order of material change than a $3D + T$ space-time arena where both the three spatial coordinates and the temporal coordinate have the same ontological physical reality. In the Minkowskian formalism of special relativity, time t obtained with clocks is only a component of the fourth coordinate x_4 of a 4D space having a physical-mathematical reality. At a fundamental level, material change runs in a three-dimensional (3D) space. Definition of time "time is numerical order of change which runs in a 3D timeless space" suggests that, in the special relativistic regime, electromagnetic waves, particles and massive bodies move exclusively in space, time is numerical order of their motion. In this chapter, after analyzing some relevant recent research regarding a 3D geometry as the fundamental arena of special relativity, we will show how, on the basis of our concept of time as a mathematical quantity measuring the numerical order of material changes, a suggestive interpretation of special relativity in a 3D space background naturally emerges which can introduce interesting perspectives in the explanation of the behavior of light clocks as well as in the study of several phenomena such as aberration of light, Jupiter's satellites occultation and radar ranging of the planets.

2.1 From Einstein and Minkowski... to the Most Recent Research of Alternative Arenas of Special Relativity

Special theory of relativity is, together with general relativity and quantum mechanics, one of the three fundamental theories of modern physics which have radically revolutionized our view of the world. In essence, special relativity, as was developed by Einstein in 1905, ironed-out the inconsistencies between Newtonian mechanics and

Maxwellian electrodynamics concerning absolute motion and the laws of physics. In order to explain the results of the famous Michelson–Morley experiment Einstein considered the following two postulates:

(1) *The laws of physics are the same for all inertial frames of reference in uniform relative motion.*
(2) *The speed of light in free space is the same for all inertial observers.*

The first postulate, known also as the principle of Relativity, dispels the notion that there is such a thing as a preferred or absolute reference system. According to the first postulate, the laws of physics must have the same form in equivalent reference systems. Inertial reference systems have the same status of motion in the sense that Newton's first law holds good in them. If the first postulate was true and Maxwell's theory was a fundamental theory of nature, then the second postulate follows immediately since Maxwell's theory predicts explicitly that the speed of light has a definite numerical value. The constancy of the speed of light predicted here lead us via Einstein's great insight to rethink our view of space and time. Time for different frames of reference runs at different rates and lengths are not absolute but depend on the observers' state of motion. From these two postulates one can derive immediately the famous Lorentz transformations of the space and time coordinates between two inertial reference frames S and S', which in the simple case in which the origin of S' moves, with respect to S, with velocity $v<c$ parallel to the first spatial axis, assume the form:

$$\begin{cases} x' = \dfrac{x - vt}{\sqrt{1 - \beta^2}}, \\ y' = y, \\ z' = z, \\ t' = \dfrac{t - \dfrac{\beta}{c}x}{\sqrt{1 - \beta^2}}. \end{cases} \quad (2.1)$$

Unlike in Lorentz's 1895 theory, inside Einstein's 1905 special relativity the derivation of the Lorentz transformations (1) does not require the existence of a stationary aether.

Special relativity as was developed by Einstein in 1905 constitutes a very successful theory. Many peculiar predictions — some of these very counterintuitive — can be derived from it. Nevertheless, when tested experimentally, they invariably turn out to be true. One of the best-known special-relativistic phenomenon is time dilation, which has been demonstrated in many experiments, including both measurements of the life time of decaying particles (muons) and direct use of atomic clocks (in this regard see, for example, the classic works of R. J. Kennedy and E. M. Thorndike [104], H. E. Ives and G. R. Stilwell [105], B. Rossi and D. B. Hall [106], M. Kaivola, O. Poulsen, E. Riis and S. A. Lee [107], S. Reinhardt et al. [108]). A famous realization of the twin paradox is the famous Hafele–Keating experiment dealing with airplane borne traveling clocks [109, 110]. The clock hypothesis (independence of time dilation of acceleration) has been verified using muons by Bailey et al. in 1977 [111]. In the light of the famous Michelson's and Morley's work as well as the more recent test of A. Brillet, J. L. Hall and D. Hils, the isotropy of the round-trip speed of light has been shown to be true with ever increasing accuracy [112–114]. Since the '60s to the present days, there have been explicit tests of the isotropy of space [115, 116] and of Lorentz invariance [117, 118] and there is a huge amount of literature on the tight limits of possible violations of Lorentz invariance [119]. The independence of the speed of light of the velocity of its source has also been explicitly addressed (see, for example, the references [120, 121]).

Today, many of the particular effects of special relativity have been demonstrated separately [119, 122, 123]. An exception seems to be Lorentz contraction, often quoted as not having been observed directly. It is however clear that even the Michelson–Morley experiment is an indirect proof of Lorentz contraction, if one accepts standard arguments as to why there can be neither contraction nor expansion in the direction perpendicular to the velocity [124]. Then length contraction parallel to it is required to explain the null result of the experiment in inertial frames where the interferometer is not at rest.

In 1908, three years after Einstein's work, Minkowski introduced the idea of a fundamental 4D space-time continuum as a conceptual framework for Einstein's theory of special relativity. Minkowski's arena fuses the three dimensions of space with an imaginary time dimension $x_4 = ict$, with the unit imaginary i producing the correct space-time distance $x^2 - c^2t^2$, and the results of Einstein's then recently developed theory of special relativity, thus providing an explanation for Einstein's theory in terms of the structure of space and time. In this picture, the space-temporal continuum of special relativity becomes pseudo-Euclidean: indeed, Minkowski's space-time constitutes a natural formal extension, in two directions, of the euclidean geometry of space; in fact, here one deals with a continuum with an adding dimension and introduces in this arena a nonpositive definite arena. Minkowski's 4D space-time continuum is so characterized by a pseudo-Pythagorean distance measure of the form $s^2 = x_1^2 + x_2^2 + x_3^2 + x_4^2$, which allows one to view space-time just as a conventional Euclidean space with no difference in treatment between the spatial coordinates x, y, z and the fourth coordinate ict, but still recovering the invariant distance measure $s^2 = x_1^2 + x_2^2 + x_3^2 - c^2t^2$. This idea was received favorably by Einstein, and by the wider scientific community at the time making the Minkowski space-time as the fundamental arena for all nongravitational interactions both in classical and quantum physics. If Minkowski's space-time continuum is endowed with geometrical, and more specifically metric, properties, however it remains opened the debate regarding the major or minor content of reality which can be ascribed to this structure, namely if it must be intended as an aspect of reality or as a useful metaphore of a "reality" which lies all and one in the formalism. In this regard, in [125] Bergia showed that, if one marries the second thesis, it turns out to be hard to build a logical path towards general relativity. On the other hand, in [126] Bergia claimed, in the light of relevant elements of information obtained by Miller, that the strong ontology provided by the Einstein–Minkowski relativity has become common patrimony in a gradual way.

Yet despite the indelible impression Minkowski's insight has left on our thinking, special relativity is a theory that does not deal fundamentally about space-time. A permissible interpretation is that this theory is foremost concerned with how different observers view the world around them, and how their views relate to each other. Namely, starting with space-time, the observer space — the space of all possible observers in the universe — is simply the space of future-directed unit time-like vectors. The space-time perspective is compelling precisely because it efficiently accounts for all these possible viewpoints.

On the other hand, it must be emphasized that, in the light of the modern research, 4D space-time is not the only possible framework for understanding observers. For example, since the '90s Selleri proposed general transformations of space and time between inertial reference frames that seem to indicate clearly that in special relativity time must be separated from space [127–132]. Given the inertial frames S_0 and S' endowed with Cartesian coordinates x_0, y_0, z_0 and x', y', z', respectively (where the origin of S', observed from S_0, is seen to move with velocity $v < c$ parallel to the $+x_0$-axis), by starting from the following two empirically based assumptions:

(1) the two-way velocity of light is the same in all directions and in all inertial systems;
(2) clock retardation takes place with the usual velocity-dependent factor when clocks move with respect to S_0,

Selleri found the following transformations for the space and time variables from S_0 to S' (called also equivalent transformations):

$$\begin{cases} x' = \dfrac{x_0 - vt_0}{\sqrt{1-\beta^2}}, \\ y' = y_0, \\ z' = z_0, \\ t' = \sqrt{1-\beta^2}\,t_0 + e_1(x_0 - vt_0), \end{cases} \quad (2.2)$$

where $\beta = \frac{v}{c}$. These transformations contain a free parameter, e_1, the coefficient of x' in the transformations of time which can be fixed by choosing a peculiar clock synchronization method in S'. The special theory of relativity (and thus the standard Lorentz transformations) is obtained for $e_1 = -\frac{\beta}{c\sqrt{1-\beta^2}}$. For all the values of e_1 except $e_1 = -\frac{\beta}{c\sqrt{1-\beta^2}}$, the system S_0 turns out to be a privileged reference frame. As it has been analyzed by Selleri, different values of e_1 determine different theories of space and time which are empirically equivalent to a large extent. In fact, Michelson-type experiments, the twin paradox experiment, radar ranging of planets and occultation of Jupiter satellites, Doppler effect and aberration of starlight, Fizeau experiment are insensitive to the choice of e_1. These results suggest that the clock synchronization in inertial systems is conventional and the assumed invariance of the one way velocity of light of the standard interpretation of the special theory of relativity has only motivations of simplicity. (No perfectly inertial reference frame exists in practice, e.g. because of Earth rotation around its axis, of orbital motion, etc. All we know about inertial systems has actually been obtained in reference frames endowed with a small but nonzero acceleration. The frame associated with Earth for many problems can be considered inertial just because its rotation, being of a fourth order, can be neglected.) But, as it has been shown by Selleri, there are some particular phenomena (the accelerating spaceships, the rotating disk and the question of superluminal signals regarding the group velocity of electromagnetic radiation) that modify the situation to the point that the condition $e_1 = 0$ becomes necessary. The adoption of $e_1 = 0$ makes Eq. (2.2) become the "inertial transformations" in which the transformation of the speed of clock does not contain the space variable:

$$\begin{cases} x' = \dfrac{x_0 - vt_0}{\sqrt{1-\beta^2}}, \\ y' = y_0, \\ z' = z_0, \\ t' = \sqrt{1-\beta^2}\, t_0. \end{cases} \quad (2.3)$$

The inertial transformations imply the absolute simultaneity: two spatially separated events happening at the same time in the inertial reference frame S_0, are simultaneous also with respect to S'.

As regards the equivalent transformations (and thus the inertial transformations), it is also interesting to remark that, on the basis of a theorem proved by Rizzi, Ruggiero and Serafini [133], Selleri's assumptions:

1. at least one inertial reference frame S_0 exists in which the velocity of light is c in all points and in all directions (S_0 is optically isotropic);
2. the two ways velocity of light is the same in all inertial reference frames and in all directions;
3. the pace of clocks in motion with respect to S_0 with velocity v slows down by the usual factor $\sqrt{1-\beta^2}$;

are equivalent to Einstein's basic assumptions of the special theory of relativity. Moreover, Rizzi, Ruggiero and Serafini showed that from this theorem the following important consequences derive:

(a) no experiment can discriminate between different values of e_1;
(b) it is impossible to detect the privileged inertial reference frame S_0 (and this means that the role of privileged inertial reference frame played by S_0 is only artificial as S_0 is the inertial reference frame in which, by convention, the Einstein synchronization procedure was adopted; any inertial reference frame S can then play the role of S_0);
(c) the transformation:

$$\begin{cases} x = \hat{x}, \\ y = \hat{y}, \\ z = \hat{z}, \\ t = \hat{t} + \dfrac{\beta + ce_1\sqrt{1-\beta^2}}{c}\hat{x} \end{cases} \quad (2.4)$$

allows one to pass, within any given inertial reference frame S, from the Einstein synchronization to Selleri's generalized synchronization.

In this way, one also passes from the Lorentz transformations to the equivalent transformations.

Selleri's theory of inertial transformations (2.3) determines an arena of special relativity in which the temporal coordinate must be clearly considered as a different entity with respect to the spatial coordinates just because the transformation of clocks' run between the two inertial systems does not depend on the spatial coordinates. Moreover, it is interesting to remark that, in the theory of the inertial transformations (2.3), the inertial reference frame S_0 is not only privileged but also physically active in the sense that clocks run slower when move with respect to S_0 and light propagates in the simplest way only in S_0. It is the motion relative to S_0 that influences the clocks' running. This means that the 3D space provides the primitive ontology of special relativity. According to the inertial transformations (2.3), the origin of S_0, observed from the inertial frame S', is seen to move with velocity $-\beta c/\left(1 - \frac{v^2}{c^2}\right)$. The latter velocity can be larger than c, but cannot be superluminal. In fact, if a particle moves with velocities v_p and u_p, relative to S_0 and S', respectively, Eq. (2.3) lead to the following transformations of velocities:

$$\begin{cases} u_{px} = \dfrac{v_{px} - v}{1 - \beta^2}, \\ u_{py} = \dfrac{1}{\sqrt{1 - \beta^2}} v_{py}, \\ u_{pz} = \dfrac{1}{\sqrt{1 - \beta^2}} v_{pz}. \end{cases} \quad (2.5)$$

Setting $u_{px} = u_p \cos \vartheta$, $u_{py} = u_p \sin \vartheta$, $u_{pz} = 0$ (where ϑ is the angle, in S', between u_p and v_p), one easily obtains

$$v_p = \sqrt{\left(v + (1 - \beta^2) u_p \cos \vartheta\right)^2 + \left((1 - \beta^2) u_p \sin \vartheta\right)^2}. \quad (2.6)$$

As a consequence of Eq. (2.6), in the approach of the inertial transformations the velocity of light is isotropic only in S_0. A luminous pulse traveling in the direction $-x'$ with respect to the velocity of S' relative to S_0 has a velocity given by

$$c_1(\vartheta) = \frac{c}{1 + \beta \cos \vartheta}, \qquad (2.7)$$

where $\vartheta = \pi$, namely

$$\tilde{c}(\pi) = \frac{c}{1 - \beta}, \qquad (2.8)$$

which is certainly larger than $\beta c/\left(1 - \frac{v^2}{c^2}\right)$ if $0 \leq \beta < 1$. Thus, in Selleri's approach of the inertial transformations, the relative velocities, in any direction ϑ, can grow without limit, but they always remain smaller than $c_1(\vartheta)$. The absolute velocities can never be larger than c. Another important result regarding Selleri's theory of inertial transformations lies in the possibility to provide a consistent description of the Compton effect which suggests a complete empirical equivalence between Einstein's special relativity and the inertial transformations approach [134].

On the other hand, it is important to mention that, on the basis of relevant current research, the arena of special relativity emerges from a fundamental 3D geometry. For example, by following the philosophy that is at the basis of Selleri's approach, Reginald T. Cahill introduced an exact mapping from Galilean time and space coordinates to Minkowski space-time coordinates, showing that Lorentz covariance and the space-time construct are consistent with the existence of a dynamical 3D-space and "absolute motion" [135–140]. While in Minkowski–Einstein space-time formalism length contraction and clock effects are transferred to the metric of the mathematical space-time, and thus appear to be merely perspective effects for different observers, in Cahill's approach they are real effects experienced by objects and clocks in motion relative to an actual 3D-space. Whereas the Minkowski–Einstein space-time construct is merely a mathematical artifact, and various observable phenomena cannot be described by that formalism, in Cahill's approach a neo-Galilean formalism defined by Galilean space and time coordinates represents the valid description of reality, and it is a deeper encompassing formalism than the Minkowski–Einstein formalism in terms of both mathematical clarity and ontology. More precisely, in Cahill's model space is a real existent fractal network of relationships of connectivities whose dynamical patterns at a coarse grained level are described by a velocity field $\vec{v}(\vec{r}, t)$ where $\vec{r}$ is the location of a small region according to some arbitrary embedding (which provides a mathematical way of characterizing the gross three-dimensionality of the network: the embedding space coordinates provide a coordinate system or frame of reference that is convenient to describe the velocity field, but which is not real). The minimal dynamics can be developed by writing down the lowest-order zero-rank tensors, of dimension $1/t^2$, that are invariant under translation and rotation. In this way, one obtains

$$\nabla \cdot \left(\frac{\partial \vec{v}}{\partial t} + (\vec{v} \cdot \nabla) \vec{v}\right) + \frac{\alpha}{8} (TrD)^2 - \frac{\alpha}{8} (TrD^2) = -4\pi G\rho, \quad (2.9)$$

where $D_{ij} = \frac{1}{2}\left(\frac{\partial v_i}{\partial x_j} + \frac{\partial v_j}{\partial x_i}\right)$, $\rho(\vec{r}, t)$ is an effective matter density (including matter and electromagnetic energy densities). In Eq. (2.9), G turns out to be Newton's gravitational constant, and describes the rate of nonconservative flow of space into matter. As regards the value of α, experiment and astrophysical data have shown that it is equal to the fine structure constant within observational errors [136, 138, 139]. As a consequence, in this picture for a quantum system with mass m one can introduce a generalized form of the Schrödinger equation [138], with the new terms required to maintain that the motion is intrinsically linked to the 3D-space and that the time evolution is unitary

$$i\hbar\frac{\partial \psi(\vec{r}, t)}{\partial t} = -\frac{\hbar^2}{2m}\nabla^2\psi(\vec{r}, t) - i\hbar\left(\vec{v} \cdot \nabla + \frac{1}{2}\nabla \cdot \vec{v}\right)\psi(\vec{r}, t).$$

(2.10)

The space and time coordinates $\{t, x, y, z\}$ in (2.9) and (2.10) ensure that the separation of a deeper and unified process into different classes of phenomena, here a dynamical three-space and a quantum system, is properly tracked and connected, and that this occurs for the same coordinates. However, it is important to realize that these coordinates, although may be used by an observer to also track the different phenomena, have no ontological significance, in the sense that they are not real. The velocities $\vec{v}$ have no ontological or absolute meaning relative to this coordinate system — that is in fact how one arrives at the form in (2.9), and so the "flow" is always relative to the internal dynamics of the 3D-space. The acceleration of the dynamical patterns in space induced by wave refraction is

$$\vec{g} = \frac{\partial \vec{v}}{\partial t} + (\vec{v} \cdot \nabla)\vec{v} + (\nabla \times \vec{v}) \times \vec{v}_R, \quad (2.11)$$

where

$$\vec{v}_R(\vec{r}_0(t), t) = \vec{v}_0(t) - \vec{v}(\vec{r}_0(t), t) \quad (2.12)$$

is the velocity of the wave packet relative to the 3D-space, $\vec{v}_0$ and $\vec{r}_0$ being the velocity and position relative to the observer, and the last term in (2.12) generates the Lense–Thirring effect as a vorticity driven effect. Together (2.10) and (2.12) amount to the derivation of gravity as a quantum effect, explaining both the equivalence principle (in the sense that g in (2.12) is independent of m) and the Lense–Thirring effect. Overall — if one neglects vorticity effects — one obtains

$$\nabla \cdot \vec{g} = -4\pi G \rho - \frac{\alpha}{8}((trD)^2 - tr(D^2)), \quad (2.13)$$

which is Newtonian gravity but with the extra dynamical term whose strength is given by α. This new dynamical effect explains the spiral galaxy flat rotation curves (and so without the need to introduce "dark matter" as a primary ontological reality), the bore hole g anomalies, the black hole "mass spectrum". On the basis of Eq. (2.13), the unknown and undetected "dark matter" density — which is invoked to account for the observed value of g — is a dynamical property of the 3D-space itself.

By starting from the dynamics defined by Eq. (2.9), Cahill realized that all special relativistic effects are dynamically and observationally relative to an ontologically real, that is, detectable dynamical 3D-space. In this picture, space and time are distinct phenomena, and are not merged into some 4D-dimensional structure. Indeed here time is seen to have a cosmic significance, and all observers can measure that time, for by measuring their local absolute speed relative to their local 3D-space they can correct the ticking rate of their clocks to remove the local time dilation effect, and so arrive at a measure of the ticking rate of cosmic time. In order to see in detail in what sense the idea of an absolute motion is not incompatible with Lorentz symmetry, namely of the idea that motion through a structured space determines actual dynamical time dilations and length contractions in agreement with the Lorentz interpretation of special relativistic effects, we follow Cahill's treatment in the recent paper *Dynamical 3-Space: A Review* [139].

Observers in uniform motion 'through' the space, equipped with standard rods (which would agree if they were brought together, and at rest with respect to space would all have length Δl_0) and accompanying standardized clocks (which behave in a similar manner), will always obtain the same numerical value c for the speed of light. An observer P and accompanying rod AB are both moving at speed v_R relative to space, with the rod longitudinal to that motion. P then measures the time Δt_R, with the clock at end A of the rod, for a light pulse to travel from end A to the other end B and back again to A. The light travels at speed c relative to space. Letting the time taken for the light pulse to travel from A→B be t_{AB} and from B→A be t_{BA}, as measured by a clock at rest with respect to space, the total travel time Δt_0 is

$$\Delta t_0 = t_{AB} + t_{BA} = \frac{\Delta l_R}{c - v_R} + \frac{\Delta l_R}{c + v_R} = \frac{2\Delta l_0}{c\sqrt{1 - \frac{v_R^2}{c^2}}}, \quad (2.14)$$

where

$$\Delta l_R = \Delta l_0 \sqrt{1 - \frac{v_R^2}{c^2}} \quad (2.15)$$

is the length of the rod moving at velocity v_R. So, for the moving observer the speed of light is defined as the distance the observer believes the light traveled $2\Delta l_0$ divided by the travel time according to the accompanying clock Δt_R, namely $2\Delta l_0/\Delta t_R = c$, from above, which is thus the same speed as seen by an observer at rest in the space, namely c. In this way, the speed v_R of the observer through space is not revealed by this procedure, and the observer is erroneously led to the conclusion that the speed of light is always c. The Einstein space-time measurement protocol actually inadvertently uses this special effect by using the radar method for assigning historical space-time coordinates to an event: the observer records the time of emission and reception of radar pulses ($t_r > t_e$) traveling through space, and then retrospectively assigns the time and distance of a distant event B according to relations

$$T_B = \frac{1}{2}(t_r + t_e), \quad D_B = \frac{c}{2}(t_r - t_e), \qquad (2.16)$$

where each observer is now using the same numerical value of c. The event B is then plotted as a point in an individual geometrical construct by each observer, known as a space-time record, with coordinates (D_B, T_B). This is the same as an historian recording events according to some agreed protocol. Cahill showed then that because of this protocol and the absolute motion dynamical effects, observers discover on comparing their historical records of the same events that the expression

$$\tau_{AB}^2 = T_{AB}^2 - \frac{1}{c^2}D_{AB}^2 \qquad (2.17)$$

is an invariant, where $T_{AB} = T_A - T_B$ and $D_{AB} = D_A - D_B$ are the differences in times and distances assigned to events A and B using the Einstein measurement protocol (2.16), as long as both are sufficiently small compared with the scale of inhomogeneities in the velocity field. Once one has established the invariance of the construct (2.17) when one of the observers is at rest in space, it emerges that for two observers P′ and P″ both in absolute motion they also agree on the invariance of (2.17). In other words, the Einstein space-time measurement protocol and the Lorentzian absolute motion effects lead to the conclusion that the quantity (2.17) is an invariant in general. This is a remarkable and subtle result. If in Einstein's theory this invariance was a fundamental assumption, in Cahill's 3D dynamical space approach it is a derived result. On the basis of the invariance of (2.17) in Cahill's approach, it follows that individual patches of space-time records may be mapped one into the other merely by a change of coordinates, and that collectively the space-time patches of all may be represented by a pseudo-Riemannian manifold, where the choice of coordinates for this manifold is arbitrary, and therefore the following invariant emerges

$$\Delta\tau^2 = g_{\mu\nu}(x)\Delta x_\mu \Delta x_\nu, \qquad (2.18)$$

with $x^\mu = \{D_1, D_2, D_3, T\}$. Equation (2.18) is invariant under the Lorentz transformations

$$x'^\mu = L^\mu_\nu x^\nu, \qquad (2.19)$$

where, for example for relative motion in the x-direction, L^μ_ν is specified by Eq. (2.1). This physically means — Cahill argues — that absolute motion and special relativity effects, and even Lorentz symmetry, are all compatible: a possible preferred frame is hidden by the Einstein measurement protocol.

As regards a 3D background space as fundamental arena of special relativity, finally, it is important to underline that, in the recent paper *Holographic special relativity*, Derek K. Wise reinterpreted special relativity, or more precisely its de Sitter deformation, in terms of a 3D conformal geometry, as opposed to 4D space-time geometry [141]. Wise's approach presents an alternative way of encoding observers by using not Lorentzian but rather conformal geometry. An inertial observer, usually described by a geodesic in space-time, becomes instead a choice of ways to reverse the conformal compactification of a Euclidean vector space up to scale. The observer's "current time" usually given by a point along the geodesic, corresponds to the choice of scale in the decompactification. In particular, Wise's approach is based on the following perspective on the de-Sitter observer space:

> "Observers live in the conformal 3D-sphere. Different observers are distinguished by their preferred Euclidean decompactification, so observer space is the space of all such decompactifications. Space-time is an auxiliary construction given by identifying all observers who share the same co-oriented unit sphere."

Moreover, Wise underlines that, on the light of the 3D conformal arena of de Sitter deformation of special relativity, the possibility is opened to extend a 3D holographic dynamic conformal geometry also to general relativity which is equivalent to the ADM formulation of general relativity under certain conditions.

In Wise's approach, the key to the relationship between conformal and Lorentzian geometry is a certain Lie group coincidence: the connected de Sitter group

$$G := SO_0(4;1) \qquad (2.20)$$

is not only the proper orthochronous isometry group of de Sitter space-time $S^{3,1}$ but also the group of symmetries of the conformal three-sphere, represented as a projective light cone P(C) in (4+1)-dimensional Minkowski space-time. The relationship between these two spaces is subtle. While both are homogeneous G-spaces, and P(C) can be thought of as the past or future boundary of $S^{3,1}$, there is no G-equivariant map between $S^{3,1}$ and P(C), and hence no direct way of mapping things happening in space-time to things happening on the boundary, or vice versa, in a way that respects the symmetries of both.

On the other hand, observers take priority over space-time. The observer space of de Sitter space-time is also a homogeneous G-space and, while there is no equivariant map between $S^{3,1}$ and P(C), there is the next best thing, a span of equivariant maps, with observer space O at the apex. Here it must be remarked that while de Sitter space-time has a conformal three-sphere as its asymptotic boundary, this boundary seems to any given observer to have more than a conformal structure: it appears to be an ordinary sphere of infinite radius, effserver's asymptotic past. Using the observer's asymptotic past as the "origin", this affine space becomes a vector space. This 'vector space in the infinite past' has an inner product only up to scale, since the magnitude diverges as we push further back into the past. However, we can renormalize the scale of this vector space by declaring the unit sphere to be at the asymptotic past of the observer's current light cone. This makes the boundary, after removing the "point at infinity", an inner product space, or Euclidean vector space.

In summary, an observer comes with a particular idea of how to put a Euclidean vector space structure on the complement of a point in the conformal sphere. This process can be referred as Euclidean decompactification, since it reverses the conformal compactification of a Euclidean vector space. Remarkably, observers in de Sitter spacetime correspond one-to-one with such Euclidean decompactifications of the sphere. Ultimately, one gets two isomorphic perspectives on the de Sitter observer space:

1. Observers live in de Sitter space-time. Different observers are distinguished by their velocity vectors, so observer space is the space of all unit future-directed time-like vectors. Conformal space is an auxiliary construction given by identifying all observers whose geodesic extensions are asymptotic in the past.
2. Observers live in the conformal three-sphere. Different observers are distinguished by their preferred Euclidean decompactification, so observer space is the space of all such decompactifications. Space-time is an auxiliary construction given by identifying all observers who share the same co-oriented unit sphere.

A co-orientation of a two-sphere embedded in P(C) is just a choice whose complementary component is 'inside' the two-sphere, here determined by the 'origin' of a given observer.

As regards the aim to provide new examples of 'observer space geometries', Wise showed that observer space geometries also arise "holographically" from 3D conformal geometry. In this regard, Wise's main mathematical result is perhaps represented by the following theorem, which gives a canonical Cartan geometry on a certain bundle over any 3D conformal manifold, modeled on the observer space of de Sitter space-time: "A Möbius geometry canonically induces an observer space geometry. Moreover, this geometry may be identified as the space of transverse three-planes in the tautological bundle corresponding to the conformal metric induced by the Möbius geometry."

2.2 Interpretation of Special Relativity in a 3D Euclid Space

In this paragraph, we will show how one can make some steps further beyond Selleri's (as well Cahill's) results by analyzing a new alternative interpretation of special relativity — recently proposed by the author of this book and Amrit Sorli — in which the fundamental arena of processes is a 3D Euclid space and time has a secondary ontological status in the sense that represents only a mathematical coordinate which measures the numerical order of material changes. We can call this approach as the "3D interpretation" of special relativity.

The starting point of the 3D interpretation of special relativity can be considered the publication of the paper *Special theory of relativity in a three-dimensional Euclid space* in Physics Essays [142]. In this paper, Amrit Sorli and the author of this book showed that Minkowski 4D space can be replaced with a 3D Euclid space with Galilean transformations

$$X' = X - v \cdot \tau,$$
$$Y' = Y, \qquad (2.21)$$
$$Z' = Z$$

for the three spatial dimensions and the following transformation

$$\tau' = \sqrt{1 - \frac{v^2}{c^2}} \cdot \tau \qquad (2.22)$$

for the rate of clocks. In these equations, X, Y, Z are the coordinates of an event measured by the observer O of a rest inertial system o and X', Y', Z' are the coordinates of the same event as measured by the observer O' of an inertial system o' which moves with respect to the inertial system o with constant velocity $v < c$ parallel to the X-axis; τ is the proper time of the observer O (namely the rate, the speed of clock of the observer O) and τ' is the proper time of the observer O' (namely the rate, the speed of clock of the observer O'). Equations (2.21) and (2.22) are valid for both observers O and O' of the inertial systems o and o'. In particular, Eq. (2.22) shows clearly that the speed of the moving clock (namely the proper time of the moving observer) does not depend on the spatial coordinates but is linked only with the speed v of the inertial system o'.

As we have shown in Sec. 2.1, a formula like Eq. (2.22) regarding the transformations of time has been found by Selleri in the context of his theory of inertial transformations which implies that space and time must be considered as distinct entities. Now, our interpretation of special relativity based on Eqs. (2.21) and (2.22) allows us to go another step further beyond Selleri's results by proposing that, at a fundamental level, the arena of processes is indeed a 3D Euclid space and time has a secondary ontological status in the sense that represents only a mathematical coordinate which measures the numerical order of material changes. In our model of special relativity, as the formalisms (2.21) and (2.22) clearly state, time and space represent two distinct entities. The transformation of the speed of clocks (2.22) derived from Selleri's formalism shows clearly that the speed of the moving clock does not depend on the spatial coordinates but is linked only with the speed v of the inertial system o'. Equations (2.21) and (2.22) determine an arena of special relativity in which the temporal coordinate must be clearly considered as a different entity with respect to the spatial coordinates just because the transformation of the speed of clocks between the two inertial systems does not depend on the spatial coordinates. This means that the three spatial coordinates of the two inertial systems turn out to have a primary ontological status, define an arena that must be considered more fundamental than the standard space-time coordinates interpreted in the sense of Einstein. On the basis of Eqs. (2.21) and (2.22) one can assume that the real arena of Special Relativity is not a mixed 4D space-time but rather a 3D space and that time does not represent a fourth coordinate of space but exists merely as a mathematical quantity measuring the numerical order of material changes. Clocks are measuring devices of the numerical order of material motion: each clock can be associated with a specific proper time of a specific observer. A clock as a measuring device of numerical order of material change in an experiment runs slower (generally, all material changes run slower) in a faster inertial system o' than in an inertial system at rest o. Experiments with clocks in a fast airplane do confirm that these relative velocities are valid for both observers O and O'.

As the author of this book and Sorli showed in the recent paper *About a new suggested interpretation of special relativity* within a three-dimensional Euclid space [143], each material motion determines a peculiar "scaling" factor of the proper time and of the spatial coordinates. By following the treatment of this paper, let us see how the concept of duration may be considered as a quantity emerging from the numerical order and how from the fundamental equations of the model (2.21) and (2.22) one can derive the standard Lorentz transformations of spatial and temporal coordinates between two inertial systems.

Equation (2.22) can be scaled by any arbitrary number which replaces τ by a new proper time $\tilde{\tau}$. By substituting this new proper time into Eq. (2.21) one obtains a different result for the transformation of the spatial coordinates. The correct results of special relativity can be then reproduced by defining the concept of physical duration of motion as a proper, "true", physical scale for the proper time τ. More precisely, the physical time — intended as duration of material change — measured by the stationary observer can be associated with a peculiar scaling physical factor of the numerical order of the following form:

$$t = \alpha(v, X, \tau), \qquad (2.23)$$

where the scaling factor α is a function also of the velocity of the moving frame with respect to the rest frame and of the position of the material object with respect to the rest frame.

The velocity of a material object can be considered as the physical entity which derives from the duration (namely from this true physical scale α of the numerical order) as well as from the position of the object. In other words, the motion of each material object and of each reference frame can be associated with the duration of material change defined as this peculiar physical scaling factor α of the numerical order as well as with the spatial coordinates. The notion of velocity derives just from the physical scale of the numerical order and from the spatial coordinates.

In particular, we assume to make the following choice for the scaling function α:

$$t = \tau\left(1 - \frac{v^2}{c^2}\right) + \frac{vX}{c^2}. \qquad (2.24)$$

The physical time t, defined by Eq. (2.24) represents the duration of material motion as measured by the stationary observer O in the rest inertial frame o. Equation (2.24) indicates that the physical time as a duration of material change emerges from a more fundamental numerical order. On the basis of Eq. (2.24), the numerical order may be considered as the internal structure of the physical, measurable time intended as duration.

The duration of material change, since is a physical scale of the numerical order, follows a transformation law analogous to Eq. (2.22). In other words, the physical time t' — representing the duration of the material motion measured by the moving observer O' in the inertial frame o' — will be linked to t by a transformation law similar to Eq. (2.22). Thus, by substituting Eq. (2.22) into Eq. (2.24) one obtains the following transformation for the duration of material motion:

$$t' \frac{1}{\sqrt{\left(1 - \frac{v^2}{c^2}\right)}} \left(1 - \frac{v^2}{c^2}\right) + \frac{vX}{c^2} = t \qquad (2.25)$$

namely

$$t' = \frac{t - \frac{vX}{c^2}}{\sqrt{\left(1 - \frac{v^2}{c^2}\right)}}, \qquad (2.26)$$

which coincides with the standard Lorentz transformation for the time coordinate.

We have thus shown that the Lorentz transformation for the time coordinate of standard Einsteinian special relativity emerges as a consequence of the more fundamental transformation for the numerical order of material changes and that the time coordinate of Einstein intended as duration of material change can be seen as a physical scaling factor of the numerical order of material change on the basis of Eq. (2.24). According to the author, this can be considered an important indication that the standard 4D space-time does not represent the fundamental arena of special relativity but rather emerges from a more fundamental 3D arena where time exists only as a numerical order of material changes (and the physical duration emerges as a physical scaling function of this numerical order).

Now, let us see how one can derive the standard Lorentz transformation for the first spatial coordinate by starting from the 3D arena described by Eqs. (2.21) and (2.22) and the duration of material changes (2.24). In this regard, before all, we write the inverse of Eq. (2.24):

$$\tau = \frac{t - \frac{vX}{c^2}}{\left(1 - \frac{v^2}{c^2}\right)}. \quad (2.27)$$

Now, the re-scaling of the numerical order determined by the material motion indirectly acts on the measurement of the position of the material object in the moving inertial frame. There is therefore a re-scaling of the position of the material object into consideration as a consequence of its motion as measured by the observer in motion. In particular, it seems permissible to assume that the effect of the duration of the motion of the material object is to originate the following scaling change of the first spatial coordinate:

$$x = \delta(\tau, t, X, v)(X - v\tau) \quad (2.28)$$

namely

$$X' = \delta(\tau, t, X, v)(X - v\tau), \quad (2.28a)$$

where the scaling function δ depends on the numerical order, the duration of material change, the velocity of the moving frame with respect to the rest frame and the position of the material object with respect to the rest frame. More precisely, the function δ may be considered as a re-scaling of the distance determined by the duration of material change (2.24) with respect to the numerical order. This function may be derived directly starting from the duration of material change (2.24), in other words may be considered as the consequence of the re-scaling of the numerical order produced by the material change into consideration. From Eq. (2.24) one has

3D Euclid Space and Special Relativity

immediately:

$$t - \frac{vX}{c^2} = \tau \left(1 - \frac{v^2}{c^2}\right). \tag{2.29}$$

By dividing (2.11) for the numerical order one has

$$\frac{1}{\tau}\left(t - \frac{vX}{c^2}\right) = \left(1 - \frac{v^2}{c^2}\right). \tag{2.30}$$

Thus, from Eq. (2.24) follows that the re-scaling factor of the distance in the first spatial coordinate determined by the material motion having duration (2.6) is of the form:

$$\delta = \sqrt{\frac{1}{\tau}\left(t - \frac{vX}{c^2}\right)} \tag{2.31}$$

and, on the basis of Eq. (2.29), this re-scaling factor is just equal to

$$\delta = \sqrt{1 - \frac{v^2}{c^2}}. \tag{2.32}$$

Therefore, by substituting Eq. 2.32 into Eq. (2.28a), one obtains the following transformation of the first spatial coordinate

$$X' = \sqrt{1 - \frac{v^2}{c^2}}\,(X - v\tau). \tag{2.33}$$

Finally, by substituting Eq. (2.24) into Eq. (2.33) one obtains

$$X' = \sqrt{1 - \frac{v^2}{c^2}}\left(X - v \cdot \frac{t - \frac{vX}{c^2}}{1 - \frac{v^2}{c^2}}\right) \tag{2.34}$$

namely

$$X' = \frac{X - vt}{\sqrt{1 - \frac{v^2}{c^2}}}, \tag{2.35}$$

which coincides just with the standard Lorentz transformation for the first spatial coordinate. The derivation (2.27)–(2.35) shows that the standard Lorentz transformation of the first spatial coordinate is not a primary physical law but emerges from a more fundamental 3D arena where time, at a fundamental level, exists only as a numerical order and the duration of material change is a scaling function of the numerical order and determines a re-scaling of the position of the object as measured by the moving observer: the behavior of the first spatial coordinate described by Eq. (2.35) is determined just by the link between the duration of the material motion and the numerical order.

In synthesis, on the basis of the treatment made in the paper [143], the suggestive perspective is opened that the standard Lorentz transformations of Einstein's special relativity concerning the temporal coordinate (2.26) and the first spatial coordinate (2.35) derive from a more fundamental 3D arena described by Eqs. (2.21) and (2.22) under the hypothesis that the duration of material change is a quantity which emerges from a more fundamental numerical order (proper time) on the basis of Eq. (2.24) and that the link between the duration of material change and the numerical order corresponding to the velocity of the moving observer with respect to the rest frame determines a re-scaling of the position of the object into consideration in the moving frame. The treatment made in this chapter suggests that the standard Einsteinian formalism of special relativity may be obtained by starting from the idea that the fundamental arena of physical processes is a 3D arena where time, at a fundamental level, is a different entity with respect to the spatial coordinates, namely exists only as a numerical order (in the sense that the transformation of the speed of clocks between the two inertial systems does not depend on the spatial coordinates) and the physical duration of material change emerges as a scaling function of the numerical order. In other words, in this approach to special relativity, Eqs. (2.21) and (2.22) define the fundamental arena of physical processes and may be considered the fundamental laws; instead, the standard Lorentz transformations (2.26) and (2.35) of Einstein's special relativity may be seen as the laws governing another level, which may be defined as the level of the measurements of the observers, and emerge from the fundamental laws (2.21) and (2.22) in the hypothesis that the measurable time as duration emerge from the numerical order on the basis of Eq. (2.24) and determines a re-scaling of the position as measured in the moving frame.

On the basis of Eqs. (2.21) and (2.22) there is no fundamental "time dilation" as it is known in Einstein's special theory of relativity, namely dilation of time as a fourth coordinate of space causes clocks to have a slower rate. What we measure in different inertial systems is the relative velocity of material change (including run of clocks), namely their numerical order.

Also the "length contraction" in the direction of motion of an inertial system along the axis X (predicted by the standard Lorentz formalism (2.35)) cannot be considered as a fundamental primary reality: it can be seen as an effect of motion determined by the link between the duration of material motion and the numerical order on the basis of Eq. (2.24), in other words by the re-scaling function, corresponding to the duration of material change, produced by the material change into consideration. At the fundamental level expressed by Eqs. (2.21) and (2.22), there is no length contraction. On the other hand, as shown in [142], length contraction along axis X would lead to a contradiction: in a moving inertial system horizontal photon clock positioned along the axis X would shrink and would have faster rate than identical photon clocks positioned vertically. Regarding "length contraction" some other research lead to the same conclusions. Since 1905 when special theory of relativity was published there has been no experimental data on "length contraction" [144] intended as a fundamental physical reality.

Moreover, in this model based on Eqs. (2.21) and (2.22), at a fundamental level, time travels into the past and into the future cannot be considered as a physical reality: in the arena described by Eqs. (2.21) and (2.22) one can travel in space only. Past, present and future exist only as a numerical order of changes which run in a 3D space [144–146].

Now, let us consider two events 1 and 2 occurring in two distinct points of space and characterized by a different numerical order (with respect to the observer O at rest). In the inertial system o, the spatial distance between these two events as measured by the observer O is

$$\Delta s^2 = \Delta X^2 + \Delta Y^2 + \Delta Z^2, \tag{2.36}$$

while the material change associated with these two events is described by the numerical order $\Delta\tau$. In the moving inertial system o′, the spatial distance between these two events as measured by the observer O′ is

$$\Delta s'^2 = \Delta X'^2 + \Delta Y'^2 + \Delta Z'^2, \qquad (2.37)$$

while the numerical order of material change associated with these two events is $\Delta\tau'$.

On the basis of Eq. (2.21), Eq. (2.37) becomes

$$\Delta s'^2 = (\Delta X - v\Delta\tau)^2 + \Delta Y^2 + \Delta Z^2 \qquad (2.38)$$

while, taking into account the transformation of the rate of clocks (2.22), the numerical order of material change associated with the two events under consideration measured with respect to the inertial system o′ is

$$\Delta\tau' = \sqrt{1 - \frac{v^2}{c^2}} \cdot \Delta\tau. \qquad (2.39)$$

In this model, Eqs. (2.38) and (2.39) come to substitute the relativistic invariant quantity represented by the square of the modulus of the four-interval in Minkowski space-time. Since the standard Lorentz transformations for the space and time coordinates emerge as a consequence of the more general transformations (2.21) and (2.22) in a more fundamental 3D arena, we can also say that the relativistic invariant quantity represented by the square of the modulus of the four-interval in Minkowski space-time can be seen as a consequence of Eqs. (2.38) and (2.39).

In synthesis, taking into account the fundamental relativistic postulate of the invariance of the velocity of light for different observers and Eqs. (2.21) and (2.22), a "3D interpretation" of special theory of relativity can be proposed which is based on the following postulates:

(1) The velocity of light has the same value in all directions and in all inertial systems (postulate of the invariance of the velocity of light in the vacuum).

(2) The fundamental arena of physical processes is a 3D Euclid space where time exists merely as a mathematical quantity measuring the numerical order of material changes (and which can be defined as the proper time of the observer under consideration). Given two inertial frames o and o', where the origin of o', observed from o, is seen to move with velocity v parallel to the X-axis, the transformations between the spatial coordinates of these two systems are given by Eq. (2.21), while the transformation of the rate of clocks, namely between the proper times of the observers of these two inertial systems, is given by Eq. (2.22). The duration of material changes satisfying the standard Lorentz transformation for the temporal coordinate is a proper, physical scaling function which emerges from the more fundamental numerical order and determines itself a re-scaling of the position measured by the moving observer expressed by the standard Lorentz transformation for the first spatial coordinate.

2.3 About Light Clocks

On the basis of the treatment made by Sorli, Klinar and the author of this book in the paper [145], the common interpretation of how an observer at rest experiences the velocity of a "light clock" in a moving inertial system seems not adequate. The common interpretation of "light clocks" in special relativity from the point of view of the observer O who is at rest is shown in Fig. 2.1. The clock o on the left is at rest, the clock o' on the right is moving with constant velocity in horizontal direction. According to the common interpretation for the observer O at the station, the moving clock o' is supposed to run ("tick") slower than the clock at rest o because the path of the photon seems longer than by the clock at rest. One "tick" of the clock means a photon passes one distance between the two mirrors [147].

The interpretation provided by Sorli and the author of this book in the papers [142, 145] is the following: the observer O experiences that photon of the moving clock o' has a longer path between mirrors only because the clock o' is moving. This "illusive experience" caused

140   The Timeless Approach

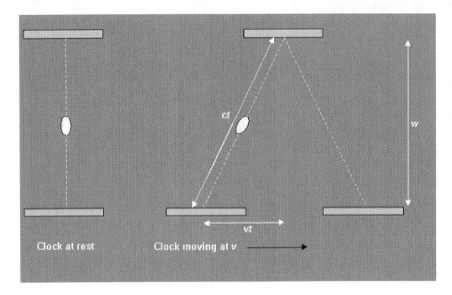

Fig. 2.1. Two identical light clocks: one at rest, one moving relative to us. The light blips in both travel at the same speed relative to us, the one in the moving clock goes further, so must take longer between clicks. t is the time from one mirror to the other.

by motion does not mean that for the observer O the moving clock o′ will have a slower rate. In the moving clock o′, photon moves in a vertical direction and has the same path between mirrors as in the clock o at rest. Out of this the following consequences can be drawn:

(a) The photons in a clock at rest o and in a moving clock o′ move in a vertical direction.
(b) The path of the photons between mirrors is equal in both clocks o and o′.
(c) The velocity c of photons is the same in both clocks o and o′.

The points (a), (b), (c) are valid for both observers O and O′. This "thought experiment" with clock at rest o and moving clock o′ seems to suggest that at the photon scale the "relative velocity" of the physical phenomena in different inertial systems is not valid any more. The "relative velocity" of material changes starts with massive particles as it is proved by different decay time of rest pi mesons and

3D Euclid Space and Special Relativity   141

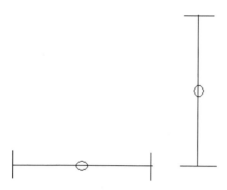

Fig. 2.2. Horizontal light clock A and vertical light clock B in a moving inertial system.

pi mesons in motion and by different lifetimes of rest muons and muons in motion [148].

Moreover, by following the treatment of Sorli and Fiscaletti in the paper [142], let us consider the following thought experiment with two "light clocks" (where, as in the example above, a photon moves between two mirrors). In a moving inertial system o', we put one light clock A horizontally in the direction of motion along the axis X and another light clock B vertically with respect to the axis X. One path of the photon between mirrors means one "tick" of the clock. According to the "length contraction" clock A should shrink and so photon would have a shorter path and clock A would "tick" faster than clock B. However, special relativity does not predict that two clocks in the same inertial system should have a different rate.

In virtue of the constancy of light speed the vertical light clock B and the horizontal light clock A should have the same rate. As shown clearly by Sorli and the author in the paper [142], the "length contraction" of horizontal light clock A creates therefore a paradox. The contradiction between the results of the vertical light clock and the horizontal light clock suggests that there should be no "length" contraction along the direction of motion (namely the axis X). To resolve this paradox, in the paper [142], Sorli and the author of this book proposed just that Minkowski's 4D space must be replaced with a 3D Euclid space with the Galilean transformations

(2.21) for the spatial coordinates and Selleri's formalism (2.22) for the transformation of the speed of clocks. On the basis of Eqs. (2.21) and (2.22), in the arena which describes electromagnetic phenomena, time is a distinct entity from space (it is only a mathematical quantity indicating the numerical order of material changes) and there is no "length contraction" of moving light clocks. Vertical light clock B and horizontal light clock A have the same rate. The 3D Euclid space described by Eqs. (2.21) and (2.22) allows us to resolve in a clear and elegant way the light clocks paradox.

2.4 Does Abolishing of Space-Time as a Fundamental Arena of the Universe Mean the Introduction of Modern Concepts of "Ether"?

According to interpretation of special relativity based on the postulates (1) and (2) enunciated in Sec. 2.1, the 3D Euclid background space described by Eqs. (2.21) and (2.22) (with the consequent abolishing of time as a primary physical reality, and thus as a fourth dimension of space and space-time, as a fundamental arena of the universe) does not require introduction of modern concepts of "ether" as proposed by Levy, Duffy and some other researchers [149]. Physical space does not behave as a medium which carries the light, but instead the propagation of light is governed by the electromagnetic properties of the quantum vacuum, its permeability and permittivity. Reintroduction of "ether" is not necessary as a fundamental quantum vacuum covers all its properties.

In his paper *Ether theory and the principle of relativity* [149], Levy has derived a set of space-time transformations which assume the existence of a preferred ether frame and the variability of the one-way speed of light in the other frames. In Levy's approach, the extended transformations can be converted into a set of equations that have a similar mathematical form to the Lorentz–Poincaré transformations, but which differ from them in the sense that they connect reference frames whose coordinates are altered by the measurement distortions due to length contraction and clock retardation

3D Euclid Space and Special Relativity 143

and by the synchronization procedures. Instead, in the view of special relativity based on the postulates (1) and (2) enunciated in Sec. 2.2, the transformations of spatial coordinates (2.21) and of time (2.22) imply the following fundamental results: on one hand, that there is no length contraction along the direction of motion of an inertial system and, on the other hand, that time is a distinct entity from space and that the idea of dilation of time can be replaced with the idea of relative velocity of material changes (including runs of clocks) in different inertial systems. The fundamental arena of special relativity is indeed a 3D Euclid space subjected to Galilean transformations for the three coordinates of space and where time exists only as a mathematical parameter measuring the numerical order of material changes: in each inertial system there is a peculiar velocity of material changes. The view here suggested presents Einstein's special relativity in an ordinary way without unnecessary introduction of the imaginary coordinate $X_4 = ict$. In the universe time intended as a fourth dimension of space does not exist and so cannot be "relative". With clocks we measure velocity and duration of material changes; what is relative is a velocity of material changes which depends on the energy density of a fundamental quantum vacuum from which universal space originates.

2.5 Aberration of Light

The 3D interpretation of special relativity here presented, on the basis of Eqs. (2.1), (2.2) (and the consequent Eqs. (2.20) and (2.21) allows us to analyze and explain in a clear and simple way several relevant phenomena inside special relativity, such as aberration of light, Doppler effect, Jupiter's satellites occultation and radar ranging of the planets.

Let us examine before all aberration. By following the treatment in [89], let us consider the propagation of a light corpuscle P on the $X - Y$ plane of the inertial system o. On the basis of the postulate (1), P propagates at the same velocity c in each inertial system and in each direction. So, with respect to the stationary observer O of the inertial system o, P is described by coordinates satisfying, at the proper time τ,

$$\begin{cases} X = c\tau \cos\vartheta, \\ Y = c\tau \sin\vartheta. \end{cases} \quad (2.40)$$

Relative to the moving inertial system o', P is described by coordinates satisfying, at the proper time τ', the following relations

$$\begin{cases} X' = c\tau' \cos\vartheta', \\ Y' = c\tau' \sin\vartheta'. \end{cases} \quad (2.41)$$

Substituting Eqs. (2.21) and (2.22) into Eq. (2.41) we obtain

$$\begin{cases} X - v\tau = c\tau\sqrt{1 - \dfrac{v^2}{c^2}} \cos\vartheta', \\ Y = c\tau\sqrt{1 - \dfrac{v^2}{c^2}} \sin\vartheta', \end{cases} \quad (2.42)$$

from which, taking account of Eq. (2.40) we have

$$\begin{cases} c\tau \cos\vartheta - v\tau = c\tau\sqrt{1 - \dfrac{v^2}{c^2}} \cos\vartheta', \\ c\tau \sin\vartheta = c\tau\sqrt{1 - \dfrac{v^2}{c^2}} \sin\vartheta', \end{cases} \quad (2.43)$$

namely

$$\begin{cases} \dfrac{c\tau \cos\vartheta - v\tau}{c\tau\sqrt{1 - \frac{v^2}{c^2}}} = \cos\vartheta', \\ \dfrac{1}{\sqrt{1 - \frac{v^2}{c^2}}} \sin\vartheta = \sin\vartheta', \end{cases} \quad (2.44)$$

which provide the following formula for the aberration of light pulse in a 3D Euclid space with transformations of the rate of clocks given by formalism (2.22):

$$tg\vartheta' = \frac{c\tau \sin\vartheta}{c\tau \cos\vartheta - v\tau}, \quad (2.45)$$

namely

$$tg\vartheta' = \frac{c\sin\vartheta}{c\cos\vartheta - v}. \qquad (2.46)$$

Equation (2.46) turns out to be in perfect agreement with the experimental evidence: it coincides perfectly with the aberration formula obtainable from a classical treatment of the problem. Therefore, the mathematical formalism of special relativity based on Eqs. (2.21) and (2.22) (with the implicit postulate of the invariance of the speed of light, namely that the speed of light in the vacuum has the same value c in all directions and in all inertial systems) leads to the same result (2.44) of the classical treatment represented by the Galilean transformations. In other words, the model of special relativity based on the postulates (1) and (2) allows us to reconcile the treatment of the aberration of light with the classical scheme.

2.6 Doppler Effect

Let us consider a plane electromagnetic wave which is propagating in the vacuum. In the stationary inertial system o, this wave is described by the wave function

$$\psi(\vec{r}, \tau) = \psi_0 \exp\left[i\omega\left(\tau - \frac{\hat{n} \cdot \vec{r}}{c}\right)\right], \qquad (2.47)$$

where ψ_0 is a constant amplitude, ω is the angular frequency, $\hat{n} = (\cos\vartheta, \sin\vartheta)$ is the unit vector normal to the wave fronts. Equation (2.47) can also be explicitly expressed as

$$\psi(\vec{r}, \tau) = \psi_0 \exp\left[i\omega\left(\tau - \frac{X\cos\vartheta + Y\sin\vartheta}{c}\right)\right]. \qquad (2.48)$$

In the moving inertial system o', this same wave is described by the wave function

$$\psi'(\vec{r}', \tau) = \psi_0' \exp\left[i\omega'\left(\tau' - \frac{\hat{n}' \cdot \vec{r}'}{c}\right)\right]. \qquad (2.49)$$

On the basis of what we have seen in the paragraph 2.5, the unit vector $\hat{n}'$ has components given by the following relation

$$\hat{n}' = \left(\cos\vartheta - \frac{v}{c}, \sin\vartheta\right). \tag{2.50}$$

By inserting Eqs. (2.21), (2.22) and (2.50) into the phase of the wave function (2.47) we obtain

$$\left[\omega'\left(\tau' - \frac{\hat{n}'\cdot\vec{r}'}{c}\right)\right]$$

$$= \omega\left[\tau\sqrt{1-\frac{v^2}{c^2}} - \frac{\left(\cos\vartheta - \frac{v}{c}, \sin\vartheta\right)\cdot(X - v\tau, Y)}{c}\right] \tag{2.51}$$

namely

$$\left[\omega'\left(\tau' - \frac{\hat{n}'\cdot\vec{r}'}{c}\right)\right]$$

$$= \omega\left[\tau\sqrt{1-\frac{v^2}{c^2}} - \frac{\left(X\cos\vartheta - \frac{vX}{c} - v\tau\cos\vartheta + \frac{v^2\tau}{c} + Y\sin\vartheta\right)}{c}\right] \tag{2.52}$$

namely

$$\left[\omega'\left(\tau' - \frac{\hat{n}'\cdot\vec{r}'}{c}\right)\right] = \omega\left[\tau\sqrt{1-\frac{v^2}{c^2}}\right.$$

$$\left. - \frac{\left(X\cos\vartheta - \frac{vX}{c} - v\tau\cos\vartheta + \frac{v^2\tau}{c} + Y\sin\vartheta\right)\sqrt{1-\frac{v^2}{c^2}}}{c\sqrt{1-\frac{v^2}{c^2}}}\right]. \tag{2.53}$$

Equation (2.53) may be rewritten as

$$\left[\omega'\left(\tau' - \frac{\hat{n}' \cdot \vec{r}'}{c}\right)\right]$$

$$= \omega \left[\frac{c\tau\left(1 - \frac{v^2}{c^2}\right) - (X\cos\vartheta + Y\sin\vartheta)\sqrt{1 - \frac{v^2}{c^2}} + v\left(\frac{X}{c} + \tau\cos\vartheta - \frac{v\tau}{c}\right)\sqrt{1 - \frac{v^2}{c^2}}}{c\sqrt{1 - \frac{v^2}{c^2}}} \right] \quad (2.54)$$

namely

$$\left[\omega'\left(\tau' - \frac{\hat{n}' \cdot \vec{r}'}{c}\right)\right]$$

$$= \omega \left[\sqrt{1 - \frac{v^2}{c^2}}\right] \left[\tau + \frac{-(X\cos\vartheta + Y\sin\vartheta) + v\left(\frac{X}{c} + \tau\cos\vartheta - \frac{v\tau}{c}\right)}{c\sqrt{1 - \frac{v^2}{c^2}}}\right]$$

$$(2.55)$$

namely

$$\left[\omega'\left(\tau' - \frac{\hat{n}' \cdot \vec{r}'}{c}\right)\right] = \omega \left[\sqrt{1 - \frac{v^2}{c^2}}\right]$$

$$\times \left[\tau'/\sqrt{1 - \frac{v^2}{c^2}} - \frac{\hat{n} \cdot \vec{r}' - v\left(\frac{X'}{c}\right)}{c\sqrt{1 - \frac{v^2}{c^2}}}\right]. \quad (2.56)$$

On the basis of Eqs. (2.51)–(2.56) we obtain thus that in the moving inertial system o', the plane wave has a frequency

$$\omega' = \omega \left[\sqrt{1 - \frac{v^2}{c^2}}\right], \quad (2.57)$$

which means that the frequency of the plane electromagnetic wave follows the same behavior of the rate of clocks namely does not depend on the spatial coordinates. Equation (2.57) is coherent with Einstein's derivation of relativistic Doppler effect — based on Lorentz transformations — in virtue of the fact that, in this approach, Lorentz transformations emerge at a secondary level from Eqs. (2.21) and (2.22) taking account of the duration of material change (2.24).

As regards the transverse Doppler effect, which occurs when the observer is displaced in a direction perpendicular to the direction of the motion of the source, assuming the objects are not accelerated, light emitted when the objects are closest together will be received some time later, at reception the amount of redshift will be given by the factor — independent of the spatial coordinates — present in Eq. (2.57). The transverse Doppler effect is a consequence of the fundamental Eq. (2.57). In the frame of the receiver, when the angle θ_0 between the direction of the emitter at emission, and the observed direction of the light at reception is equal to $\pi/2$, the light was emitted at the moment of closest approach, and one obtains the transverse redshift expressed by Eq. (2.57).

2.7 Jupiter's Satellites Occultation

Let a Jupiter's satellite (for example, Io) be in a state of motion on the X-axis of the stationary inertial system o. Io sends a light signal (occultation) characterized by a numerical order T, with respect to this inertial system, when it is in the position

$$X_1 = -L. \qquad (2.58)$$

The equation of motion of the signal relative to the inertial system o is:

$$X = -L + c(\tau - T). \qquad (2.59)$$

Let us suppose that the Earth is moving with constant speed v and thus constitute instantaneously a moving inertial system o'. The equation of motion of Earth as seen from the stationary system o is the following:

$$X_E = v\tau. \tag{2.60}$$

The event represented by the arriving of the signal on Earth is associated with a numerical order τ_a which, on the basis of Eqs. (2.59) and (2.60) satisfies the following condition

$$v\tau_a = -L + c(\tau_a - T), \tag{2.61}$$

whence

$$\tau_a = \frac{L + cT}{c - v}. \tag{2.62}$$

The Earth position corresponding to the numerical order (2.62) is

$$X_a = v\tau_a. \tag{2.63}$$

Our problem is to determine the numerical order τ'_a marked by a clock on Earth when the signal is received. Between the stationary system o and the Earth system o' the transformation (2.22) applies and one obtains

$$\tau'_a = \sqrt{1 - \frac{v^2}{c^2}} \cdot \tau_a, \tag{2.64}$$

whence, using Eq. (2.62), we have

$$\tau'_a = \sqrt{1 - \frac{v^2}{c^2}} \cdot \frac{L + cT}{c - v} \tag{2.65}$$

which is just the result obtained inside the standard Einsteinian special theory of relativity. The 3D interpretation based on postulates (1) and (2) predict therefore exactly the same occultation time of Jupiter's satellites to be observed on Earth.

2.8 Radar Ranging of the Planets

Let the equations of motion, written in the stationary inertial system o, of Earth, of Venus, and of a radar signal sent from Earth towards Venus respectively be:

$$X_1 = v\tau, \quad X_2 = v_2\tau + d, \quad X = c\tau. \tag{2.66}$$

The numerical order τ_R corresponding to the reflection of the radar pulse on the Venus surface must thus satisfy the following condition [150]

$$c\tau_R = v_2\tau_R + d, \qquad (2.67)$$

whence

$$\tau_R = \frac{d}{c - v_2}. \qquad (2.68)$$

During the return journey of the radar pulse from Venus to Earth the latter still obeys the first equation of (2.66), while the pulse must satisfy relation

$$X_2 = X_R - c(\tau - \tau_R), \qquad (2.69)$$

where

$$X_R = c\tau_R = \frac{dc}{c - v_2} \qquad (2.70)$$

is the position occupied jointly by Venus and the pulse for the numerical order τ_R. Therefore, the arrival of the pulse on Earth can be described by a numerical order τ_A which must satisfy the following condition:

$$v\tau_A = X_R - c(\tau_A - \tau_R). \qquad (2.71)$$

By using (2.68) and (2.69), Eq. (2.70) becomes:

$$\tau_A = \frac{2dc}{(c - v_2)(c + v)}. \qquad (2.72)$$

Now we want to study the phenomenon in the moving inertial system o′ associated with Earth. In this regard, by using Eq. (2.22) we obtain:

$$\tau'_A = \sqrt{1 - \frac{v^2}{c^2}} \cdot \tau_A, \qquad (2.73)$$

whence

$$\tau'_A = \sqrt{1 - \frac{v^2}{c^2}} \cdot \frac{2dc}{(c - v_2)(c + v)}, \qquad (2.74)$$

which is the same result obtained in standard Einsteinian special theory of relativity.

Chapter 3

Three-Dimensional non-Euclid Space as a Direct Information Medium and Quantum Phenomena

Quantum mechanics is not just a theory describing the behavior of microobjects: it is our current fundamental theory of motion. It can provide a deeper explanation and understanding of natural processes than classical mechanics. The empirical success of quantum mechanics is immense. Its physical obscurity is undeniable. The interpretation of what the quantum theory actually tells us about the physical world has raised a lively debate, which has continued from the early days of the theory to date. There are many interpretations of quantum mechanics described in some excellent books and articles. No consensus about which one is "valid", if any, has been established so far. Indeed each interpretation seems to have its own merits and elucidates certain aspects of quantum mechanics.

As is known, in quantum mechanics each physical system is described by a wave function. The wavefunction of an isolated microsystem evolves freely according to the Schrödinger evolution, that is certainly one of the most important equations of physics, as it allows us to understand the behavior of many materials and physical systems, such as for example semiconductors and lasers. Quantum mechanics was however originally formulated as a theory of quantum microsystems that interact with classical macrosystems. In the original formulation of the theory (known as Copenhagen interpretation or standard interpretation), the interaction of a quantum microsystem S with a classical macrosystem O is described in terms of "quantum measurements" [151, 152]. Once the microsystem S under consideration interacts with its surroundings, they become entangled and they are in a quantum mechanical superposition. If the macrosystem O interacts with the variable q of the microsystem S, and S is in a superposition of states of different values of q, then the macrosystem O measures only one of the values of q, and the interaction modifies the state of S by projecting it into a state with that value: in every measurement the wavefunction of a microsystem collapses into the state specified by the outcome of the measurement.

Therefore, there are two kinds of dynamics of physical systems in the standard interpretation of quantum mechanics: the well understood and uncontroversial continuous, unitary evolution of quantum states according to the Schrödinger equation; and the ill-understood and controversial discontinuous, nonunitary evolution of the same states during quantum measurements. Whenever a measurement is performed the wavefunction ceases to evolve according to the Schrödinger equation and collapses into one of its eigenstates: in every measurement the wavefunction of a microsystem collapses into the state specified by the outcome of the measurement. Therefore, according to the standard interpretation, on the basis of the so-called reduction postulate of the wavefunction, when a measurement is performed on a system which at the beginning is in a superposition of macroscopically different states, after the measurement its wavefunction collapses into one of these states. This final state is not predictable with certainty but only probabilistically. The absolute square of the scalar product of the wavefunction with its eigenfunctions are the probabilities (or probability densities) of the occurrence of these particular eigenvalues in the measurement process [151, 153]. As a consequence of this, according to the Copenhagen interpretation of quantum mechanics, it is not possible to provide a causal description of microscopic processes: the wavefunction carries only the information about possible outcomes of a measurement process and the reduction of wavefunction is caused by the interference, by the interaction between the observer and the measured system [154].

3D non-Euclid Space as a Direct Information Medium   153

Although the formal apparatus of standard quantum theory functions perfectly under the point of view of the empirical predictions, it is characterized by inner contradictions and therefore cannot be considered completely self-consistent. In synthesis: many authors do not find it satisfactory to resort to two different postulates as regards the modality of evolution of a system whether it is subjected to observation or not; that, in virtue of the unlimited validity of the superposition principle, superpositions of macroscopic states such as alive cat–dead cat, according to the famous Schrödinger's mental experiment, exist; that a boundary between the microscopic world (governed by the superposition principle) and the macroscopic world (in which we have well-defined perceptions as regards the properties of physical systems) cannot be defined in a precise way [155]. The failure of orthodox quantum theory to offer any sort of coherent resolution to concerns of this sort is largely the reason for which it has continually remained so ambiguous and obscure.

In this chapter, after a review of the role of time in the interpretations of quantum mechanics (also in the light of some current research), our purpose is to analyse an approach to quantum mechanics, recently proposed by the author, according to which the fundamental background of quantum phenomena is a timeless three-dimensional (3D) non-Euclid space which acts as an immediate information medium between subatomic particles.

3.1 About Time in the Interpretations of Quantum Theory

In the debate regarding the interpretation of quantum theory, the most important matters probably are quantum nonlocality and the quantum measurement problem, of which the second is generally regarded as more fundamental. The resolutions of the debate between these two problems have significant repercussions for our conception of time. Here, we will focus our attention on the following three well-known interpretations, or rather families of more precise interpretations: the collapse of the wavefunction (which covers several approaches such as the Copenhagen interpretation and the dynamical reduction programme of Ghirardi, Pearle, Penrose and others), the pilot-wave interpretation (which covers not only de Broglie's and Bohm's original theories with point-particles always having a definite position, but also more elaborated theories, including field theories), and Everett's many-worlds interpretation. Our aim is to describe how each of these approaches has distinctive consequences about the nature of time.

3.1.1 The collapse of the wavefunction

The collapse of the wavefunction refers to an irreducibly indeterministic, abrupt (discontinuous) and irreversible (non-unitary) change in the state of an isolated quantum system, contravening the deterministic, continuous and unitary evolution prescribed by the quantum theory's fundamental equation of motion (the Schrödinger equation). The collapse is postulated in order to solve the measurement problem: viz. by assuring that measurements have definite outcomes, albeit ones that are not determined by the previous quantum state. Anyone who advocates such a collapse is supposed to face several questions. Three of the most pressing ones are as follows. Under exactly what conditions does the collapse occur? What determines the physical quantity (in the formalism: the basis) with respect to which it occurs? How can the collapse mesh with relativity?

Though these questions are relevant for the foundations of physics, however, the irreducible indeterminism associated with the collapse of the wavefunction raises, for philosophy, the question how exactly to understand the various alternative futures. In particular, the following question emerges naturally: is indeterminism compatible with the idea of a single actual future, i.e. the 'block universe' or 'B-theory of time'? In this regard, some authors answer in affirmative way (e.g., Lewis [156] and Earman [157]). Broadly speaking, the idea at the basis of the approach of these authors is the following: an indeterministic event requires that two or more possible histories (possible worlds) match utterly up to the event in question, but thereafter fail to match (namely 'diverge'), reflecting the differences in the event's various possible outcomes and in these outcomes' later causal consequences. On the other hand, other authors answer in negative way (e.g., McCall [158]). In this second kind of approaches, the fundamental idea is the following: an indeterministic event requires that a single possible history (possible world) splits at the event in question, with future branches incorporating the different outcomes and their respective causal consequences.

In the light of the problems emerging from the collapse of the wavefunction, the topic of time in quantum physics therefore runs up against notoriously hard questions in the metaphysics of modality. How exactly should we conceive of possible histories? We need to answer this question in order to settle how we should understand their matching: as a matter of some kind of isomorphism or counterparthood, or as fully-fledged identity? In this regard, in [159] Earman discusses how a wide variety of theorems about spacetime structures make it very hard to make the branching view mesh with relativity and how the branching view relates to the work of the 'Pittsburgh–Krakow' school of Belnap, Placek and others, on what they call 'branching space-times' (developed with an eye on quantum theory, especially quantum non-locality, e.g. by Placek [160]). The same Earman analyses, again in the light of modern physics, the proposal that over time, reality 'grows by the accretion of facts'.

3.1.2 The pilot-wave

The pilot-wave interpretation adds to quantum theory's deterministic evolution of the orthodox quantum state (wavefunction), the postulate that certain preferred quantities have at all times a definite value: these quantities are assumed to evolve deterministically in a manner governed by the quantum state/wavefunction.

The original and best-studied version of this theory postulates that, in nonrelativistic quantum theory, the preferred quantity is the position of point-particles. These positions evolve deterministically according to a guidance equation that requires the particle's momentum at any time to be proportional to the gradient of the phase of the wavefunction. In other more elaborated approaches, other quantities, such as a field quantity like the magnetic field, are preferred, i.e. postulated to have definite values that evolve deterministically 'guided' by the quantum state.

But as the phrase 'at any time' hints, many current versions of the pilot-wave interpretation utilize an absolute time structure (frame-independent simultaneity), as in Newtonian physics. Some of these approaches, particularly in field theory, guarantee the Lorentz symmetry characteristic of special relativity. However, it must be remarked that here the Lorentz symmetry is an approximate and emergent symmetry, governing a certain regime or sector of the theory, not a fundamental one. (In a somewhat similar way, the pilot-wave interpretation recovers the apparent indeterminism of orthodox quantum theory — "the collapse of the wavefunction" — as an emergent feature due to averaging over the unknown, but deterministically evolving, definite values.)

Therefore, if we base ourselves on the lead of these current versions, the consequences for the nature of time are striking: and they are crucial for physics, not "just" philosophy. In fact, the following result emerges natural: these approaches resurrect the absolute time structure of Newtonian physics.

3.1.3 *Everett's many worlds*

The Everett or many-worlds interpretation proposes to reconcile quantum theory's deterministic evolution of the orthodox quantum state with the collapse of the wavefunction, i.e. measurements having definite outcomes with various frequencies, by assuming that measurement processes involve a splitting of the universe into different branches. Obviously, this returns us to the deep and disconcerting questions, both physical and philosophical, mentioned in Sec. 3.1.1: about the conditions under which a branching occurs, how branching can mesh with relativity, and how we should understand branching. So, it is hardly surprising that this interpretation has traditionally been regarded as vague and more controversial than others. In this context, for example Bell wrote that it "is surely the most bizarre of all [quantum theory's possible interpretations]" and seems "an extravagant, and above all extravagantly vague, hypothesis. I could almost dismiss it as silly" [161].

But since the quotation of Bell, the followers of the Everett interpretation have made major improvements to this approach. According to Butterfield [162], the two main improvements have been the following:

(i) To combine the physics of decoherence with the philosophical ("functionalist") idea that objects in 'higher-level' ontology, e.g. a cat, are not some kind of aggregate (e.g. a mereological fusion) of lower-level objects, but rather dynamically stable patterns of them. This suggests that the proverbial Schrödinger's cat measurement involves an approximate and emergent splitting, after which there really are two cats (or two broad kinds of cat), since the total wavefunction is peaked over two distinctive patterns in the classical configuration space. More precisely, it is peaked over two kinds of pattern: the legs, tail and body when they are all horizontal, still and cool ('dead'), and the legs and tail when vertical, moving and warm ('alive').

(ii) To develop various arguments justifying, from an Everettian perspective, the orthodox (Born-rule) form of quantum probabilities.

Both these improvements mentioned by Butterfield, which have been developed in detail by many papers over the last 20 years, lead to relevant open questions; in particular, as regards the topic of time in quantum physics, the point (i) is the most fundamental one.

In this regard, the important point is how, in the many-worlds interpretation, decoherence processes yield a continual, but approximate and emergent, splitting of the universe that

(a) meshes fundamentally with relativity, so that there is no absolute time structure; and
(b) is to be combined with almost all objects — not just macroscopic objects like cats, tables and stars, but anything that classical physics successfully describes as having a spatial trajectory,

etc.: for example, large molecules — being treated as patterns (or better: being treated as the quantum state being peaked above such patterns in an abstract classical configuration space).

About the above listed points (a) and (b), the Everett's vision turns out to be of course hard and dizzying, but this does not mean that it is less alluring. Indeed, Everettians admit the difficulty of their approach, as well as the attraction. And some think there is important technical work to be done here. For example, Deutsch recently suggests to explore the new physics that the Everett interpretation promises to contain; and in the course of this, he admits that the exploration will be very challenging, since no one has yet given a precise mathematical description of (even toy models of) this branching structure [163, 164].

Thus, there are open questions about what more is required to articulate the Everettian vision. But however those further details go, the consequences for the philosophy of time will obviously be radical. A fundamentally deterministic (and relativistic) evolution of the universe's state will be "overlaid" by an approximate and emergent branching structure for time (and also by an apparent indeterminism).

Moreover, for philosophers of time interested in the Everett interpretation, Butterfield recommends an interesting analogy proposed by Wallace [165, 166] between times as understood on the 'block universe' or 'B-theory' of time, and worlds or branches as understood by the Everettian view. Butterfield writes: "Recall that the B-theorist says:

(i) reality is 4D, and slices across it are in principle arbitrary and artefactual (above all in relativity theory, with no absolute simultaneity); and, at the same time,
(ii) for describing the history of the universe — and in particular, in physics, for doing dynamics — only a small subset of slices will be useful; though the criteria to select that subset will be a bit approximate.

Then the proposed analogy is as follows. Similarly, the Everettian view says:

(i′) the 'slicing' of reality by choosing a basis in Hilbert space is in principle arbitrary and artefactual; but also

(ii′) for describing the history of the universe — and in physics, for doing the approximate and emergent dynamics of a world — only a small subset of bases will be useful; although the criteria to select that subset will be a bit approximate (since decoherence gives no absolute criterion for a system-environment split, or for when interference terms are small enough)."

3.2 About Recent Research on the Role of Time in Quantum Theory

Interesting considerations about the role of time in quantum theory have been made by Busch [167, 168], who has suggested a trichotomy of roles (or senses) of time. Busch's distinction of three roles of time in quantum theory provides an invaluable element for organizing discussion (i.e. preventing confusion!) about time in quantum physics (besides being equally useful for discussing time in classical physics, albeit with some obvious adjustments). This distinction has also been emphasized by Hilgevoord [169–172] who has rightly underlined that much of the historical confusion about time in quantum theory (including such figures as von Neumann and Pauli) can be dispelled by respecting this distinction, and the corresponding distinction about space, or position: viz. between (a) position as a coordinate of space-time, and (b) position as a dynamical variable, e.g. the position of a point-particle. According to Hilgevoord's view, armed with these distinctions, there is no temptation to say that quantum physics marks a break from classical physics, by making position but not time into a dynamical variable. On the contrary: both classical and quantum physics treat both time and position, either as a coordinate, or as a dynamical variable.

By following Busch's treatment in [167, 168], let us see now what are these three roles of time in quantum theory. First, Busch defines 'external time' (or also 'pragmatic time' in [167]) as the time measured by clocks that are not coupled to the objects studied in the experiment. As regards this first role, time specifies practically a parameter or parameters of the experiment: e.g. an instant or duration of preparation or of measurement, or the time-interval between preparation and measurement. In this first role, there thus seems to be no scope for uncertainty about time. But, as Busch discusses, there is a consolidated tradition (which derives from the founding fathers of quantum theory) of an uncertainty principle between (i) the duration of an energy measurement, and (ii) either the range of an uncontrollable change of the measured system's energy or the resolution of the energy measurement or the statistical spread of the system's energy. Busch argues that Aharonov and Bohm [173] refuted, however, this tradition [174, 168]. Aharonov and Bohm provided a simple model of an arbitrarily accurate and arbitrarily rapid energy measurement. In Aharonov's and Bohm's model, two particles are confined to a line and are both free, except for an impulsive measurement of the momentum and so energy of the first by the second, with the momentum of the second being the pointer-quantity. A proper analysis and vindication of Aharonov's and Bohm's refutation uses positive-operator-valued measures (POVMs), which provide a generalized representation of quantum observables, determining the most general correspondence between quantum states, represented by a density matrix ρ and the statistics $P(x) = \text{tr}[\rho \Delta(x)]$ of a measuring arrangement, where $\Delta(x)$ is a family of linear operators.

Then, Busch defines 'intrinsic time' (or also 'dynamical time' in [167]) for a dynamical variable of the studied system, that functions to measure the time. For example: the position of a clock's dial relative to the background, i.e. the clock's face. Another example is the position of a classical free particle: it also registers the time, if only one could (i) 'read off' its position relative to the background space and (ii) keep up with its motion, so as to be able to read off that position. Busch suggests that in line of principle every nonstationary quantity A defines for any quantum state ρ a characteristic time $\tau_P(A)$ in which $\langle A \rangle$ changes 'significantly'. This implies that one

3D non-Euclid Space as a Direct Information Medium 161

should expect there will be various uncertainty principles for various definitions of intrinsic times. In this regard, in [163] Butterfield considers 1) the well-known principle due to Mandelstam and Tamm

$$\tau_P(A)\Delta_\rho(H) \geq \frac{1}{2}\hbar, \quad (3.1)$$

which combines the Heisenberg equation of motion of an arbitrary quantity A

$$i\hbar\frac{dA}{dt} = [A, H], \quad (3.2)$$

H being the Hamiltonian, with the Heisenberg–Robertson uncertainty principle

$$\Delta_\rho A \Delta_\rho B \geq \frac{1}{2}|\langle[A, B]\rangle_\rho| \quad (3.3)$$

for any quantities A, B and quantum state (density matrix) ρ and the definition of a characteristic time

$$\tau_P(A) = \Delta_\rho A / |d\langle A\rangle_\rho/dt|. \quad (3.4)$$

and 2) a principle due to Hilgerwood and Uffink, which, by applying the idea of discrimination of states to the translation of a quantum state in time instead of in space, can define the widths for energy E and momentum p_x according to relations

$$\tau_r W_\alpha(E) \geq C(\alpha, r)\hbar, \quad \xi_r W_\alpha(p_x) \geq C(\alpha, r)\hbar, \quad (3.5)$$

where

$$\int_{W_\alpha(E)} |\langle E|\psi\rangle|^2 dE = \alpha; \quad \int_{W_\alpha(p_x)} |\langle p_x|\psi\rangle|^2 dp_x = \alpha, \quad (3.6)$$

and the constant is

$$C(\alpha, r) = 2\arccos\frac{2 - r - \alpha}{\alpha} \quad (3.7)$$

for $r > 2(1 - \alpha)$, and thus implies that, in a many-particle system, the width in the total energy can control the temporal spread, in the sense of temporal translation width, of each component particle to be small.

Finally, Busch defines 'observable time' (also called, in [167], 'event time') for the sort of case, where the measured time has a physical meaning 'as a time': e.g. a time of arrival, or of sojourn, in a given spatial region, or a time of flight between two given places, or a time of decay. In quantum theory, such cases prompt the idea of a time operator, i.e. a self-adjoint operator whose spectrum is the possible values of the time in question. And so here one may meet Pauli's one-liner 'proof' that there cannot be such an operator. In this regard, in [163] Butterfield mentions that one can indeed derive the following two different scenarios:

(1) The first point expresses Pauli's idea that a time operator would imply that the Hamiltonian has the entire real line as its spectrum: forbidding discrete eigenvalues, and also making the Hamiltonian unphysical because unbounded below. Thus: a self-adjoint operator T generating translations in energy according to

$$\exp(i\tau T/\hbar) H \exp(-i\tau T/\hbar) = H + \tau I, \quad \forall \tau \in R, \quad (3.8)$$

would imply that the spectrum of H is R. The reason is simply that unitary transformations are spectrum-preserving, namely the left-hand side has the same spectrum as the H. So H has the same spectrum as $H + \tau I$ for all τ; so the spectrum of H is R. Here, we should note that Eq. (3.8) would imply that in a dense domain

$$[H, T] = i\hbar I, \quad (3.9)$$

and this would imply our opening 'prototype form' of the time-energy uncertainty principle for any state ρ

$$\Delta_\rho T \Delta_\rho H \geq \frac{1}{2}\hbar. \quad (3.10)$$

(2) But the converse fails. That is: Eq. (3.9) does not imply Eq. (3.8). So the door is open for some realistic Hamiltonians (namely bounded below and/or with discrete spectrum; i.e. without the whole of R as spectrum), to have a T satisfying Eq. (3.2), which Busch calls a canonical time operator [168]. Busch points out that although there is little general theory of which Hamiltonians

have such a T (even when we generalize the concept of quantity to POVMs), there are many examples, including familiar systems like the harmonic oscillator.

Finally, another important aspect linked with the wavefunction collapse which is worth to be analyzed is the appearance of particles from the wave description of matter. As it can be deduced from various interpretations of quantum theory, the particles can be considered as distinct among manifestations of the collapse in the sense that are localized entities with permanent properties. The properties of particles are at odds with their quantum mechanical description as spread-out waves evolving according to the Schrödinger equation, and their appearance as localized objects presents a problem. Already Schrödinger recognized this problem and underlined that "emerging particle from decaying nuclei is described as a spherical wave that impinges continuously on a surrounding luminescent screen over its full expanse [...] the screen, however, does not show a more or less constant uniform surface glow, but rather lights up at one instant at one spot" [175]. Schrödinger found this idea of abrupt, instantaneous and spatially nonlocal collapsing of the entire spread-out wavefunction into a function of point support at the moment of the appearance of the particle "ridiculous" and altogether denied the existence of particles [176].

However, such collapse of the wavefunction seems inevitable if we need to reconcile quantum description with our experience with measuring instruments. And starting from the inevitability of the collapse a question arises: is the collapse of the wavefunction on the appearance of a particle fundamental? That is, does the collapse occur at the moment of the appearance of a particle and hence cannot be broken down into a series of casually separated processes? In this regard, the point of view of standard quantum mechanics is that the collapse of the wavefunction is fundamental.

Several authors have analyzed the problem of the appearance of a particle in the light of the quantum time of arrival problem (QTPOAP) inside standard quantum mechanics [177–193]. The appearance of a particle, as already underlined by Schrödinger, is clearly a quantum arrival problem: in fact, it means to find a quantum mechanical mechanism for the localization of a unitarily evolving wavefunction at a definite point in space at a definite time at the registration of the particle. Any solution then to the QTOAP should not only be able to predict the time-of-arrival distributions but must also explain how the particle appears at the moment of arrival. The QTOAP, however, is affected with controversy. If, on one hand there are some researchers who tried to find solutions to this problem (see, for example, Muga's and Leavens's paper [185]), on the other hand there are other authors who remarked that this problem is not meaningful or possesses no solution at all [194–200]. Moreover, many available QTOA theories have only addressed the time-of-arrival distribution aspect of the problem [177–180, 182–189, 193, 201].

However, since 2005 Galapon has proposed an interesting model about the connection between the QTOAP and the problem of the appearance of particles in one dimension within the confines of standard quantum mechanics. Galapon's approach provides a generalization of solution to the one-dimensional free QTOAP [192] for arbitrary arrival point, for arbitrary interaction potential via spatial confinement followed by a limiting procedure for arbitrarily large confining lengths. In particular, in the recent article *Theory of quantum arrival and spatial wavefunction collapse on the appearance of particle*, Galapon found that the resulting quantum time-of-arrival theory suggests that the collapse of the wavefunction on the appearance of the particle is not fundamental: the collapse occurs much earlier than the appearance of the particle and the subsequent localization of the wavefunction on the appearance of the particle arises from the unitary Schrödinger equation [202].

In Galapon's model, the time-of-arrival operator T, that reduces to the classical time of arrival

$$T_x(q,p) = -\text{sgn}(p)\sqrt{\mu/2} \int_x^q (H(q,p) - V(q'))^{-1/2} dq' \quad (3.11)$$

(where H is the Hamiltonian, V is the interaction potential, μ is the mass of the particle and (q, p) are, respectively, the position and the

3D non-Euclid Space as a Direct Information Medium   165

momentum of the particle at $t = 0$) is the integral operator

$$(T\varphi)(q) = \int_{-\infty}^{\infty} \langle q|T|q'\rangle \varphi(q')dq' \qquad (3.12)$$

with the kernel given by the following relation

$$\langle q|T|q'\rangle = \frac{\mu}{i\hbar} T(q,q') \text{sgn}(q - q') \qquad (3.13)$$

in which the kernel factor $T(q, q')$ is determined by the interaction potential $V(q)$. By considering arrival at the origin $x = 0$, the kernel factor is the solution to equation

$$-\frac{\partial^2 T(q,q')}{\partial q^2} + \frac{\partial^2 T(q,q')}{\partial q'^2} + \frac{2\mu}{\hbar^2}(V(q) - V(q'))T(q,q') = 0 \qquad (3.14)$$

subjected to the conditions $T(q, q) = q/2$ and $T(q, -q) = 0$. The classical time of arrival derives from T via the Weyl–Wigner transform of its kernel:

$$T_\hbar(q,p) = \int_{-\infty}^{+\infty} \langle q + (v/2)|T|q - (v/2)\rangle \exp(-i(vp/\hbar))dv. \qquad (3.15)$$

The physical content of the time-of-arrival operator T can be explored by successive coarse graining, namely by successive approximation of T with discrete observables. By writing T in terms of the position and momentum operators, in Galapon's approach a coarse graining of T is constructed by confining the system in a large box of length $2l$ centred at the arrival point and then by projecting its explicit operator form in the Hilbert space $H_l = L^2[-l, l]$, under the condition that the Hamiltonian is purely kinetic for vanishing interaction potential. Galapon showed that a coarse graining of T is given by the confined-time-of-arrival (CTOA) operators for a given interaction potential for a given confining length. In particular, for continuous potentials, the CTOA operators turn out to be compact nondegenerate self-adjoint operators. Their compactness implies that they have pure discrete spectrum (with corresponding complete square integrable eigenfunctions). It is just this compactness that makes these operators a coarse graining of the operator T. On the basis of Galapon's treatment, the spectral properties of the CTOA operators are intimately tied with the internal unitary dynamics of the system: the eigenfunctions evolve according to Schrödinger's equation such that the event of the position expectation value assuming the arrival point, and the event of the position uncertainty being minimum occur at the same instant of time equal to their corresponding eigenvalues, with the uncertainty decreasing with the magnitude of the eigenvalue. This property can be referred as the unitary arrival of the eigenfunctions at the arrival point [190, 203]. According to Galapon's results, the unitary arrival of the eigenfunctions at their respective eigenvalues is consistent with the interpretation that T is a first time-of-arrival operator. This property already hints that T may account for the localization of the wavefunction in space and time on the appearance of the particle [202].

Moreover, Galapon found that computations on different potentials show the same dynamical behavior for arbitrarily large confining lengths. In particular, by solving Eq. (3.14) in the case of the harmonic oscillator CTOA operator for the arrival point $x = 0$ in the limit of arbitrarily large confining lengths, one finds that the corresponding kernel factor is

$$T(q,q') = \hbar \sinh(\mu\omega(q^2 - q'^2)/2\hbar) \frac{1}{2\mu\omega(q-q')}. \qquad (3.16)$$

In virtue of Eq. (3.16), the CTOA eigenfunctions evolve to a singular support at the arrival point at their respective eigenvalues when the confining lengths tend to infinity; in particular, the modulus square of the wavefunctions for a given fixed time tends to the position eigenfunction at the arrival point (and computations on other potentials, including the quartic oscillator, show the same dynamical behavior for arbitrarily large confining lengths). These results suggest that the behaviors of the CTOA operators in the limit are the same.

On the basis of Galapon's model, the QTOAP within the scheme of standard quantum mechanics implies an unitarily evolving CTOA eigenfunction to a localized support at the arrival point. By assuming that particles are wave packets of singular support, hence it can be concluded that the CTOA eigenfunctions for the arbitrarily large values of confining lengths are particles at their respective eigenvalues. This supports a mechanism for the localization of the wavefunction in space and in time at the registration of the particle. To be more explicit, let us consider a quantum particle prepared in some initial state, ψ_0. Without loss of generality, we can assume that ψ_0 has a compact support. We can then enclose the system by a box of very large length (for all practical purposes infinite), with the support of ψ_0 laid completely in the box. Since the CTOA eigenfunctions are complete we can decompose ψ_0 in terms of these eigenfunctions. Now, if we initially assume that particle detectors somehow respond only to a localized wave packet or localized energy, then we can say that registration or arrival of the particle at a certain arrival point at time τ can be interpreted as the detection of the component eigenfunction whose eigenvalue is τ that is unitarily arriving (essentially collapsing for arbitrarily large confining length) at the arrival point. This implies that the appearance or arrival of the particle is a combination of a collapse of the initial wavefunction into one of the eigenfunctions of the time-of-arrival operator right after the preparation of the initial state followed by a unitary evolution of the eigenfunction. That is, the collapse of the wavefunction on the appearance of particle is not fundamental but decomposable into a series of casually separated processes.

On the other hand, it must be emphasized that Galapon's interpretation contrasts with the standard interpretation of the collapse of the spatial wavefunction on the appearance of the particle. In standard quantum mechanics, when a quantum object is prepared in some initial state ψ_0 and when an observable of the object is measured at a later time T, then the state at the moment of measurement, which is $\psi(T) = U_T \psi_0$, where U_T is the time-evolution operator, collapses randomly into one of the eigenfunctions of the observable. Moreover, in standard quantum mechanics the appearance of particle derives from a position measurement, so that the appearance at point q_0 at some time T is the projection of the evolved wavefunction $\psi(q,T) = U_T\psi_0(q)$ to the eigenfunction $\delta(q - q_0)$ of the position operator. Instead, according to Galapon's model of the quantum arrival description, the collapse occurs much earlier than the appearance of the particle, with the initial state collapsing into one of the eigenfunctions of the time-of-arrival operator right after the preparation and with the particle appearing later at the moment the eigenfunction has evolved to a state of localized support at the arrival point. In the quantum arrival picture proposed by Galapon, therefore, the appearance of particles then does not emerge from position measurement but rather from time measurement.

In synthesis, Galapon's model can be considered an interesting attempt to treat the QTOAP in one dimension for arbitrary arrival point, for arbitrary interaction potential. It is a theory based on a self-adjoint and canonical coarse graining of a time-of-arrival operator that reduces to the classical time of arrival observable in the classical limit; moreover, it is a theory based on the collapse-supplemented Schrödinger equation, with probabilities computed in the standard way. According to Galapon's results, the appearance of a particle arises as a combination of the collapse of the initial wavefunction into one of the eigenfunctions of the time-of-arrival operator, followed by the unitary Schrödinger evolution of the eigenfunction. In other words, particles do not arise out of position measurements but out of time-of-arrival measurements, and the collapse of the wavefunction on the appearance of particle is decomposable into a series of casually separated processes.

3.3 Bohm's Quantum Potential and EPR-type Experiments: A 3D non-Euclid Space as a Direct Information Medium Between Subatomic Particles

Bohm's quantum potential and the nonlocal correlations characterizing many-body quantum processes can be considered the starting-points in order to introduce the view of a timeless 3D non-Euclid space which acts as a direct information medium for the transmission of information.

As is known, in his classic works of 1952 and 1953 [177, 204, 205], rediscovering an interpretation already provided by de Broglie at the 1927 Solvay Congress [206–208], Bohm showed that if we interpret each individual physical system as composed by a corpuscle and a wave guiding it, by writing its wavefunction in polar form and decomposing the Schrödinger equation into real and imaginary parts, the movement of the corpuscle under the guide of the wave happens in agreement with a law of motion which assumes the following form

$$\frac{\partial S}{\partial t} + \frac{|\nabla S|^2}{2m} - \frac{\hbar^2}{2m}\frac{\nabla^2 R}{R} + V = 0 \qquad (3.17)$$

(where R is the absolute value and S is the phase of the wavefunction, $\hbar$ is Planck's reduced constant, m is the mass of the particle and V is the classical potential). This equation is equal to the classical equation of Hamilton–Jacobi except for the appearance of the additional term

$$Q = -\frac{\hbar^2}{2m}\frac{\nabla^2 R}{R}, \qquad (3.18)$$

having the dimension of an energy and containing Planck's constant and therefore appropriately defined quantum potential.

The treatment provided by relations (3.17) and (3.18) can be extended in a simple way to many-body systems. If we consider a wavefunction $\psi = R(\mathbf{x}_1, \ldots, \mathbf{x}_N, t)e^{iS(\mathbf{x}_1,\ldots,\mathbf{x}_N,t)/\hbar}$, defined on the configuration space R^{3N} of a system of N particles, the movement of this system under the action of the wave ψ happens in agreement to the law of motion

$$\frac{\partial S}{\partial t} + \sum_{i=1}^{N}\frac{|\nabla_i S|^2}{2m_i} + Q + V = 0, \qquad (3.19)$$

where

$$Q = \sum_{i=1}^{N} -\frac{\hbar^2}{2m_i}\frac{\nabla_i^2 R}{R} \qquad (3.20)$$

is the many-body quantum potential. The equation of motion of the ith in particle, in the limit of big separations, can also be written in the following form

$$m_i \frac{\partial^2 \mathbf{x}_i}{\partial t^2} = -[\nabla_i Q(\mathbf{x}_1, \mathbf{x}_2, \ldots, \mathbf{x}_n) + \nabla_i V_i(\mathbf{x}_i)], \qquad (3.21)$$

which is a quantum Newton law for a many-body system. Equation (3.21) shows that the contribution to the total force acting on the ith particle coming from the quantum potential, i.e. $\nabla_i Q$, is a function of the positions of all the other particles and thus in general does not decrease with growing distance.

The quantum potential is the crucial entity which allows us to understand the features of the quantum world determined by Bohm's version of quantum mechanics. In virtue of its mathematical expression, the quantum potential determines a nonlocal, instantaneous action onto the particles under consideration. In Bohm's view, the quantum potential implies, at a fundamental level, a universal interconnection of things that could no longer be questioned. In relations (3.18) and (3.20) the appearance of the absolute value of the wavefunction in the denominator explains why the quantum potential can produce strong long-range effects that do not necessarily fall off with distance and so the typical properties of entangled wavefunctions. Thus even though the wavefunction spreads out, the effects of the quantum potential need not necessarily decrease (as the equation of motion (3.21) of the many-body systems shows clearly, the total force acting on the ith particle coming from the quantum potential, i.e. $\nabla_i Q$, does not necessarily fall off as distance grows and indeed the forces between two particles of a many-body system may become stronger, even if $|\psi|$ may decrease in this limit). This is just the type of behavior required to explain EPR-type correlations.

Moreover, the quantum potential has a geometric, contextual nature, a global information on the process and its environment, and at the same time has an active, dynamical information in the sense that it modifies the behavior of the particle. In a double-slit experiment, for example, if one of the two slits is closed the quantum potential changes, and this information arrives instantaneously to the particle, which behaves as a consequence. As the author pointed out in the recent paper *"The geometrodynamic nature of the quantum potential"* [209], if the quantum potential produces a space-like and an active information, then it cannot be seen as an external entity in space but as an entity which contains a spatial information, i.e. as an entity which represents space. The author suggested that the quantum potential, in virtue of its features, can be considered a geometrodynamic entity, namely that the information determined by the quantum potential is a type of geometrodynamic information "woven" into the 3D space. If the quantum potential has an instantaneous action and contains an active information about the environment, one can say that it is the 3D space the medium responsible of the behavior of quantum particles. In the geometrodynamic picture provided by the author in [209], one can say that the quantum potential contains the geometric properties of the 3D space from which the quantum force, and thus the behavior of quantum particles are derived. And the presence of Laplace operator (and of the absolute value of the wavefunction in the denominator) indicates just that the geometrodynamic information contained in quantum potential determine a nonlocal, instantaneous action. As regards the geometrodynamic nature of the quantum potential and the nonlocal nature of the interactions in physical space, one can also say, by paraphrasing J. A. Wheeler's famous saying about general relativity, that the evolution of the state of a quantum system changes active global information, and this in turn influences the state of the quantum system, redesigning the nonlocal geometry of the universe.

In the authors view, the crucial point regarding the quantum potential is "its space-like action and its active geometrodynamic nature lead to consider the 3D space as the fundamental arena which determines the motion and the behavior of subatomic particles". In this picture, one can say that Eqs. (3.18) and (3.20) defining the quantum potential contain the idea of a 3D space as an immediate information medium in an implicit way [210].

In particular, if one considers a many-body quantum process (such as for example the case of an EPR-type experiment, of two subatomic particles, before joining and after separated and carried away at big distances from one another), one can say that the 3D physical space assumes the special "state" represented by quantum potential (3.20), and this allows an instantaneous communication between the particles into consideration [211]. By means of quantum entanglement, in EPR-type experiments the 3D timeless space (where time exists only as a numerical order of material changes) acts as an immediate medium of information transfer between the systems under consideration through its special state represented by Bohm's quantum potential. According to this view — proposed by the author and Sorli in the papers [210–214] — the reading of EPR-type experiments provided by Bohm's quantum potential can be considered as the fundamental element which suggests the introduction of a 3D timeless space which acts as an immediate information medium, as the fundamental arena, of the quantum domain.

Let us consider the classic example of EPR-type experiment given by Bohm [215] in 1951. We have a physical system given by a molecule of total spin 0 composed by two spin 1/2 atoms in a singlet state:

$$\psi(\mathbf{x}_1, \mathbf{x}_2) = f_1(\mathbf{x}_1) f_2(\mathbf{x}_2) \frac{1}{\sqrt{2}} (u_+ v_- - u_- v_+), \qquad (3.22)$$

where $f_1(\mathbf{x}_1)$, $f_2(\mathbf{x}_2)$ are nonoverlapping packet functions, $u_\pm$ are the eigenfunctions of the spin operator $\hat{s}_{z_1}$ in the z-direction pertaining to particle 1 and $v_\pm$ are the eigenfunctions of the spin operator $\hat{s}_{z_2}$ in the z-direction pertaining to particle 2: $\hat{s}_{z_1} u_\pm = \pm \frac{\hbar}{2} u_\pm$, $\hat{s}_{z_2} v_\pm = \pm \frac{\hbar}{2} v_\pm$. Let us suppose to perform a spin measurement on the particle 1 in the z-direction when the molecule is in such a state. And let us suppose moreover that we obtain the result spin up for this particle 1. Then, according to the usual quantum theory, the wavefunction (3.22) reduces to the first of its summands:

$$\psi \to f_1 f_2 u_+ v_-. \qquad (3.23)$$

The result of the measurement carried out on the particle 1 leads us to have knowledge about the state of the unmeasured system 2: if the particle 1 is found in the state of spin up, we know immediately that the particle 2 is in the state v_- which indicates that the particle 2 has spin down. But this outcome regarding particle 2 depends on the kind of measurement carried out on particle 1. In fact, by performing different types of measurement on particle 1 we will bring about distinct states of the particle 2. This means that as regards spin measurements there are correlations between the two particles. By considering the particles 1 and 2 separately one can think about strange influence of one particle onto the other. A measurement on one of the two particles automatically fixes also the state of the other particle, independently from the distance between them. Although the two partial systems (the particle 1 and the particle 2) are clearly separated in space (in the conventional sense that the outcomes of position measurements on the two systems are widely separated), indeed they cannot be considered physically separated because the state of the particle 2 is indeed instantaneously influenced by the kind of measurements made on the particle 1. Bohm's example shows, therefore, clearly that entanglement in spin space implies nonlocality and nonseparability in Euclidean 3D space: this comes about because the spin measurements couple the spin and space variables.[1]

As we have illustrated through Bohm's example, the surprising fact regarding the quantum entanglement lies in the fact that the results of the measurement of the spin of two particles are 100% correlated, if for both particles we measure the spin along the same direction. But then how can we interpret and reproduce the fact that the measurement of a particle defines instantaneously in what state the other particle is found, independently from the distance? According to the previsions of quantum theory, in EPR-type experiments the transmission of the information has zero numerical order. The time of information transfer from one particle to the other in EPR-type experiments is zero [216]. If the state of the second particle changes instantaneously after the measurement on the first particle, this fact does not imply a transmission of the information at a higher speed than light speed because one cannot influence the outcome of the first measurement. According to the approach developed by the author of this book and Sorli in the papers [210–214], a 3D timeless space associated with the geometrodynamic,

---

[1] It is also important to underline that if one assumes the quantum interference as a fundamental property, independent of space separation of quantum mechanical systems, the nonlocality becomes a natural phenomenon.

nonlocal action of Bohm's quantum potential is the fundamental arena responsible of the transmission of the information in EPR-type experiments. The information between the two particles in EPR-type experiments is instantaneous thanks to the medium of a 3D timeless space (where time is not a primary physical reality but exists only as a numerical order of material changes) which is associated with the geometrodynamic, nonlocal action of Bohm's quantum potential. If one takes under examination the situation considered by Bohm in 1951 (illustrated before) one can say that it is the state of space in the form of the quantum potential (3.20) which produces an instantaneous connection between the two particles as regards the spin measurements. In synthesis, one can say that in EPR-type experiments the quantum potential (3.20) makes the 3D physical space an "immediate information medium" that informs particle 1 about the behavior of particle 2 and opposite. In EPR-type experiments the behavior of a subatomic particle is influenced instantaneously by the other particle thanks to the 3D space which functions as an immediate information medium in virtue of the geometric properties represented by the quantum potential (3.20). EPR-type experiments demonstrate that the fundamental arena of quantum processes is a timeless 3D space which acts as an immediate information medium between the particles under consideration. Therefore, the following interesting perspective emerges in the theory proposed by the author and Sorli: the same timeless 3D space can be considered as the background for both the quantum processes and the processes of the special relativistic domain.

The fundamental difference of the 3D timeless background space of quantum processes with respect to the 3D timeless space of special relativity lies in the fact that it is non-Euclidean. In this regard, it is important to underline that according to current research, quantum phenomena can be interpreted as a modification of the geometrical properties of physical space. In particular, in the recent article *On a geometrical description of quantum mechanics*, Novello, Salim and Falciano showed that quantum mechanics can be seen as a manifestation of the non-Euclidean structure of the 3D space: by starting from a variational principle which defines the non-Euclidean structure of space, they provided a geometrical interpretation to quantum phenomena in the picture of the Weyl geometry, thus remarking a close connection between the de Broglie–Bohm interpretation of quantum mechanics and the Weyl integrable space [217].

In Novello's, Salim's and Falciano's approach, the geometrical structure of space is obtained by starting from the action

$$I = \int dt d^3 x \sqrt{g} \Omega^2 \left(\lambda^2 R - \frac{\partial S}{\partial t} - H_m \right), \quad (3.24)$$

where $g = \det g_{ij}$, $R \equiv g^{ij} R_{ij}$, $R_{ij} = \Gamma^m_{mi,j} - \Gamma^m_{ij,m} + \Gamma^l_{mi}\Gamma^m_{jl} - \Gamma^l_{ij}\Gamma^m_{lm}$ is the Ricci curvature tensor, λ^2 is a constant having dimension of energy multiplied for length squared. The connection of the 3D-space Γ^i_{jk}, the Hamilton's principal function S and the scalar function Ω are the independent variables of the approach.

In the case of a point-like particle the matter Hamiltonian is $H_m = \frac{1}{2m}\nabla S \cdot \nabla S + V$. Variation of the action (3.24) with respect to the independent variables gives

$$g_{ij;k} = -4(\ln \Omega)_{,k} g_{ij}, \quad (3.25)$$

where ";" denotes covariant derivative and "," simple spatial derivative. Variation with respect to Ω gives

$$\lambda^2 R = \frac{\partial S}{\partial t} + \frac{1}{2m}\nabla S \cdot \nabla S + V. \quad (3.26)$$

Equation (3.26) describes the affine properties of the physical space. On the other hand, by setting $\lambda^2 = \frac{\hbar^2}{16m}$ and taking into account the expression of the curvature in terms of the scalar function Ω, $R = 8\frac{\nabla^2 \Omega}{\Omega}$, Eq. (3.26) becomes

$$\frac{\partial S}{\partial t} + \frac{1}{2m}\nabla S \cdot \nabla S + V - \frac{\hbar^2}{2m}\frac{\nabla^2 \Omega}{\Omega} = 0. \quad (3.27)$$

Equation (3.27) is analogous to the quantum Hamilton–Jacobi equation (3.17) of the de Broglie–Bohm interpretation of quantum mechanics, if one identifies the scalar function Ω with the amplitude of the wavefunction. In Novello's, Salim's and Falciano's approach based on Weyl integrable space, the quantum potential can be thus merely identified with the curvature scalar which characterizes this geometry. Moreover, the inverse square root of the curvature scalar defines a typical length (Weyl length) that can be used to evaluate the strength of quantum effects

$$L_W = \frac{1}{\sqrt{R}}. \quad (3.28)$$

Inside this picture, the classical behavior is recovered when the length defined by the Weyl curvature scalar is small compared to the typical length scale of the system. Once the Weyl curvature becomes nonnegligible the system goes into a quantum regime. It is also interesting to mention that in this approach, as long as one accepts that quantum mechanics is a manifestation of a non-Euclidean geometry, all theoretical issues related to quantum effects receive a pure geometrical meaning. In particular, in this picture, the identification of Weyl integrable space's curvature scalar as the ultimate origin of quantum effects leads to a geometrical interpretation of Heisenberg's uncertainty principle. This geometrical description considers the uncertainty principle as a break down of the classical notion of standard rulers. The uncertainty principle derives from the fact that we are unable to perform a classical measurement to distances smaller than the Weyl curvature length. In other words, the size of a measurement has to be bigger than the Weyl length

$$\Delta L \geq L_W = \frac{1}{\sqrt{R}}. \quad (3.29)$$

The quantum regime is extreme when the Weyl curvature term dominates. Novello's, Salim's and Falciano's interpretation of Heisenberg's uncertainty principle resembles Bohr's complementary principle inasmuch as the impossibility of applying the classical definitions of measurements. However, while Bohr's complementary principle is based on the uncontrolled interference of a classical apparatus of measurement, Novello, Salim and Falciano argue that the notion of a classical standard ruler breaks down because its meaning is intrinsically dependent on the validity of Euclidean geometry. In Novello's, Salim's and Falciano's model, once it becomes necessary to include the Weyl curvature, one can no longer perform a classical measurement of distance.

Moreover, as regards the interpretation of the geometrodynamic, nonlocal information of the quantum potential as a deformation of the geometrical properties of space, the author and Licata recently proposed a dynamical explanation of Novello's, Salim's and Falciano's approach by deriving Bohm's potential from a minimum condition of Fisher information: in this model, Bohm's quantum potential emerges as an information channel indicating the modification of the geometry of physical space in a picture based on a superposition of Boltzmann entropies [218]. By following the references [218, 219], let us outline here the fundamental features of this approach.

On the basis of an extension of the tensor calculus to operators represented by nonquadratic matrices a new interesting geometrical reading of de Broglie–Bohm interpretation of quantum mechanics can be provided in which the quantum potential emerges as an information medium determined by the vector of the superposition of entropies

$$\begin{cases} S_1 = k \log W_1(\theta_1, \theta_2, \ldots, \theta_p) \\ S_2 = k \log W_2(\theta_1, \theta_2, \ldots, \theta_p) \\ \ldots \\ S_n = k \log W_n(\theta_1, \theta_2, \ldots, \theta_p) \end{cases}, \quad (3.30)$$

where k is the Boltzmann constant, W are the number of the microstates for the same parameters θ as temperatures, pressures, etc. In this picture, quantum effects are equivalent to a geometry which is described by the following equation

$$\frac{\partial}{\partial x^k} + \frac{\partial^2 S_j}{\partial x^k \partial x^p} \frac{\partial x^i}{\partial S_j} = \frac{\partial}{\partial x^k} + \frac{\partial \log W_j}{\partial x_h} = \frac{\partial}{\partial x^k} + B_{j,h}, \quad (3.31)$$

where $B_{j,h} = \frac{\partial S_j}{\partial x_h} = \frac{\partial \log W_j}{\partial x_h}$ is a Weyl-like gauge potential. This Weyl-like gauge potential determines a deformation of the moments for the change of the geometry stated by the following expression of the action:

$$A = \int \rho \left[\frac{\partial A}{\partial t} + \frac{1}{2m}(p_i + B_{k,i})(p_j + B_{k,j}) + V \right] dt d^n x, \quad (3.32)$$

which gives

$$A = \int \rho \left[\frac{\partial A}{\partial t} + \frac{1}{2m}(p_i p_j + B_{k,i} B_{k,j}) + V \right] dt d^n x \qquad (3.33)$$

namely

$$A = \int \rho \left[\frac{\partial A}{\partial t} + \frac{1}{2m}\left(p_i p_j + \frac{\partial \log W_k}{\partial x_i} \frac{\partial \log W_k}{\partial x_j} \right) + V \right] dt d^n x \qquad (3.34)$$

namely

$$A = \int \rho \left[\frac{\partial A}{\partial t} + \frac{1}{2m} p_i p_j + V + \frac{1}{2m}\left(\frac{\partial \log W_k}{\partial x_i} \frac{\partial \log W_k}{\partial x_j} \right) \right] dt d^n x. \qquad (3.35)$$

The quantum action assumes the minimum value when

$$\delta A = 0 \qquad (3.36)$$

namely

$$\delta \int \rho \left[\frac{\partial A}{\partial t} + \frac{1}{2m} p_i p_j + V \right] dt d^n x$$

$$+ \delta \int \frac{\rho}{2m} \frac{\partial \log W_k}{\partial x_i} \frac{\partial \log W_k}{\partial x_j} dt d^n x = 0 \qquad (3.37)$$

and thus

$$\frac{\partial A}{\partial t} + \frac{1}{2m} p_i p_j + V + \frac{1}{2m}\left(\frac{1}{W_k^2} \frac{\partial W_k}{\partial x_i} \frac{\partial W_k}{\partial x_j} - \frac{2}{W_k} \frac{\partial^2 W_k}{\partial x_i \partial x_j} \right)$$

$$= \frac{\partial S_k}{\partial t} + \frac{1}{2m} p_i p_j + V + Q, \qquad (3.38)$$

where Q is Bohm's quantum potential. Equation (3.38) states clearly that the quantum potential — and thus its geometrodynamic features — can be derived as a consequence for the extreme condition of Fisher information. On the basis of Eq. (3.38), one can interpret Bohm's quantum potential as an information channel determined by the functions W_k defining the number of microstates of the physical system under consideration, which depend on the parameters θ of the distribution probability (and thus, for example, on the space-temporal distribution of an ensemble of particles, namely the density

3D non-Euclid Space as a Direct Information Medium 179

of particles in the element of volume d^3x around a point **x** at time t) and which correspond to the vector of the superpose Boltzmann entropies (3.30). In other words, the distribution probability of the wavefunction determines the functions W_k defining the number of microstates of the physical system under consideration, a vector of superposition of Boltzmann entropies emerges from these functions W_k given by Eqs. (3.30), and these functions W_k, and therefore the vector of the superpose entropies, can be considered as the fundamental physical entities which determine the action of the quantum potential (in the extreme condition of the Fisher information) on the basis of the following equation[2]

$$Q = \frac{1}{2m}\left(\frac{1}{W^2}\frac{\partial W}{\partial x_i}\frac{\partial W}{\partial x_j} - \frac{2}{W}\frac{\partial^2 W}{\partial x_i \partial x_j}\right). \quad (3.39)$$

In this picture, in the extreme condition of Fisher information, the Boltzmann entropies defined by Eqs. (3.30) (and thus the functions W_k) emerge as "informational lines" of the quantum potential.

Moreover, on the basis of Eqs. (3.31) and (3.38), one can say that the quantum phenomena corresponds to a change of the geometry, which is expressed by a Weyl-like gauge potential and is determined by the functions W and thus by the vector of the superpose entropies. According to this model, it becomes so permissible for the following reading of the mathematical formalism in nonrelativistic bohmian quantum mechanics: the distribution probability of the wavefunction determines the functions W defining the number of microstates of the physical system under consideration, a vector of superposition of entropies emerges from these functions W given by Eqs. (3.30), and these functions W (and thus also the vector of the superpose entropies given by Eqs. (3.30)) determine a change of the geometry expressed by a Weyl-like gauge potential and characterized by a deformation of the moments given by Eq. (3.35). The quantum potential, in the extreme condition of Fisher information, can be

---

[2]In the next equations of this paragraph, for simplicity we are going to denote the generic function W_k (defined by Eqs. (3.30)) with W.

therefore considered as an information channel describing the change of the geometry of the physical space in the presence of quantum effects.

Moreover, by introducing the definition (3.39) of the quantum potential inside the quantum Hamilton–Jacobi equation (3.17), we obtain

$$\frac{|\nabla S|^2}{2m} + V + \frac{1}{2m}\left(\frac{1}{W^2}\frac{\partial W}{\partial x_i}\frac{\partial W}{\partial x_j} - \frac{2}{W}\frac{\partial^2 W}{\partial x_i \partial x_j}\right) = -\frac{\partial S}{\partial t}, \quad (3.40)$$

which provides a new alternative way (in the regime of Fisher information) to read the energy conservation law in quantum mechanics. In Eq. (3.40) two quantum corrector terms appear in the energy of the system, which are owed to the functions W linked with the vector of the superpose entropies, and which thus describe the change of the geometry in the presence of quantum effects. On the basis of Eq. (3.40), we can say that the distribution probability of the wavefunction determines the functions W defining the number of microstates of the physical system under consideration, a vector of superpose entropies emerges from these functions W given by Eqs. (3.30), these functions W (and thus also the vector of the superpose entropies given by Eqs. (3.30)) determine a change of the geometry of the physical space, and this change of the geometry produces two quantum corrector terms in the energy of the system. These two quantum corrector terms can thus be interpreted as a sort of modification of the physical features of the background space determined by the ensemble of particles associated with the wavefunction under consideration. According to the approach suggested by the author and Licata in the paper [218], it is just the functions W and thus the vector of superposition of entropies (3.30) the fundamental entities which, by expressing the deformation of the geometry of the background space determined by the ensemble of particles associated with the wavefunction under consideration, represent what are the geometric properties of space from which the quantum force, and thus the behavior of quantum particles, are derived. It is also interesting to observe that, in analogy to the quantum lengths of Novello's, Salim's and Falciano's model, also in this approach the inverse square root of the quantity

$$L_{\text{quantum}} = \frac{1}{\sqrt{\frac{1}{\hbar^2}\left(\frac{2}{W}\frac{\partial^2 W}{\partial x_i \partial x_j} - \frac{1}{W^2}\frac{\partial W}{\partial x_i}\frac{\partial W}{\partial x_j}\right)}} \qquad (3.41)$$

defines a typical quantum-entropic length that can be used to evaluate the strength of quantum effects. Once the quantum-entropic length (3.41) becomes non-negligible the system goes into a quantum regime. If in Novello's, Salim's and Falciano's model it is possible to provide a geometrical interpretation to quantum mechanics by identifying Bohm's quantum potential with the curvature scalar of Weyl integrable space, in analogous way, in the approach suggested by the author and Licata in the reference [218], one obtains a geometrodynamic (but less radical) picture of Bohm's quantum potential in the context of a Fisher geometry which allows us to characterize the deformations of physical space in the presence of quantum effects in the space of parameters and to express Bohm's quantum potential as the information channel indicating the modification of the geometry of physical space determined by the microstates characterizing the system under consideration and thus by the superpose vector of the Boltzmann entropies.

3.4 The Quantum Entropy as the Fundamental Entity Which Determines the non-Euclid and Nonlocal Features of the 3D Timeless Space in the Quantum Domain

According to the model proposed by the author in the papers [210, 214] a fundamental entity describing the properties of the background space can be introduced which determines the action of the 3D space as an immediate information medium in EPR-type experiments and which makes non-Euclidean the 3D space responsible of quantum processes. In this view, the non-Euclid feature and, at the same time, the nonlocal feature of the 3D timeless space characterizing the quantum domain associated with the quantum potential are determined by the quantum entropy, which can be considered as the ultimate visiting card describing the geometry of quantum processes.

The idea of a quantum entropy from which the action of the quantum potential derives was originally proposed by Sbitnev in the paper *Bohmian split of the Schrödinger equation onto two equations describing evolution of real functions* [220]. In this paper, Sbitnev showed that the quantum potential can be interpreted as an information channel into the movement of the particles as a consequence of the fact that it determines two quantum correctors into the energy of the particle depending on a more fundamental physical quantity which can be appropriately called "quantum entropy". This new way of reading bohmian mechanics can be called as the "entropic version" of bohmian mechanics or, more briefly, "entropic bohmian mechanics". In the recent papers [221, 222], the author of this book then extended the approach based on the quantum entropy also to many-body systems and provided a geometrodynamic reading of the theory, in which the quantum entropy can be considered as the ultimate visiting card of quantum processes.

For a one-body system, the quantum entropy can be defined by the logarithmic function

$$S_Q = -\frac{1}{2}\ln\rho, \quad (3.42a)$$

where $\rho = |\psi(\mathbf{x},t)|^2$ is the probability density (describing the space-temporal distribution of the ensemble of particles, namely the density of particles in the element of volume d^3x around a point $\mathbf{x}$ at time t) associated with the wavefunction $\psi(\mathbf{x},t)$ of an individual physical system. In the case of a many-body system, the quantum entropy can be always defined by the logarithmic function

$$S_Q = -\frac{1}{2}\ln\rho, \quad (3.42b)$$

where here $\rho = |\psi(\mathbf{x}_1,\mathbf{x}_2,\ldots,\mathbf{x}_N,t)|^2$ is the probability density (describing the space-temporal distribution of the ensemble of particles, namely the density of particles in the element of volume d^3x around a point $\mathbf{x}$ at time t) associated with the wavefunction $\psi(\mathbf{x}_1,\mathbf{x}_2,\ldots,\mathbf{x}_N,t)$ of the many-body system under consideration. In the entropic version of bohmian quantum mechanics, one can assume that the space-temporal distribution of the ensemble of particles describing the physical system under consideration generates a modification of the background space characterized by the quantity given by Eq. (3.42a) (or (3.42b)). The quantum entropy ((3.42a) and (3.42b)) can be interpreted as the physical entity that, in the quantum domain, characterizes the degree of order and chaos of the vacuum — a storage of virtual trajectories supplying optimal ones for particle movement — which supports the density ρ describing the space-temporal distribution of the ensemble of particles associated with the wavefunction under consideration.

In the recent articles *Bohmian split of the Schrödinger equation onto two equations describing evolution of real functions* and *Bohmian trajectories and the path integral paradigm. Complexified lagrangian mechanics* Sbitnev [220, 223] showed that the quantum potential can be determined as an information channel into the movement of the particles as a consequence of the fact that it determines two quantum correctors into the energy of the particle depending on the "quantum entropy". By introducing the quantum entropy, for one-body systems the quantum potential can be expressed in the following convenient way

$$Q = -\frac{\hbar^2}{2m}(\nabla S_Q)^2 + \frac{\hbar^2}{2m}(\nabla^2 S_Q), \qquad (3.43)$$

and one obtains the following equation of motion for the corpuscle associated with the wavefunction $\psi(\mathbf{x},t)$:

$$\frac{|\nabla S|^2}{2m} - \frac{\hbar^2}{2m}(\nabla S_Q)^2 + V + \frac{\hbar^2}{2m}(\nabla^2 S_Q) = -\frac{\partial S}{\partial t}, \qquad (3.44)$$

which provides an energy conservation law where the term $-\frac{\hbar^2}{2m}(\nabla S_Q)^2$ can be interpreted as the quantum corrector of the kinetic energy $\frac{|\nabla S|^2}{2m}$ of the particle while the term $\frac{\hbar^2}{2m}(\nabla^2 S_Q)$ can be interpreted as the quantum corrector of the potential energy V.

In the case of many-body systems, the quantum potential is given by the following expression

$$Q = \sum_{i=1}^{N}\left[-\frac{\hbar^2}{2m_i}(\nabla_i S_Q)^2 + \frac{\hbar^2}{2m_i}(\nabla_i^2 S_Q)\right], \qquad (3.45)$$

and the equation of motion is

$$\sum_{i=1}^{N}\frac{|\nabla_i S|^2}{2m_i}-\sum_{i=1}^{N}\frac{\hbar^2}{2m_i}(\nabla_i S_Q)^2+V+\sum_{i=1}^{N}\frac{\hbar^2}{2m_i}(\nabla_i^2 S_Q)=-\frac{\partial S}{\partial t}, \quad (3.46)$$

which provides an energy conservation law where the term $-\sum_{i=1}^{N}\frac{\hbar^2}{2m_i}(\nabla_i S_Q)^2$ can be interpreted as the quantum corrector of the kinetic energy of the many-body system while the term $\sum_{i=1}^{N}\frac{\hbar^2}{2m_i}(\nabla_i^2 S_Q)$ can be interpreted as the quantum corrector of the potential energy.

On the basis of the treatment provided by the author in the papers [221, 222], in the entropic bohmian mechanics the following reading of the quantum potential and of the energy conservation law in quantum mechanics becomes permissible. The quantum potential can be considered as the special state of the 3D timeless space which derives from the quantum entropy describing the modification of the geometrical properties of the background space from Euclidean to non-Euclidean produced by the density of the ensemble of particles associated with the wavefunction under consideration. In the entropic version of nonrelativistic bohmian quantum mechanics, one can say that the probability density associated with the wavefunction under consideration determines a quantum entropy which describes the degree of order and chaos of the vacuum supporting this probability density and thus corresponds to a deformation of the geometry of the physical space. And, on the basis of Eqs. (3.44) and (3.46), we can say that the quantum entropy (and thus the non-Euclidean features of the 3D space in the quantum domain) determines two quantum correctors in the energy of the physical system under consideration (of the kinetic energy and of the potential energy respectively) and without these two quantum correctors (linked just with the quantum entropy) the total energy of the system would not be conserved.

Moreover, in this entropic approach to bohmian quantum mechanics, the classical limit can be expressed by the conditions

$$(\nabla S_Q)^2 \to (\nabla^2 S_Q) \quad (3.47)$$

for one-body systems and

$$(\nabla_i S_Q)^2 \to (\nabla_i^2 S_Q) \tag{3.48}$$

for many-body systems. The quantum dynamics will approach the classical dynamics, and thus the non-Euclid geometry of space will approach the Euclid geometry when the quantum entropy satisfies conditions (3.47) (for one-body systems) or (3.48) (for many-body systems) which can be considered as the expression of a correspondence principle in quantum mechanics.

It is also interesting to observe that, in this entropic picture, the inverse square root of the quantity

$$L_{\text{quantum}} = \frac{1}{\sqrt{(\nabla S_Q)^2 - \nabla^2 S_Q}} \tag{3.49}$$

for one-body systems and the inverse square root of the quantity

$$L_{\text{quantum}} = \frac{1}{\sqrt{\sum_{i=1}^{N}((\nabla_i S_Q)^2 - \nabla_i^2 S_Q)}} \tag{3.50}$$

for many-body systems, define typical quantum-entropic lengths that can be used to evaluate the strength of quantum effects and, therefore, the modification of the geometry with respect to the Euclidean geometry of the 3D space characteristic of classical physics. The quantum entropic-lengths (3.49) and (3.50) can be introduced as a result of an equivalence with Novello's, Salim's and Falciano's approach. In fact, taking into account that Bohm's potential can be identified with the curvature scalar of the Weyl integrable space, namely is $Q = -\frac{\hbar^2}{2m}\frac{\nabla^2 \Omega}{\Omega}$, where the scalar function Ω is linked with the curvature through relation $R = 8\frac{\nabla^2 \Omega}{\Omega}$ and that, in this picture, the inverse square root of the curvature scalar defines a typical length (Weyl length) that can be used to evaluate the strength of quantum effects, in other words the quantity $L_W = \frac{1}{\sqrt{\frac{\nabla^2 \Omega}{\Omega}}}$ can be defined as the quantum length, the entropic quantum potential for one-body systems (3.43) leads to define the quantum-entropic length given by (3.49) (and, in analogous way, the entropic quantum potential for many-body systems (3.45) leads to define the quantum-entropic length (3.50)).

Once the quantum-entropic lengths ((3.49) or (3.50)) become nonnegligible the system goes into a quantum regime. In this picture, Heisenberg's uncertainty principle derives from the fact that we are unable to perform a classical measurement to distances smaller than the quantum-entropic lengths. In other words, the size of a measurement has to be bigger than the quantum-entropic lengths

$$\Delta L \geq L_{\text{quantum}} = \frac{1}{\sqrt{(\nabla S_Q)^2 - \nabla^2 S_Q}} \qquad (3.51)$$

(for one-body systems)

$$\Delta L \geq L_{\text{quantum}} = \frac{1}{\sqrt{\sum_{i=1}^{N}((\nabla_i S_Q)^2 - \nabla_i^2 S_Q)}} \qquad (3.52)$$

(for many-body systems). The quantum regime is entered when the quantum-entropic lengths must be taken under consideration.

It is also interesting to remark that, in the entropic bohmian mechanics here presented, the presence of the two quantum correctors of the energy seems to suggest that the quantum-entropic lengths (3.49) and (3.50) provide an indicator of nonlocal correlations in quantum systems. In the recent papers [224, 225], the author of this book and Licata, by building an analogy between the quantum treatment of a rigid rotator and a typical system with spin $1/2$ and by studying the behavior of the quantum potential, analyzed spin–spin correlations in entangled qubit pairs characterized by the wavefunction

$$|\psi\rangle = \cos\frac{\vartheta}{2}|\uparrow\downarrow\rangle + e^{i\varphi}\sin\frac{\vartheta}{2}|\uparrow\downarrow\rangle \qquad (3.53)$$

(where $|\uparrow\downarrow\rangle$ corresponds to the state of the system when the first qubit is in the "up" state, i.e., in the direction of the z-axis, and the second qubit is in the "down" state, while $|\downarrow\uparrow\rangle$ corresponds to the state of the system when the first qubit is in the "down" state and the second qubit is in the "up" state), in terms of a quantum-entropic length which assumes the form

$$L_{\text{quantum}} = \frac{1}{\sqrt{\frac{1}{2I}((\hat{\mathbf{M}}_1^2 S_Q) - (\hat{\mathbf{M}}_1 S_Q)^2 + (\hat{\mathbf{M}}_2^2 S_Q) - (\hat{\mathbf{M}}_2 S_Q)^2)}}, \qquad (3.54)$$

where $\mathbf{M}_1 = i\hat{\mathbf{M}}_1 S$ and $\mathbf{M}_2 = i\hat{\mathbf{M}}_2 S$ are the angular momenta of the two qubits 1 and 2 respectively, S is the phase of the wavefunction (3.53) of the system. In the approach developed in the papers [224, 225], in two-qubit systems in the general state (3.53), such as the ones of EPR-type experiments, the maximum correlation is determined by the limit value 1 of the quantum-entropic length (3.50) and, in this regard, a generalization of Bell inequalities, which takes explicit account of the physical complexity of the interplay between information and entropy in a quantum system, can be easily introduced which has the following form

$$CHSH_{entropic} = \langle AB \rangle + \langle AB' \rangle + \langle A'B \rangle - \langle A'B' \rangle - \frac{4}{2 - L_{quantum}^{max}} \leq 0. \quad (3.55)$$

The violation of the entropic length Bell inequality (3.55) essentially coincides with the one of the standard CHSH inequality

$$CHSH \equiv \langle AB \rangle + \langle AB' \rangle + \langle A'B \rangle - \langle A'B' \rangle \leq 2. \quad (3.56)$$

As regards the entangled states of a two qubit pair of the form (3.53), the maximal violation of the entropic length Bell inequality (3.55) which corresponds to $L_{quantum}^{max} = 1$, is obtained for $\vartheta = \frac{\pi}{2}$, on which one gets $CHSH_{entropic} \approx +0,237$. For other values of ϑ, the maximal violation of (3.53) when optimized over the measurements, follows the exact same profile as for the standard Bell-CHSH inequality (3.56). However, the measurements that maximize the violation of CHSH are not the ones which give the maximal violation of $CHSH_{entropic}$. In general, for the standard Bell-CHSH scenario, the violation of the standard inequality (3.56) is a necessary but not sufficient condition for the violation of (3.55). The advantage of the entropic length Bell inequality (3.57) lies in the fact that the results depend directly on the quantum-entropic length which emerges directly as correlation degree in quantum systems (for this reason, the quantum-entropic length has been appropriately called by the author and Licata as Bell length).

While in Novello's, Salim's and Falciano's model Bohm's quantum potential is identified with the curvature scalar of the Weyl integrable space, in analogous way in the approach of a 3D timeless non-Euclidean space as fundamental arena of quantum processes proposed by the author in the papers [221, 222, 214] and illustrated in this chapter, the quantum effects are owed to the quantum entropy. In analogy with Novello's, Salim's and Falciano's approach that implies that the quantum effects are the manifestations of the modification of the structure of the 3D physical space from Euclidean to a non-Euclidean Weyl integrable space, inside the entropic approach to Bohm's quantum mechanics proposed by the author, the presence of quantum effects is linked with a change in the geometry of the physical space, but here the change of the geometry, and thus the length that can be used to evaluate the strength of quantum effects are determined by the quantum entropy.

Moreover, the quantum entropy allows us to throw new fundamental light into the interpretation of quantum information. Inside this approach, since the quantum potential is an information channel which describes the deformation of the geometry of space in the presence of quantum effects and is determined by the quantum entropy, the real ultimate grid of quantum information can be considered the deformation of the background space associated with the quantum entropy and described by the quantum-entropic lengths (3.49) and (3.50). The quantum entropy indicating the non-Euclid and nonlocal geometrical properties of the 3D space emerges thus as the most fundamental source of quantum information.

In order to illustrate in major detail the concept of quantum information as a measure of the deformation of the background space associated with the quantum entropy, let us take under consideration the classic double-slit experiment. Here, we have a wavefunction characterized by n probability densities $h_1, h_2, \ldots, h_n$:

$$|\psi\rangle = |h_1\rangle + |h_2\rangle + \cdots + |h_n\rangle, \quad (3.57)$$

where $h_1 = \alpha_1 + i\beta_1$, $h_2 = \alpha_2 + i\beta_2, \ldots, h_n = \alpha_n + i\beta_n$.

The probability for the interference is

$$P(x) = \langle\psi|\psi\rangle = \sum_{i,j} g_{ij} \xi^i(x) \xi^j(x), \quad (3.58)$$

where

$$\xi^i = \sqrt{I^i}, \quad \xi^j = \sqrt{I^j}, \quad (3.59)$$

where $I_i = \langle h_i | h_i \rangle$

3D non-Euclid Space as a Direct Information Medium

and

$$g = \begin{bmatrix} 1 & \cos(\alpha_1 - \alpha_2) & \cdots & \cos(\alpha_1 - \alpha_n) \\ \cos(\alpha_2 - \alpha_1) & 1 & \cdots & \cos(\alpha_2 - \alpha_n) \\ \cdots & \cdots & \cdots & \cdots \\ \cos(\alpha_n - \alpha_1) & \cos(\alpha_n - \alpha_2) & \cdots & 1 \end{bmatrix}.$$
(3.60)

On the basis of the metric (3.60), the square-distance between the end-points of two vectors ξ^i and η^i is

$$s^2(\theta_k) = \sum_{i,j} g_{i,j}(\xi^i(\theta_k) - \eta^i(\theta_k))(\xi^j(\theta_k) - \eta^j(\theta_k)). \qquad (3.61)$$

Moreover, when $\eta^i = \xi^i + \frac{\partial \xi^i}{\partial \theta_k}$ we obtain

$$ds^2 = \sum_{i,j} g_{i,j} \left(\frac{\partial \xi^i}{\partial \theta_k} \partial \theta_k \right) \left(\frac{\partial \xi^i}{\partial \theta_h} \partial \theta_h \right) = G_{h,k} \partial \theta^h \partial \theta^k, \qquad (3.62)$$

where

$$G = A^T g A \qquad (3.63)$$

and

$$A = \begin{bmatrix} \frac{\partial \xi^1}{\partial \theta_1} & \frac{\partial \xi^1}{\partial \theta_2} & \cdots & \frac{\partial \xi^1}{\partial \theta_q} \\ \frac{\partial \xi^2}{\partial \theta_1} & \frac{\partial \xi^2}{\partial \theta_2} & \cdots & \frac{\partial \xi^2}{\partial \theta_q} \\ \cdots & \cdots & \cdots & \cdots \\ \frac{\partial \xi^n}{\partial \theta_1} & \frac{\partial \xi^n}{\partial \theta_2} & \cdots & \frac{\partial \xi^n}{\partial \theta_q} \end{bmatrix}. \qquad (3.64)$$

In analogy with Resconi's and Licata's treatment in the reference [226], here Eqs. (3.62), (3.63) and (3.64) indicate that we have n quantum states and n parameters of the states in order to describe the process of interference of n probability densities.

In the picture of a 3D non-Euclidean arena described by the quantum entropy, one can also introduce the quantum-entropic length given by Eq. (3.50) inside Eq. (3.62):

$$ds^2 = G_{h,k}\partial\theta^h\partial\theta^k = \frac{1}{\sum_{i=1}^{N}((\nabla_i S_Q)^2 - \nabla_i^2 S_Q)}. \qquad (3.65)$$

Equation (3.65) indicates clearly that the parameter states characterizing the quantum states in the process of the interference are determined by the quantum entropy. It becomes so permissible that the following reading of the formalism of the double-slit interference (in the case of a wavefunction characterized by n probabilities densities) inside the picture based on the quantum entropy: the probability density associated with the wavefunction under consideration determines a quantum entropy characterizing the vacuum associated with the beams of electrons during the process, and this quantum entropy corresponds to a change of the geometry of space described by a quantum-entropic length given by Eq. (3.61). The interference process is associated with the quantum entropic length (3.61) and thus is determined by the quantum entropy. The quantum entropy can be thus considered as the origin and the ultimate visiting card of the processes in the experiment of the double-slit interference. The quantum-entropic distance given by Eq. (3.61) and determined by the quantum entropy can be considered as the ultimate source of quantum information as regards the double-slit interference.

Finally — and this is a fundamental result obtained inside the approach proposed by the author in the papers [221, 222, 214] — with the introduction of the quantum entropy ((3.42a) for one-body systems and (3.42b) for many body systems) which leads to the energy conservation law (Eq. (3.44) for one-body system and Eq. (3.46) for many-body system), it is also possible to throw new light on the interpretation of the action of the 3D space as an immediate information medium in EPR-type correlations. In fact, for example, for many-body systems, on the basis of Eq. (3.46), one can say that the action of the 3D space as an immediate information medium derives just from the two quantum correctors to the energy of the system under consideration, namely from the quantum corrector to the potential energy $\sum_{i=1}^{N} \frac{\hbar^2}{2m_i}(\nabla_i^2 S_Q)$ and the quantum corrector to the kinetic energy $-\sum_{i=1}^{N} \frac{\hbar^2}{2m_i}(\nabla_i S_Q)^2$ (while the other two terms $\sum_{i=1}^{N} \frac{|\nabla_i S|^2}{2m_i}$ and V determine a local feature of space). On the basis of the two quantum correctors of the energy appearing in Eq. (3.46), the quantum-entropic length (3.50) — determined by the quantum entropy — can be interpreted as an indicator of the nonlocal correlations in many-body systems (and thus provides a direct measure of the degree of departure from the Euclidean geometry characteristic of classical physics). The maximum value of (3.50) is obtained for $L_{\text{quantum}}^{\max} = 1$, which corresponds to the maximum degree of nonlocal correlation of a quantum system. The feature of the quantum potential to make the 3D space an immediate information channel into the behavior of quantum particles derives just from the quantum-entropic length (3.50) and thus from the quantum entropy. In other words, one can see that by introducing the quantum entropy given by Eq. (3.42b), it is just the two quantum correctors to the energy of the system under consideration, which are determined by the quantum entropy describing the change of the geometry of the 3D space from Euclidean to non-Euclidean the fundamental elements which, at a fundamental level, produce an immediate information medium in the behavior of the particles in EPR-type experiments. The space we perceive seems to be characterized by local features because in our macroscopic domain the quantum entropy satisfies conditions (3.47) or (3.48) which imply a Euclid nature of space.

In synthesis, according to the view suggested by the author, the introduction of the quantum entropy (given by Eq. (3.42a) or (3.42b)) as the fundamental entity that determines the behavior of quantum particles leads to an energy conservation law in quantum mechanics (expressed by Eqs. (3.44) and (3.46)) which lets us realize what makes indeed the 3D space an immediate information medium in EPR-type correlations, which lets us realize how the property of quantum potential to determine the action of the 3D space as an immediate information medium arises. The ultimate source, the ultimate visiting card which determines the action of the 3D space as an immediate information medium between quantum particles is a fundamental vacuum defined by the quantum entropy ((3.42a) or (3.42b)) and whose geometry is defined by the quantum-entropic length ((3.49) or (3.50)). The quantum entropy, by producing two quantum corrector terms in the energy, can be considered the fundamental element which gives origin to the nonlocal action of the quantum potential [214].

3.5 The Symmetrized Quantum Potential and the Symmetrized Quantum Entropy

The nonlocal geometry of the 3D space which acts as an immediate information medium between quantum particles in EPR-type experiments is characterized by a fundamental feature: if one imagines to exchange, to invert the roles of the two particles what happens is always the same type of process, namely an instantaneous communication between the two particles. In other words, the nonlocal geometry regarding the instantaneous communication between two quantum particles in EPR-type experiments is characterized by a symmetry: it occurs in either case i.e., if one intervenes on one and if one intervenes on the other, in both cases the same type of process happens and — we can say — always thanks to the 3D space which functions as an immediate information medium. According to the author, this symmetric property of the instantaneous communication between two quantum particles in EPR-type experiments is not well explained by the original Bohm's quantum potential. Although Bohm's quantum potential ((3.18) for one-body systems and (3.20) for many-body systems) has a space-like, instantaneous action, however it comes from the Schrödinger equation which is not time-symmetric and therefore its expression cannot be considered completely satisfactory just because it meets problems if one inverts the sign of time: thus it cannot reproduce in a coherent way the symmetric feature of the nonlocal geometry of space as regards the instantaneous communication of quantum particles in EPR-type experiments if one imagines to see the process backwards.

In the light of these considerations, in order to reproduce and explain this symmetric property of the nonlocal geometry of space in EPR-type experiments, with the publication of the paper *Nonlocality and the symmetrized quantum potential* [227], the author and Sorli introduced a research line based on a symmetrized extension of the quantum potential. The symmetrized quantum potential reproduces, from the mathematical point of view, the fact that, if one imagines to film the process of an instantaneous communication between two subatomic particles in EPR-type experiments backwards, namely inverting the sign of time, one should expect to see what really happened. In other words, the symmetrized quantum potential can explain a symmetric and instantaneous communication between subatomic particles and thus can be considered as an interesting attempt to develop a mathematical formalism for the state of the 3D space as an immediate information medium in EPR-type experiments. In order to analyse the most important features of the symmetrized quantum potential approach, we base our discussion on the papers [210, 213, 214, 227–229].

In the symmetrized approach developed by the author, the postulates of de Broglie–Bohm pilot wave theory can be generalized in four more fundamental postulates of a symmetrized bohmian mechanics:

- An individual physical system comprises a wave propagating in space and time together with a point particle which moves under the guidance of the wave.
- The wave is mathematically described by $|C(t)\rangle = \begin{pmatrix} \psi(t) \\ \phi(t) \end{pmatrix} = \begin{pmatrix} R_1 e^{iS_1/\hbar} \\ R_2 e^{iS_2/\hbar} \end{pmatrix}$, a solution to the Schrödinger symmetrized equation

$$\begin{pmatrix} H & 0 \\ 0 & -H \end{pmatrix} C(t)\rangle = i\hbar \frac{\partial}{\partial t} |C(t)\rangle, \qquad (3.66)$$

where $|C(t)\rangle = \begin{pmatrix} \psi(t) \\ \phi(t) \end{pmatrix}$, $\psi(t)$ is the wavefunction describing the forward-time process (solution to the standard Schrödinger equation), $\phi(t)$ is the wavefunction describing the time-reverse process (solution to the time-reversed Schrödinger equation).

- The particle motion is obtained as the solution to the symmetrized quantum Hamilton–Jacobi equation

$$\frac{\partial}{\partial t}\begin{pmatrix}S_1\\S_2\end{pmatrix} + \frac{1}{2m}\begin{pmatrix}(\nabla S_1)^2\\(\nabla S_2)^2\end{pmatrix} - \frac{\hbar^2}{2m}\begin{pmatrix}\frac{\nabla^2 R_1}{R_1}\\-\frac{\nabla^2 R_2}{R_2}\end{pmatrix} + \begin{pmatrix}V\\-V\end{pmatrix} = 0 \tag{3.67}$$

or to the equivalent symmetrized quantum Newton's equation of motion

$$m\frac{d^2\mathbf{x}}{dt^2} = -\nabla\begin{pmatrix}V - \frac{\hbar^2}{2m}\frac{\nabla^2 R_1}{R_1}\\-V + \frac{\hbar^2}{2m}\frac{\nabla^2 R_2}{R_2}\end{pmatrix}. \tag{3.68}$$

To solve Eqs. (3.67) and (3.68) one has to specify the initial condition $\mathbf{x}_0$. An ensemble of possible motions associated with the same wave $|C(t)\rangle = \begin{pmatrix}\psi(t)\\\phi(t)\end{pmatrix} = \begin{pmatrix}R_1 e^{iS_1/\hbar}\\R_2 e^{iS_2/\hbar}\end{pmatrix}$ is generated by varying $\mathbf{x}_0$.

- The probability that a particle in the ensemble lies between the points $\mathbf{x}$ and $\mathbf{x} + d\mathbf{x}$ at time t is given by $\begin{pmatrix}R_1^2(\mathbf{x},t)\\R_2^2(\mathbf{x},t)\end{pmatrix}d^3x$. This postulate has the effect of selecting among all the possible motions implied by the laws (3.67) and (3.68) those that are compatible with an initial distribution $\begin{pmatrix}R_{10}^2(\mathbf{x},t)\\R_{20}^2(\mathbf{x},t)\end{pmatrix}$.

We obtain in this way a symmetrized extension of Bohmian mechanics which is characterized by a symmetrized quantum potential at two components of the form

$$Q = -\frac{\hbar^2}{2m}\begin{pmatrix}\frac{\nabla^2 R_1}{R_1}\\-\frac{\nabla^2 R_2}{R_2}\end{pmatrix}, \tag{3.69}$$

where R_1 is the amplitude function of the forward-time wavefunction ψ and R_2 is the amplitude function of the time-reverse wavefunction ϕ.

Let us examine now in more detail the form of the symmetrized quantum potential (3.69). As one can easily see, just like the quantum potential of the original Bohm theory, also the symmetrized quantum potential (3.69) has an action which is stronger when the mass is more comparable with Planck constant, and Laplace operator indicates that the action of this potential is like-space, nonlocal, instantaneous. The difference from the original bohmian mechanics lies in the fact that (3.69) has two components, namely depends also on the wavefunction concerning the time-reverse process, and therefore its space-like, nonlocal, instantaneous action is predicted not only by the forward-time process but also by the time-reverse process (and this implies therefore that the process of the instantaneous action between two subatomic particles can be interpreted in the correct way also exchanging t in $-t$).

With the introduction of the symmetrized quantum potential, one can say that, in the presence of quantum particles, space assumes a special "state" at two components that determines the following facts. The first component of this special state $Q_1 = -\frac{\hbar^2}{2m}\frac{\nabla^2 R_1}{R_1}$ regards the forward-time process: it practically coincides with the original Bohm's quantum potential, allows us to explain quantum processes in terms of well-defined motions of particles and determines a nonlocal, instantaneous communication on the particle into consideration. Instead the second component $Q_2 = \frac{\hbar^2}{2m}\frac{\nabla^2 R_2}{R_2}$ regards the time-reverse process, reproduces what physically happens if one would imagine to film a quantum process backwards: it allows us to explain quantum phenomena in the most correct and complete way from the point of view of the symmetry in time. Moreover, the opposed sign of the second component with respect to the first component translates from the mathematical point of view the idea that, in a quantum process, time exists only as a measuring system of the numerical order of material changes: the sign of the second component seems to indicate that it is not possible to go backwards in the physical time intended as a numerical order.

In order to illustrate the symmetrized extension of bohmian mechanics better, let us consider the simple example of the one-dimensional harmonic oscillator, subjected thus to the potential

$V = \frac{1}{2}m\omega^2 x^2$. In this particular case, Eq. (3.69) assumes the form

$$\frac{\partial}{\partial t}\begin{pmatrix} S_1 \\ S_2 \end{pmatrix} + \frac{1}{2m}\begin{pmatrix} (\nabla S_1)^2 \\ (\nabla S_2)^2 \end{pmatrix} - \frac{\hbar^2}{2m}\begin{pmatrix} \frac{\nabla^2 R_1}{R_1} \\ -\frac{\nabla^2 R_2}{R_2} \end{pmatrix}$$

$$+ \begin{pmatrix} \frac{1}{2}m\omega^2 x^2 \\ -\frac{1}{2}m\omega^2 x^2 \end{pmatrix} = 0, \qquad (3.70)$$

the stationary states are given by $C(t) = \begin{pmatrix} u_n(x)e^{-iE_n t/\hbar} \\ u_n(x)e^{iE_n t/\hbar} \end{pmatrix}$ (where $u_n(x)$ are real functions proportional to Hermite polynomials and $E_n = (n+\frac{1}{2})\hbar\omega$, $n = 0,1,2,\ldots$ is the quantum number associated with each stationary state) and the quantum potential which derives from (3.70) is

$$Q = \begin{pmatrix} \left(n+\frac{1}{2}\right)\hbar\omega - \frac{1}{2}m\omega^2 x^2 \\ -\left(n+\frac{1}{2}\right)\hbar\omega + \frac{1}{2}m\omega^2 x^2 \end{pmatrix}. \qquad (3.71)$$

The first component of the symmetrized quantum potential (3.71) is the real physical component which can explain the forward-time process of the instantaneous action on the particle subjected to the potential $V = \frac{1}{2}m\omega^2 x^2$ while the second component of the symmetrized quantum potential allows us to recover the symmetry in time in quantum processes regarding the harmonic oscillator and thus to interpret in the correct way the behavior of the harmonic oscillator if one would imagine to film the process backwards.

In analogy to what happens in the original bohmian theory, also in this symmetrized extension the quantum potential (3.69) must not be considered a term ad hoc. It plays a fundamental role in the symmetrized quantum formalism: in the formal plant of the symmetrized Bohm's theory it emerges directly from the symmetrized Schrödinger equation. Without the term (3.69) the total energy of the physical system would not be conserved. In fact, Eq. (3.67) can also be written in the equivalent form

$$\frac{1}{2m}\begin{pmatrix}(\nabla S_1)^2\\(\nabla S_2)^2\end{pmatrix} - \frac{\hbar^2}{2m}\begin{pmatrix}\dfrac{\nabla^2 R_1}{R_1}\\-\dfrac{\nabla^2 R_2}{R_2}\end{pmatrix} + \begin{pmatrix}V\\-V\end{pmatrix} = -\frac{\partial}{\partial t}\begin{pmatrix}S_1\\S_2\end{pmatrix}, \quad (3.72)$$

which can be seen as a real energy conservation law for the forward-time and the reverse-time process in symmetrized quantum mechanics: here one can easily see that without the symmetrized quantum potential (3.69) energy would not be conserved. Equation (3.72) tells us also that the reverse-time of a physical process is characterized by a classical potential and a quantum potential which are endowed with an opposed sign with respect to the corresponding potentials characterizing the forward-time process.

It is also interesting to observe that inside this time-symmetric extension of bohmian mechanics the correspondence principle becomes

$$-\frac{\hbar^2}{2m}\begin{pmatrix}\dfrac{\nabla^2 R_1}{R_1}\\-\dfrac{\nabla^2 R_2}{R_2}\end{pmatrix} \to \begin{pmatrix}0\\0\end{pmatrix}. \quad (3.73)$$

In this classical limit we have the classical Hamilton–Jacobi equation at two components:

$$\frac{\partial}{\partial t}\begin{pmatrix}S_1\\S_2\end{pmatrix} + \frac{1}{2m}\begin{pmatrix}(\nabla S_1)^2\\(\nabla S_2)^2\end{pmatrix} + \begin{pmatrix}V\\-V\end{pmatrix} = 0, \quad (3.74)$$

which shows just that the time-reverse of the classical process involves a classical potential which is endowed with an opposed sign with respect to the classical potential characterizing the forward-time process.

As regards the symmetrized quantum potential, the fundamental result lies in the fact that, for many-body systems, it constitutes a significant mathematical entity which can describe the state of the 3D space as an immediate information medium between subatomic particles in EPR-type experiments, in the sense that it allows us to reproduce in a coherent way the symmetric feature of the nonlocal geometry of space in these experiments. In the case of a system of N particles the symmetrized quantum potential assumes the form

$$Q = \sum_{i=1}^{N} -\frac{\hbar^2}{2m_i} \left(\frac{\nabla_i^2 R_1}{R_1} - \frac{\nabla_i^2 R_2}{R_2} \right), \qquad (3.75)$$

where R_1 is the absolute value of the wavefunction $\psi = R_1 e^{iS_1/\hbar}$ describing the forward-time process (solution of the standard Schrödinger equation of a many-body system) and R_2 is the absolute value of the wavefunction $\phi = R_2 e^{iS_2/\hbar}$ describing the time-reverse process (solution of the time-reversed Schrödinger equation of a many-body system). On the basis of Eq. (3.75), one can say that the 3D space which acts as a direct information medium and which reproduces the symmetric features of the nonlocal geometry in quantum processes regarding many-body systems gets a mathematical reality: one can explain nonlocal correlations in many-body systems — and thus EPR experiments — from the mathematical point of view in the correct way (namely also if one would imagine to film back the process of these correlations). The first component of the many-body symmetrized quantum potential,

$$Q_1 = \sum_{i=1}^{N} -\frac{\hbar^2}{2m_i} \frac{\nabla_i^2 R_1}{R_1}, \qquad (3.76)$$

regards the nonlocal geometry of the quantum world for forward-time process and coincides with the original Bohm's quantum potential: it is the primary physical component which produces observable effects in the quantum world and expresses the nonlocality of many-body systems. The second component

$$Q_2 = \sum_{i=1}^{N} \frac{\hbar^2}{2m_i} \frac{\nabla_i^2 R_2}{R_2} \qquad (3.77)$$

is introduced to reproduce in the correct way the time-reverse process of the instantaneous action and thus it guarantees the symmetric nature of the nonlocal geometry of the quantum world as regards the instantaneous communication between quantum particles in EPR-type experiments. Thus, it allows us to reproduce mathematically the instantaneous communication between quantum particles in EPR-type experiments with the idea of space as an immediate information medium if one would imagine to film the process backwards. Just like in the one-body symmetrized quantum potential (3.69), also in the many-body symmetrized quantum potential (3.75) the opposed sign of the second component with respect to the first component is the mathematical translation of the idea of the measurable time as a measuring system of the numerical order of material changes.

Now, on the basis of the mathematical formalism illustrated in chapter 3.2, at a fundamental level, the action of the 3D space as an immediate information channel into the behavior of quantum particles in EPR-type experiments derives from the quantum entropy describing the deformation of the geometry of the 3D space from Euclidean to non-Euclidean as well as from a quantum-entropic length — associated with the quantum entropy — providing a direct correlation degree. As a consequence, in the symmetrized extension of bohmian mechanics also the one-body symmetrized quantum potential and the many-body symmetrized quantum potential can be seen as physical entities which derive from an opportune quantum entropy and an opportune quantum-entropic length.

To explore in detail this point, we base our treatment on the paper [229]. Let us consider the many-body symmetrized quantum potential (3.75). Its first component derives from a quantum entropy $S_{Q1} = -\frac{1}{2}\ln\rho_1$ defining the degree of order and chaos of the vacuum for the forward-time processes (where $\rho_1 = |\psi(\mathbf{x}_1, \mathbf{x}_2, \ldots, \mathbf{x}_N, t)|^2$, $\psi(\mathbf{x}_1, \mathbf{x}_2, \ldots, \mathbf{x}_N, t) = R_1 e^{-iS_1/\hbar}$ being the forward-time many-body wavefunction, solution of the standard Schrödinger equation) and thus can be expressed as

$$Q_1 = \sum_{i=1}^{N} \left[-\frac{\hbar^2}{2m_i}(\nabla_i S_{Q1})^2 + \frac{\hbar^2}{2m_i}(\nabla_i^2 S_{Q1}) \right]. \quad (3.78)$$

Its second component derives from a quantum entropy $S_{Q2} = -\frac{1}{2}\ln\rho_2$ defining the degree of order and chaos of the vacuum for the time-reverse processes (where $\rho_2 = |\phi(\mathbf{x}_1, \mathbf{x}_2, \ldots, \mathbf{x}_N, t)|^2$,

$\phi(\mathbf{x}_1, \mathbf{x}_2, \ldots, \mathbf{x}_N, t) = R_2 e^{-iS_2/\hbar}$ being the time-reverse many-body wavefunction, solution to the time-reversed Schrödinger equation) and thus can be expressed as

$$Q_2 = \sum_{i=1}^{N} \left[\frac{\hbar^2}{2m_i}(\nabla_i S_{Q2})^2 - \frac{\hbar^2}{2m_i}(\nabla_i^2 S_{Q2}) \right]. \tag{3.79}$$

Therefore, the many-body symmetrized quantum potential (3.75) reads as

$$Q = \begin{pmatrix} \sum_{i=1}^{N} \left[-\frac{\hbar^2}{2m_i}(\nabla_i S_{Q1})^2 + \frac{\hbar^2}{2m_i}(\nabla_i^2 S_{Q1}) \right] \\ \sum_{i=1}^{N} \left[\frac{\hbar^2}{2m_i}(\nabla_i S_{Q2})^2 - \frac{\hbar^2}{2m_i}(\nabla_i^2 S_{Q2}) \right] \end{pmatrix}. \tag{3.80}$$

Thus, the nonlocal geometry of the quantum world as regards the instantaneous communication of quantum particles in EPR-type experiments, in virtue of its symmetric features, can be defined by a symmetrized quantum entropy at two components of the form

$$S_Q = \begin{pmatrix} -\frac{1}{2}\ln\rho_1 \\ -\frac{1}{2}\ln\rho_2 \end{pmatrix}, \tag{3.81}$$

where the first component $S_{Q1} = -\frac{1}{2}\ln\rho_1$ corresponds to the change of the geometric properties of physical space for the forward-time processes while the second component $S_{Q2} = -\frac{1}{2}\ln\rho_2$ corresponds to the change of the geometric properties of physical space for the reversed-time processes.

The modification of the geometrical properties of space expressed by the symmetrized quantum entropy (3.81) implies two appropriate energy conservation laws. The energy conservation law for the forward-time process is

$$\sum_{i=1}^{N} \frac{|\nabla_i S_1|^2}{2m_i} - \sum_{i=1}^{N} \frac{\hbar^2}{2m_i}(\nabla_i S_{Q1})^2 + V + \sum_{i=1}^{N} \frac{\hbar^2}{2m_i}(\nabla_i^2 S_{Q1}) = -\frac{\partial S_1}{\partial t} \tag{3.82}$$

while the energy conservation law for the reversed-time process is

$$\sum_{i=1}^{N} \frac{|\nabla_i S_2|^2}{2m_i} + \sum_{i=1}^{N} \frac{\hbar^2}{2m_i}(\nabla_i S_{Q2})^2 - V - \sum_{i=1}^{N} \frac{\hbar^2}{2m_i}(\nabla_i^2 S_{Q2}) = -\frac{\partial S_2}{\partial t}.$$
(3.83)

Equations (3.82) and (3.83) may be written as a symmetrized energy conservation law in the following form

$$\begin{pmatrix} \sum_{i=1}^{N} \frac{|\nabla_i S_1|^2}{2m_i} - \sum_{i=1}^{N} \frac{\hbar^2}{2m_i}(\nabla_i S_{Q1})^2 + V + \sum_{i=1}^{N} \frac{\hbar^2}{2m_i}(\nabla_i^2 S_{Q1}) = -\frac{\partial S_1}{\partial t} \\ \sum_{i=1}^{N} \frac{|\nabla_i S_2|^2}{2m_i} + \sum_{i=1}^{N} \frac{\hbar^2}{2m_i}(\nabla_i S_{Q2})^2 - V - \sum_{i=1}^{N} \frac{\hbar^2}{2m_i}(\nabla_i^2 S_{Q2}) = -\frac{\partial S_2}{\partial t} \end{pmatrix}.$$
(3.84)

The symmetrized quantum entropy can be thus considered as the fundamental entity which determines the nonlocal features of the quantum geometry. On the basis of the symmetrized energy conservation law (3.84), one can say that the nonlocal quantum geometry characterized by a symmetric feature as regards the instantaneous correlation between subatomic particles in EPR-type experiments is expressed by the action of the 3D space as an immediate information medium which derives just from the two quantum correctors to the energy of the system under consideration, both for the forward-time processes and for the reversed-time processes, namely from the quantum corrector to the potential energy ($\sum_{i=1}^{N} \frac{\hbar^2}{2m_i}(\nabla_i^2 S_{Q1})$) for the forward-time processes and $-\sum_{i=1}^{N} \frac{\hbar^2}{2m_i}(\nabla_i^2 S_{Q2})$ for the reversed-time processes) and the quantum corrector to the kinetic energy ($-\sum_{i=1}^{N} \frac{\hbar^2}{2m_i}(\nabla_i S_{Q1})^2$ for the forward-time processes and $\sum_{i=1}^{N} \frac{\hbar^2}{2m_i}(\nabla_i S_{Q2})^2$ for the reversed-time processes), while the other two terms $\sum_{i=1}^{N} \frac{|\nabla_i S|^2}{2m_i}$ and V on the left-hand of the two equations (3.84) determine a local feature of space. In other words, the symmetrized quantum entropy given by Eq. (3.81), by determining two quantum correctors to the energy of the system under consideration (where if one imagines to film the process backwards, one would see what physically happens) can be considered the fundamental entity which, at a fundamental level, produces an immediate information medium in the behaviour of the particles in EPR-type experiments. The space we perceive seems to be characterized by local features because in our macroscopic domain the symmetrized quantum entropy satisfies the following condition

$$\begin{pmatrix} (\nabla_i S_{Q1})^2 \to (\nabla_i^2 S_{Q1}) \\ (\nabla_i S_{Q2})^2 \to (\nabla_i^2 S_{Q2}) \end{pmatrix}, \qquad (3.85)$$

which can be seen as a correspondence principle inside the symmetrized quantum potential approach based on the symmetrized quantum entropy. The symmetrized correspondence principle (3.85) of the symmetrized quantum entropy can be considered the fundamental mathematical formalism which expresses the local features of the space we perceive in our everyday life. Moreover, as we will see in chapter 3.4, the nonlocal geometry of the 3D space background in the symmetrized quantum potential approach can be described by introducing opportune symmetrized quantum-entropic lengths which provide a direct measure of the correlation degree (and thus of the degree of the departure from the Euclidean geometry characteristic of classical physics) in a quantum system in a time-symmetric picture.

On the basis of its mathematical features, the symmetrized quantum potential (deriving from the symmetrized quantum entropy) implies that in the quantum domain a timeless 3D space has a crucial role in determining the motion of a subatomic particle in the sense that the symmetrized quantum potential produces a likespace, instantaneous action on the particles under consideration (which is characterized by a symmetric feature) and contains an active information about the environment and, on the other hand, implies the concept of time as a numerical order of material change. In EPR-type experiments a 3D timeless space acts as an immediate information medium in the sense that, as a consequence of the symmetrized quantum entropy, the first component of the symmetrized quantum potential makes physical space an "immediate information medium" which keeps two elementary particles in an immediate contact (while the second component of the symmetrized quantum potential reproduces, from the mathematical point of view, the symmetry in time of this communication and the fact that time exists only as a numerical order of material change). This peculiar interpretation of the nonlocal quantum geometry (which derives from the symmetrized quantum potential approach) can be called as the "immediate symmetric interpretation" of nonlocal quantum geometry.

3.6 The Quantum Geometry in the Symmetrized Quantum Potential Approach

While in the classical domain the concept of space is linked with our common experience of translating material objects and from the possible configurations and states they can occupy, from the quantum mechanical point of view, it is not clear what is meant by a material object being "immersed" in space. The material object consists of electrons, protons and neutrons (or the quarks and gluons that compose the protons and neutrons), and the states of these particles belong to the corresponding Hilbert spaces different from the physical space or the phase space of classical physics. If all the particles constituting the material object into consideration move together in some approximate sense, one can introduce universal translation group elements — represented by operators that act on each Hilbert space — that provide us a concept of "space" that is independent of the particular system that partakes in it.

In the paper [230], Anandan defined a geometry for quantum theory by starting from the relations determined by a universal group S, which is the generalization of the translation group. This group is universal in the sense that the same S has a representation in each Hilbert space. An object may be displaced by any $s \in S$, which means that s acts on each of the Hilbert spaces of the particles or fields constituting the object under consideration through the corresponding representation of S. Each wavefunction ψ in each of these Hilbert spaces is mapped to a corresponding ψ_s by this action of s, and the group element s that determines the relation between ψ and ψ_s is independent of the Hilbert space and is therefore universal.

In Anandan's model, the geometry of quantum theory is based on an invariant geometrical "distance", which is associated with a gauge field which is the generalization of the translation group. As regards the translation of a material object that may be performed by acting on all the quantum states of the particles constituting the object, one can consider the universal group element $\exp(-\frac{i}{\hbar}\hat{p}l)$ where $\hat{p}$ is a generator of translation and define the gauge-covariant transformation $\psi_l(x) = f_l(x)\psi(x)$ where x stands for x^μ and

$$f_l(x) = \exp\left(-\frac{i}{\hbar}\hat{\vec{p}}\cdot\vec{l}\right)\exp\left\{i\frac{q}{\hbar}\int_x^{x+l}\vec{A}(\vec{y},t)\cdot d\vec{y}\right\}, \qquad (3.86)$$

where the integral is taken along the straight line joining $x = (\vec{x},t)$ and $(\vec{x}+\vec{l},t)$ for simplicity. Under a gauge transformation, $\psi'(x) = u(x)\psi(x)$, where $u(x) = \exp\{i\frac{q}{\hbar}\Lambda(x)\}$ and $A'_\mu(x) = A_\mu(x) - \partial_\mu\Lambda(x)$, the quantity (3.86) transforms to

$$f'_l(x) = \exp\left(-\frac{i}{\hbar}\hat{\vec{p}}\cdot\vec{l}\right)u(\vec{x}+\vec{l},t)\exp\left\{i\frac{q}{\hbar}\int_x^{x+l}\vec{A}(\vec{y},t)\cdot d\vec{y}\right\}u(\vec{x},t). \qquad (3.87)$$

More generally, for an arbitrary gauge field in the previous equations eA_μ must be replaced by $g_0 A_\mu^k T_k$, where T_k generates the gauge group, and $u(x)$ is the corresponding local gauge transformation. The gauge field exponential (parallel transport) operator is along an arbitrary piecewise-differentiable curve $\hat{\gamma}$ in R^4. The energy-momentum operator is defined by $\hat{p}_0 = H$, $\hat{p}_i = i\hbar\frac{\partial}{\partial x^i}$. The gauge field $g_0 A_\mu^k T_k$ corresponds to a geometry defined by a transformation g_γ on the Hilbert space by $\psi_\gamma(x) = g_\gamma(x)\psi(x)$, where

$$g_\gamma(x) = P\exp\left(-\frac{i}{\hbar}\int_{\bar{\gamma}}\hat{p}_\mu dy^\mu\right)P\exp\left\{-i\frac{g_0}{\hbar}\int_\gamma A_\mu^k(y)T_k dy^\mu\right\}, \qquad (3.88)$$

with P denoting path ordering, and γ is a curve in space-time that is congruent to $\hat{\gamma}$ while $\bar{\gamma}$ is the curve γ traversed in the reversed order. In Eq. (3.88), the curve γ begins at x and ends at $x+l$, where l^μ is a fixed vector (independent of x^μ), and $\bar{\gamma}$ begins at $x+l$ and ends in x. In Anandan's model, the quantity g_γ given by Eq. (3.88) can be considered as a quantum distance that replaces the classical space-time distance along the curve γ [230].

Now, in the recent paper [229] the author and Sorli generalized Anandan's geometric approach based on the geometrical distance (3.88) by introducing the symmetrized quantum distance given by the following relation:

$$g_\gamma(x) = \begin{pmatrix} P\exp\left(-\frac{i}{\hbar}\int_{\bar\gamma}\hat{p}_\mu dy^\mu\right) P\exp\left\{-i\frac{g_0}{\hbar}\int_\gamma A_\mu^k(y)T_k dy^\mu\right\} \\ P\exp\left(\frac{i}{\hbar}\int_{\bar\gamma}\hat{p}_\mu dy^\mu\right) P\exp\left\{i\frac{g_0}{\hbar}\int_\gamma A_\mu^k(y)T_k dy^\mu\right\} \end{pmatrix}, \tag{3.89}$$

where the second component

$$g_\gamma(x) = P\exp\left(\frac{i}{\hbar}\int_{\bar\gamma}\hat{p}_\mu dy^\mu\right) P\exp\left\{i\frac{g_0}{\hbar}\int_\gamma A_\mu^k(y)T_k dy^\mu\right\} \tag{3.90}$$

regards the time-reverse processes, namely is introduced to assure a symmetry in time of the processes of translation if one would imagine to see the process backwards.

Under a local gauge transformation the symmetrized quantum distance (3.89) transforms as

$$g'_\gamma(x)$$
$$= \begin{pmatrix} P\exp\left(-\frac{i}{\eta}\int_{\bar\gamma}\hat{p}'_\mu dy^\mu\right)u(x+l)P\exp\left\{-i\frac{g_0}{\eta}\int_\gamma A_\mu^k(y^\mu)T_k dy^\mu\right\}u^+(x) \\ P\exp\left(\frac{i}{\eta}\int_{\bar\gamma}\hat{p}'_\mu dy^\mu\right)u(x+l)P\exp\left\{i\frac{g_0}{\eta}\int_\gamma A_\mu^k(y)T_k dy^\mu\right\}u^+(x) \end{pmatrix}, \tag{3.91}$$

where $\hat{p}'_0 = H'$, $\hat{p}'_i = \hat{p}_i = i\hbar\frac{\partial}{\partial x^i}$. Moreover, taking into account that

$$P\exp\left(-\frac{i}{\hbar}\int_{\bar\gamma}\hat{p}_\mu dy^\mu\right) = \left\{P\exp\left(-\frac{i}{\hbar}\int_\gamma \hat{p}_\mu dy^\mu\right)\right\}^+ \tag{3.92}$$

one gets

$$\langle\psi|g_\gamma|\psi\rangle = \begin{pmatrix} \int d^3x \left[P\exp\left(-\frac{i}{\hbar}\int_\gamma \hat{p}_\mu dy^\mu\right)\psi(x)\right]^+ \\ P\exp(-ig_0\int_\gamma A_\mu^k T_k dy^\mu)\psi(x) \\ \int d^3x \left[P\exp\left(\frac{i}{\hbar}\int_\gamma \hat{p}_\mu dy^\mu\right)\psi(x)\right]^+ \\ P\exp(ig_0\int_\gamma A_\mu^k T_k dy^\mu)\psi(x) \end{pmatrix}.$$

(3.93)

The expectation value (3.93) is gauge invariant (for both the components) because the integrand is gauge invariant.

The symmetrized quantum distance (3.91) can be considered as the fundamental notion of distance in the geometry of the quantum world which is associated with the gauge field which is the generalization of the translation group. Now, and this is the fundamental point, in the symmetrized quantum potential approach the fundamental entity which describes the geometrical properties of the background space of quantum processes is the quantum entropy: the deformation of the geometrical properties of space with respect to the Euclidean geometry characteristic of classical physics is determined by the quantum entropy. Therefore, one can say that the action of the gauge field $g_0 A_\mu^k T_k$ — which is the generalization of the group of translation and determines the symmetrized quantum distance (3.91) — derives from the quantum entropy. So, as the author showed in the paper [229], in the symmetrized quantum potential approach, the nonlocal quantum geometry associated with the 3D timeless space which acts as an immediate information medium can be described by introducing the following notions of quantum distances:

$$L_{\text{quantum}} = \begin{pmatrix} \dfrac{1}{\sqrt{(\nabla S_{Q1})^2 - \nabla^2 S_{Q1}}} \\ \dfrac{1}{\sqrt{-(\nabla S_{Q2})^2 + \nabla^2 S_{Q2}}} \end{pmatrix}$$

$$= \begin{pmatrix} P\exp\left(-\frac{i}{\hbar}\int_{\tilde{\gamma}}\hat{p}_\mu dy^\mu\right) P\exp\left\{-i\frac{g_0}{\hbar}\int_\gamma A_\mu^k(y)T_k dy^\mu\right\} \\ P\exp\left(\frac{i}{\hbar}\int_{\tilde{\gamma}}\hat{p}_\mu dy^\mu\right) P\exp\left\{i\frac{g_0}{\hbar}\int_\gamma A_\mu^k(y)T_k dy^\mu\right\} \end{pmatrix}$$
(3.94)

for one-body systems and

$$L_{\text{quantum}} = \begin{pmatrix} \frac{1}{\sqrt{\sum_{i=1}^N((\nabla_i S_{Q1})^2 - \nabla_i^2 S_{Q1})}} \\ \frac{1}{\sqrt{\sum_{i=1}^N(-(\nabla_i S_{Q2})^2 + \nabla_i^2 S_{Q2})}} \end{pmatrix}$$

$$= \begin{pmatrix} P\exp\left(-\frac{i}{\hbar}\int_{\tilde{\gamma}}\hat{p}_\mu dy^\mu\right) P\exp\left\{-i\frac{g_0}{\hbar}\int_\gamma A_\mu^k(y)T_k dy^\mu\right\} \\ P\exp\left(\frac{i}{\hbar}\int_{\tilde{\gamma}}\hat{p}_\mu dy^\mu\right) P\exp\left\{i\frac{g_0}{\hbar}\int_\gamma A_\mu^k(y)T_k dy^\mu\right\} \end{pmatrix}$$
(3.95)

for many-body systems. Inside the symmetrized quantum potential approach, Eqs. (3.94) and (3.95) define typical quantum-entropic lengths that, by providing a direct correlation degree in quantum systems, can be used to evaluate the strength of quantum effects and, therefore, the modification of the geometry with respect to the Euclidean geometry characteristic of classical physics. In particular, the first component of (3.94) and (3.95) regards the forward-time processes while the second component of (3.94) and (3.95) regards the time-reverse processes. Once the quantum-entropic lengths (3.94) or (3.95) become nonnegligible the system into consideration goes into a quantum regime, characterized by a nonlocal geometry. In this picture, Heisenberg's uncertainty principle derives from the fact that we are unable to perform a classical measurement to distances smaller than the quantum-entropic lengths. In other words, the size of a measurement has to be bigger than the quantum-entropic lengths

$$\Delta L \geq L_{\text{quantum}} = \begin{pmatrix} \frac{1}{\sqrt{(\nabla S_{Q1})^2 - \nabla^2 S_{Q1}}} \\ \frac{1}{\sqrt{-(\nabla S_{Q2})^2 + \nabla^2 S_{Q2}}} \end{pmatrix}$$
(3.96)

(for one-body systems)

$$\Delta L \geq L_{\text{quantum}} = \left(\frac{1}{\sqrt{\sum_{i=1}^{N} ((\nabla_i S_{Q1})^2 - \nabla_i^2 S_{Q1})}} \frac{1}{\sqrt{\sum_{i=1}^{N} (-(\nabla_i S_{Q2})^2 + \nabla_i^2 S_{Q2})}} \right) \quad (3.97)$$

(for many-body systems). The quantum regime is entered when the quantum-entropic lengths must be taken under consideration. In the nonlocal quantum geometry corresponding to the 3D space which acts as a direct information medium, Eqs. (3.94) and (3.95) may be considered as the fundamental definition of distances: they replace the concept of distance of classical space-time in such a way to reproduce, understand and explain in a clear and direct way the nonlocal correlations of quantum systems. Moreover, on the basis of Eqs. (3.94) and (3.95), there is an equivalence between the effect of the gauge field associated with $g_0 A_\mu^k T_k$, and the effect of the quantum entropy. On one hand, one can say that the effect of the gauge field associated with $g_0 A_\mu^k T_k$, where T_k generates the gauge group of translation, P is the path ordering and γ is a curve in space-time that is congruent to $\hat{\gamma}$ while $\bar{\gamma}$ is the curve γ traversed in the reversed order, is to determine a deformation of the geometrical properties of the background space, with respect to the Euclidean geometry characteristic of classical physics, expressed by a symmetrized quantum entropy. In other words, under this point of view, Eqs. (3.94) and (3.95) indicate in what sense the action of the gauge field associated with $g_0 A_\mu^k T_k$, where T_k generates the gauge group, determines a 3D space which acts as an immediate information medium whose geometrical properties are defined by a symmetrized non-Euclidean quantum distance which is associated with the symmetrized quantum entropy. On the other hand, since the quantum entropy is the ultimate visiting card expressing the geometry of the quantum world inside the symmetrized quantum potential approach, one can say that the action of the gauge field of determining a deformation of the geometry in the quantum regime derives just from the symmetrized quantum entropy.

3.7 Some Examples

In this section, we are going to analyse two relevant applications of the symmetrized quantum potential approach, namely the double-slit interference and the Aharonov–Bohm effect, showing in what sense the symmetrized quantum entropy is the ultimate entity describing the background of the processes. In this regard we follow the treatment provided in the paper [229].

3.7.1 The double slit-interference

In the symmetrized quantum potential approach, the double-slit interference characterized by n probability densities is described by a wavefunction at two components $|C(t)\rangle = \begin{pmatrix} \psi(t) \\ \phi(t) \end{pmatrix}$, where $\psi(t) = h_1(t) + h_2(t) + \cdots h_n(t)$ is the usual quantum wavefunction describing the forward-time process (solution to the standard Schrödinger equation), $\phi(t) = k_1(t) + k_2(t) + \cdots + k_n(t)$ is the wavefunction describing the time-reverse process (solution to the time-reversed Schrödinger equation), where $h_1 = \alpha_1 + i\beta_1$, $h_2 = \alpha_2 + i\beta_2, \ldots, h_n = \alpha_n + i\beta_n$, $k_1 = \gamma_1 + i\delta_1$, $k_2 = \gamma_2 + i\delta_2, \ldots, k_n = \gamma_n + i\delta_n$.

The probability for the interference may be written as

$$P(x) = \begin{pmatrix} \langle \psi | \psi \rangle \\ \langle \phi | \phi \rangle \end{pmatrix} = \begin{pmatrix} \sum_{i,j} g_{ij} \xi^i(x) \xi^j(x) \\ \sum_{i,j} f_{ij} \zeta^i(x) \zeta^j(x) \end{pmatrix}, \quad (3.98)$$

where

$$\xi^i = \sqrt{I^i}, \quad \xi^j = \sqrt{I^j}, \quad (3.99)$$

$$g = \begin{bmatrix} 1 & \cos(\alpha_1 - \alpha_2) & \cdots & \cos(\alpha_1 - \alpha_n) \\ \cos(\alpha_2 - \alpha_1) & 1 & \cdots & \cos(\alpha_2 - \alpha_n) \\ \cdots & \cdots & \cdots & \cdots \\ \cos(\alpha_n - \alpha_1) & \cos(\alpha_n - \alpha_2) & \cdots & 1 \end{bmatrix}$$

(3.100)

for the forward-time process;

$$\zeta^i = \sqrt{J^i}, \quad \zeta^j = \sqrt{J^j}, \quad (3.101)$$

$$f = \begin{bmatrix} 1 & \cos(\gamma_1 - \gamma_2) & \cdots & \cos(\gamma_1 - \gamma_n) \\ \cos(\gamma_2 - \gamma_1) & 1 & \cdots & \cos(\gamma_2 - \gamma_n) \\ \cdots & \cdots & \cdots & \cdots \\ \cos(\gamma_n - \gamma_1) & \cos(\gamma_n - \gamma_2) & \cdots & 1 \end{bmatrix}$$

(3.102)

for the time-reverse process.

On the basis of the metrics (3.100) and (3.102), the square-distance between the end-points of two vectors $\begin{pmatrix} \xi^i \\ \zeta^i \end{pmatrix}$ and $\begin{pmatrix} \eta^i \\ \lambda^i \end{pmatrix}$ is

$$s^2(\theta_k) = \begin{pmatrix} \sum_{i,j} g_{i,j}(\xi^i(\theta_k) - \eta^i(\theta_k))(\xi^j(\theta_k) - \eta^j(\theta_k)) \\ \sum_{i,j} f_{i,j}(\zeta^i(\theta_k) - \lambda^i(\theta_k))(\zeta^j(\theta_k) - \lambda^j(\theta_k)) \end{pmatrix}. \quad (3.103)$$

Moreover, when $\begin{pmatrix} \eta^i \\ \lambda^i \end{pmatrix} = \begin{pmatrix} \xi^i + \frac{\partial \xi^i}{\partial \theta_k} \\ \zeta^i + \frac{\partial \zeta^i}{\partial \theta_k} \end{pmatrix}$ one obtains

$$ds^2 = \begin{pmatrix} \sum_{i,j} g_{i,j} \left(\frac{\partial \xi^i}{\partial \theta_k} \partial \theta_k \right) \left(\frac{\partial \xi^i}{\partial \theta_h} \partial \theta_h \right) \\ \sum_{i,j} f_{i,j} \left(\frac{\partial \zeta^i}{\partial \theta_k} \partial \theta_k \right) \left(\frac{\partial \zeta^i}{\partial \theta_h} \partial \theta_h \right) \end{pmatrix} = \begin{pmatrix} G_{h,k} \partial \theta^h \partial \theta^k \\ F_{h,k} \partial \theta^h \partial \theta^k \end{pmatrix}$$

(3.104)

where

$$G = A^T g A, \quad (3.105)$$
$$F = B^T f B, \quad (3.106)$$

and

$$A = \begin{bmatrix} \frac{\partial \xi^1}{\partial \theta_1} & \frac{\partial \xi^1}{\partial \theta_2} & \cdots & \frac{\partial \xi^1}{\partial \theta_q} \\ \frac{\partial \xi^2}{\partial \theta_1} & \frac{\partial \xi^2}{\partial \theta_2} & \cdots & \frac{\partial \xi^2}{\partial \theta_q} \\ \cdots & \cdots & \cdots & \cdots \\ \frac{\partial \xi^n}{\partial \theta_1} & \frac{\partial \xi^n}{\partial \theta_2} & \cdots & \frac{\partial \xi^n}{\partial \theta_q} \end{bmatrix}, \quad (3.107)$$

$$B = \begin{bmatrix} \dfrac{\partial \zeta^1}{\partial \theta_1} & \dfrac{\partial \zeta^1}{\partial \theta_2} & \cdots & \dfrac{\partial \zeta^1}{\partial \theta_q} \\ \dfrac{\partial \zeta^2}{\partial \theta_1} & \dfrac{\partial \zeta^2}{\partial \theta_2} & \cdots & \dfrac{\partial \zeta^2}{\partial \theta_q} \\ \cdots & \cdots & \cdots & \cdots \\ \dfrac{\partial \zeta^n}{\partial \theta_1} & \dfrac{\partial \zeta^n}{\partial \theta_2} & \cdots & \dfrac{\partial \zeta^n}{\partial \theta_q} \end{bmatrix}. \quad (3.108)$$

Here, inside this formalism one can define a symmetrized quantum-entropic length for the double-slit interference given by the following relation:

$$ds^2 = \begin{pmatrix} G_{h,k} \partial \theta^h \partial \theta^k \\ F_{h,k} \partial \theta^h \partial \theta^k \end{pmatrix} = \begin{pmatrix} \dfrac{1}{\sum_{i=1}^{N} ((\nabla_i S_{Q1})^2 - \nabla_i^2 S_{Q1})} \\ \dfrac{1}{\sum_{i=1}^{N} (-(\nabla_i S_{Q2})^2 + \nabla_i^2 S_{Q2})} \end{pmatrix}.$$
(3.109)

Equation (3.109) indicates clearly that the parameter states characterizing the quantum states in the process of the interference are determined by the symmetrized quantum entropy. The following reading of the formalism of the double-slit interference becomes permissible (in the case of a wavefunction characterized by N probability densities) inside the symmetrized quantum potential approach: the probability density associated with the wavefunction under consideration determines a symmetrized quantum entropy, a deformation of the geometry of the process of double-slit interference emerges which is expressed by a quantum-entropic length given by Eq. (3.109) (providing a direct measure of the correlation degree in the double-slit interference). In the symmetrized quantum potential approach, the interference process is associated with the quantum entropic length (3.109) and thus is determined by the symmetrized quantum entropy. The symmetrized quantum-entropic distance given by Eq. (3.109) can be considered as the ultimate source of quantum information as regards the double-slit interference inside the symmetrized quantum potential approach.

The symmetrized quantum-entropic length given by Eq. (3.109) is the fundamental entity which allows us to explain the results in the double-slit experiment in agreement with a principle of causality: in this picture, one can say that it is the symmetrized quantum-entropic length (3.109) determined by the quantum entropy the ultimate visiting card of the geometry of this process which is responsible of the interference figure and of the well-defined trajectories of the particles that determine it. As a consequence of the geometrical properties of the background defined by the symmetrized quantum-entropic distance (3.109) and which derive from the symmetrized quantum entropy, one obtains the following results. On one hand, that each single particle follows a precise and calculable trajectory which comes from one slit or the other and this aggregate of trajectories produces the requested interference figure and, at the same time, shows that the final position of the particle on the screen allows us to deduce through which slit it has passed; on the other hand, that if one imagines to watch the process backwards, one can reproduce in a coherent way what physically happens. The trajectory described by each subatomic particle in the double-slit interference can be interpreted as a consequence of the geometry associated with the symmetrized quantum-entropic length (3.109) and thus with the quantum entropy. As a consequence of the geometry described by this symmetrized quantum-entropic length, the region of space of the double-slit experiment assumes the special state represented by symmetrized quantum potential. While in Bohm's original theory the trajectory described by each particle in the double-slit experiment is tied to quantum potential, in the symmetrized quantum potential approach developed by the author one can explain and reproduce coherently the results both for the forward-time process and for the time-reverse process with a symmetrized quantum potential, and the ultimate entity which is responsible of these results is the geometry of the background expressed by the symmetrized quantum-entropic length (3.109) and thus by the symmetrized quantum entropy.

3.7.2 About the Aharonov–Bohm effect

The second process we take under consideration is the Aharonov–Bohm effect due to a magnetic field. The Aharonov–Bohm effect regards the quantum mechanical scattering of electrons in the presence of a classical magnetic vector potential determined by a current-carrying solenoid. Electrons are prevented from entering the region where the magnetic field itself is nonzero so there is no classical force on them. Nevertheless, information about the flux can be obtained.

As Aharonov and Bohm [231] showed, information about the magnetic flux (modulo a constant) through a solenoid can be measured via its effect on the interference pattern of electrons in a double-slit experiment. A particle of charge q traversing a path $C_\pm$ in the presence of a magnetic vector potential $\vec{A}$ can be described by a wavefunction $\psi_\pm(x)e^{\frac{iq}{\hbar c}\int_{C_\pm}\vec{A}\cdot d\vec{l}}$ where $\psi_\pm(x)$ is the wavefunction in the absence of a vector potential and the integral is taken along the corresponding path $C_\pm$. We define the flux parameter ϕ as

$$\phi = \frac{q}{\hbar c}\left(\int_{C_+}\vec{A}\cdot d\vec{l} - \int_{C_-}\vec{A}\cdot d\vec{l}\right) = \frac{q}{\hbar c}\oint_C \vec{A}\cdot d\vec{l} = \frac{q}{\hbar c}\Phi, \quad (3.110)$$

where C is a closed curve created by following C_+ and returning along C_- and Φ is the flux through the solenoid, and thus through any closed surface bounded by C. In terms of these variables, the total wavefunction at the screen is

$$\psi(x,\phi) = e^{\frac{iq}{\hbar c}\int_{C_+}\vec{A}\cdot d\vec{l}}(\psi_+(x) + \psi_-(x)e^{-i\phi}), \quad (3.111)$$

which is gauge invariant up to an overall phase.

To be more explicit, let us consider, for example, two wave packets A and B of an electron that passes on the two sides of a solenoid containing a magnetic flux Φ. On the basis of Eq. (3.110), when the line AB passes the solenoid, the wave packet A acquires a phase difference $\alpha = \frac{e}{\hbar c}\Phi$ with respect to the wave packet B. As a consequence, the expectation value of the modular momentum s, given by relation

$$\langle\psi|\exp\left(-i\frac{pl}{\hbar}\right)\psi\rangle = \frac{\exp(i\alpha)}{2}, \qquad (3.112)$$

which was $1/2$ before the line AB passed the solenoid, becomes $\exp(i\frac{e}{\hbar c}\Phi)/2$ and this change in the expectation value of the modular momentum by the factor $\exp(i\frac{e}{\hbar c}\Phi)/2$ as the line AB joining the two wave packets crosses the solenoid is the same in every gauge. The quantum distance corresponding with the general gauge-covariant symmetrized operator g_γ given by Eq. (3.89) in this case of the vector Aharonov–Bohm effect becomes

$$g_\gamma(x) = \begin{pmatrix} \exp\left(-\frac{i}{\hbar}\hat{\vec{p}}\cdot\vec{l}\right)\exp\left\{i\frac{e}{\hbar}\int_{\gamma\vec{x}}^{\vec{x}+\vec{l}} A(\vec{y},t)\cdot d\vec{y}\right\} \\ \exp\left(\frac{i}{\hbar}\hat{\vec{p}}\cdot\vec{l}\right)\exp\left\{-i\frac{e}{\hbar}\int_{\gamma\vec{x}}^{\vec{x}+\vec{l}} A(\vec{y},t)\cdot d\vec{y}\right\} \end{pmatrix}, \qquad (3.113)$$

where the integrals are from $\vec{x}$ to $\vec{x}+\vec{l}$ along γ.

In the recent article *Quantum Measurement and the Aharonov–Bohm Effect with Superposed Magnetic Fluxes*, Bradonijc and Swain considered what happens to the Aharonov–Bohm effect if the flux is in superposition of two states with different classical values [232]. These two authors found that the interference pattern in the Aharonov–Bohm effect contains information about the nature of the superposition, allowing information about the state of the flux to be extracted without disturbing it. Bradonijc's and Swain's approach is novel in that the information is obtained by a nonlocal operation involving the vector potential without transfer of energy or momentum.

By assuming that the current in the solenoid is in a superposition of two macroscopic states corresponding to equal and opposite currents, resulting in a superposition of positive and negative magnetic flux inside the solenoid, since the probability amplitude for a particle to be found at position x on the screen flux up due to vector potential

$\vec{A}_\uparrow$ is

$$\psi_\uparrow(x,\phi) = e^{\frac{iq}{\hbar c}\int_{C_+}\vec{A}_\uparrow \cdot \vec{dl}}(\psi_+(x) + \psi_-(x)e^{i|\phi|}), \quad (3.114)$$

and for the same flux down due to a vector potential $\vec{A}_\downarrow$ is

$$\psi_\downarrow(x,\phi) = e^{\frac{iq}{\hbar c}\int_{C_+}\vec{A}_\downarrow \cdot \vec{dl}}(\psi_+(x) + \psi_-(x)e^{-i|\phi|}). \quad (3.115)$$

The general wavefunction of the particle is

$$|\psi(x,\phi)\rangle = \cos\frac{\vartheta}{2}|\psi_\uparrow\rangle + \sin\frac{\vartheta}{2}e^{i\omega}|\psi_\downarrow\rangle \quad (3.116)$$

where $0 \leq \vartheta \leq \pi$ and $0 \leq \omega \leq 2\pi$. On the basis of the wavefunction (3.116), the total probability density at the screen is

$$|\psi(x,\phi)|^2 = |\psi_+|^2 + |\psi_-|^2 + 2R(\psi_+^*\psi_-)\cos|\phi|$$
$$-2J(\psi_+^*\psi_-)\sin|\phi|\left(\cos^2\left(\frac{\vartheta}{2}\right) - \sin^2\left(\frac{\vartheta}{2}\right)\right).$$
$$(3.117)$$

It is interesting to remark that Bradonijc's and Swain's approach can be considered as a toy model for the quantum mechanical propagation of a particle in a background space-time which is a superposition of different classical geometries. Bradonijc's and Swain's model provides a situation which is somewhat equivalent to a superposition of two space-times with oppositely spinning cosmic strings [233, 234] which are outside the string.

Now, the quantum mechanical propagation of a particle as it is expected by Bradonijc's and Swain's reading of Aharonov–Bohm effect (with the flux being in a superposition of two states with different classical values) can be indeed seen, inside the symmetrized quantum potential approach proposed by the author, as a consequence of the nonlocal geometry of the background space corresponding with the symmetrized quantum entropy. The symmetrized quantum entropy can be indeed seen as the fundamental entity which determines the quantum mechanical propagation of a particle in a background, given by a superposition of different classical geometries, characteristic of the Aharonov–Bohm effect by determining an opportune symmetrized quantum-entropic distance which provides a direct measure of the correlation degree, characterized by a coherent symmetry in time. This result can be shown by appropriately including the symmetrized quantum potential inside Bradonijc's and Swain's approach of Aharonov–Bohm effect.

In de Broglie–Bohm pilot wave theory, the Aharonov–Bohm effect is explained through the local but indirect action of the vector potential on the particle via the quantum force [235]. The Hamilton–Jacobi equation and the force law for this problem are

$$\frac{\partial S}{\partial t} + \frac{|\nabla S|^2}{2m} + eA_0 + Q = 0, \tag{3.118}$$

$$m\frac{d\vec{v}}{dt} = -\nabla Q + \vec{F}, \tag{3.119}$$

where $\vec{F}$ is the Lorentz force, and the physical momentum $m\vec{v} = \nabla S - \frac{e}{c}\vec{A}$ and the quantum potential Q are gauge invariant quantities. Even when $\vec{F} = 0$ the quantum potential is modified by $\vec{A}$. The latter may therefore be expected to cause a redistribution of the trajectories in the two-slit experiment. In the Aharonov–Bohm effect, the quantum potential may be expressed as

$$Q(\vec{x}, \Phi) = Q_0(\vec{x}) + f(\vec{x}, \Phi), \tag{3.120}$$

where $Q_0(\vec{x})$ is independent of the flux.

In order to analyse in detail the Aharonov–Bohm effect and its geometry inside the symmetrized quantum potential approach, we base our discussion on the paper [229]. By following the treatment of the paper [229], the geometry of the Aharonov–Bohm effect inside the symmetrized quantum potential approach starts with a generalization of the quantum potential (3.120) with a symmetrized entity at two components of the form

$$Q_0(\vec{x}, \Phi) = \begin{pmatrix} Q_{01}(\vec{x}) + f_1(\vec{x}, \Phi) \\ Q_{02}(\vec{x}) + f_2(\vec{x}, \Phi) \end{pmatrix}, \tag{3.121}$$

where

$$Q_{01}(\vec{x}) = -\frac{\hbar^2}{2m}(\nabla S_{Q1})^2 + \frac{\hbar^2}{2m}(\nabla^2 S_{Q1}), \tag{3.122}$$

$$Q_{02}(\vec{x}) = \frac{\hbar^2}{2m}(\nabla S_{Q2})^2 - \frac{\hbar^2}{2m}(\nabla^2 S_{Q2}), \quad (3.123)$$

where $S_{Q1} = -\frac{1}{2}\ln\rho_1$ is the quantum entropy defining the degree of order and chaos of the vacuum for the forward-time processes (here $\rho_1 = |\psi(\vec{x}_1,\vec{x}_2,\ldots,\vec{x}_N,t)|^2$, $\psi(\vec{x}_1,\vec{x}_2,\ldots,\vec{x}_N,t) = R_1 e^{-iS_1/\hbar}$ being the forward-time many-body wavefunction, solution of the standard Schrödinger equation) and $S_{Q2} = -\frac{1}{2}\ln\rho_2$ is the quantum entropy defining the degree of order and chaos of the vacuum for the time-reverse processes (where here $\rho_2 = |\phi(\vec{x}_1,\vec{x}_2,\ldots,\vec{x}_N,t)|^2$, $\phi(\vec{x}_1,\vec{x}_2,\ldots,\vec{x}_N,t) = R_2 e^{-iS_2/\hbar}$ being the time-reverse many-body wavefunction, solution of the time-reversed Schrödinger equation). On the basis of Eqs. (3.122) and (3.123), the term $Q_0(\vec{x})$ is just determined by the quantum entropy defining the degree of order and chaos of the vacuum for the forward-time processes and by the quantum entropy defining the degree of order and chaos of the vacuum for the time-reverse processes. By introducing the quantum potential (3.121) in the picture analysed by Bradonijc and Swain, one can rewrite the wavefunction of a particle to be found at position x on the screen flux up due to a symmetrized vector potential $\vec{A}_\uparrow = \begin{pmatrix}\vec{A}_{\uparrow 1}\\\vec{A}_{\uparrow 2}\end{pmatrix}$ (namely the Eq. (3.114)) as a symmetrized wavefunction at two components of the following form

$$\begin{pmatrix}\psi_\uparrow(x,\phi)\\\varphi_\uparrow(x,\phi)\end{pmatrix} = \begin{pmatrix} e^{\frac{iq}{\hbar c}\int_{C_+}(\vec{A}_{\uparrow 1}+Q_{01}+f_1(\vec{x},\Phi))\cdot d\vec{l}}(\psi_+(x)+\psi_-(x)e^{i|\phi|}) \\ e^{\frac{iq}{\hbar c}\int_{C_+}(\vec{A}_{\uparrow 2}+Q_{02}+f_2(\vec{x},\Phi))\cdot d\vec{l}}(\varphi_+(x)+\varphi_-(x)e^{i|\phi|}) \end{pmatrix}$$
(3.124)

and analogously the wavefunction for the same flux down due to a symmetrized vector potential $\vec{A}_\downarrow = \begin{pmatrix}\vec{A}_{\downarrow 1}\\\vec{A}_{\downarrow 2}\end{pmatrix}$ (Eq. (3.115)) as the following symmetrized wavefunction

$$\begin{pmatrix}\psi_\downarrow(x,\phi)\\\varphi_\downarrow(x,\phi)\end{pmatrix} = \begin{pmatrix} e^{\frac{iq}{\hbar c}\int_{C_+}(\vec{A}_{\downarrow 1}+Q_{01}+f_1(\vec{x},\Phi))\cdot d\vec{l}}(\psi_+(x)+\psi_-(x)e^{-i|\phi|}) \\ e^{\frac{iq}{\hbar c}\int_{C_+}(\vec{A}_{\downarrow 2}+Q_{02}+f_2(\vec{x},\Phi))\cdot d\vec{l}}(\varphi_+(x)+\varphi_-(x)e^{-i|\phi|}) \end{pmatrix}$$
(3.125)

and thus the general wavefunction of the particle as

$$\begin{pmatrix} |\psi(x,\phi)\rangle \\ |\varphi(x,t)\rangle \end{pmatrix} = \cos\frac{\vartheta}{2}\begin{pmatrix} |\psi_\uparrow\rangle \\ |\varphi_\uparrow\rangle \end{pmatrix} + \sin\frac{\vartheta}{2}e^{i\omega}\begin{pmatrix} |\psi_\downarrow\rangle \\ |\varphi_\downarrow\rangle \end{pmatrix}. \quad (3.126)$$

In Eqs. (3.124), (3.125) and (3.126), the origin of the quantum potential $Q_0(\vec{x})$ is given just by the quantum entropy defining the degree of order and chaos of the vacuum for the forward-time processes and by the quantum entropy defining the degree of order and chaos of the vacuum for the time-reverse processes. Equations (3.124), (3.125) and (3.126) suggest thus the following reading of the mathematical formalism of the Aharonov–Bohm effect: if the flux is in superposition of two states with different classical values, and thus the background is a superposition of different classical geometries, the symmetrized quantum potential is modified by the symmetrized vector potential in the sense that it contains, besides to the term associated with the symmetrized quantum entropy defining the degree of order and chaos of the vacuum (characterized by two components, the first regarding the forward-time processes, the second regarding the time-reverse processes), a symmetrized term depending itself on the flux (where, of course, the first component regards the forward-time processes, the second component assures that if one would imagine the process backwards, would see what physically happens). The symmetrized wavefunctions (3.124), (3.125) and (3.126) indicate that the change of the geometry of physical space determined by the Aharonov–Bohm effect is expressed by the quantities

$$\begin{pmatrix} A_{\uparrow 1} + Q_{01} + f_1(\vec{x},\Phi) \\ A_{\uparrow 2} + Q_{02} + f_2(\vec{x},\Phi) \end{pmatrix} \quad (3.127)$$

and

$$\begin{pmatrix} A_{\downarrow 1} + Q_{01} + f_1(\vec{x},\Phi) \\ A_{\downarrow 2} + Q_{02} + f_2(\vec{x},\Phi) \end{pmatrix}. \quad (3.128)$$

On the basis of the quantities (3.127) and (3.128) describing the deformation of the geometry of physical space in the presence of the Aharonov–Bohm effect, the total probability density at the screen, given by equation

$$\begin{pmatrix} |\psi(x,\phi)|^2 \\ |\varphi(x,\phi)|^2 \end{pmatrix}$$
$$= \begin{pmatrix} |\psi_+|^2 + |\psi_-|^2 + 2\mathrm{Re}(\psi_+^*\psi_-)\cos|\phi| - 2\mathrm{Im}(\psi_+^*\psi_-)\sin|\phi|\left(\cos^2\left(\frac{\vartheta}{2}\right) - \sin^2\left(\frac{\vartheta}{2}\right)\right) \\ |\varphi_+|^2 + |\varphi_-|^2 + 2\mathrm{Re}(\varphi_+^*\psi_-)\cos|\phi| - 2\mathrm{Im}(\varphi_+^*\varphi_-)\sin|\phi|\left(\cos^2\left(\frac{\vartheta}{2}\right) - \sin^2\left(\frac{\vartheta}{2}\right)\right) \end{pmatrix}$$

(3.129)

where ψ_+, ψ_-, φ_+ and φ_- are derived from Eqs. (3.124) and (3.125), turns out to be determined just by these quantities (3.127) and (3.128).

Let us analyse now some interesting results which derive from Eq. (3.129) for various choices of ϑ.

1. For $\vartheta = 0$, $\begin{pmatrix} |\psi(x,\phi)|^2 \\ |\varphi(x,\phi)|^2 \end{pmatrix}$ goes to regular Aharonov–Bohm effect with an "up" flux.
2. For $\vartheta = \pi$, $\begin{pmatrix} |\psi(x,\phi)|^2 \\ |\varphi(x,\phi)|^2 \end{pmatrix}$ goes to regular Aharonov–Bohm effect with a "down" flux.
3. For $\vartheta = \pi/2$, the last term of both Eqs. (3.129) goes to zero. There is still an interference pattern, but it is different from the interference patterns in cases 1 and 2 above.

The first two limits indicate that that Eq. (3.129) reduces to the expected expressions for two classical flux states. More interesting is the third limit, which indicates that it is in principle possible to extract information about the quantum mechanical state of magnetic flux in the Aharonov–Bohm experiment from the electron diffraction pattern without disturbing the state of the flux. This information emerges from a fundamentally nonlocal operation which involves the deformation of physical space determined by the symmetrized quantum potential and the symmetrized vector potential (associated with the quantities (3.127) and (3.128)) and over an extended region of space-time, and without any interaction in the region where the (superposition of) classical magnetic fields is present (i.e. the excluded region inside the solenoid). In other words, according to the symmetrized approach, the deformation of the background expressed by the quantities (3.127) and (3.128) can be seen as the ultimate source of information in the Aharonov–Bohm effect.

Moreover, in order to describe the deformation of the geometry of the background, also as regards the Aharonov–Bohm effect one can introduce a quantum-entropic length characterizing this quantum phenomenon given by the following equations:

$$L_{Aharonov-Bohm\uparrow} = \begin{pmatrix} \dfrac{1}{\sqrt{(\nabla S_{Q1})^2 - \nabla^2 S_{Q1} - 2m(A_{\uparrow 1} + f_1(\vec{x}, \Phi))}} \\ \dfrac{1}{\sqrt{-(\nabla S_{Q2})^2 + \nabla^2 S_{Q2} - 2m(A_{\uparrow 2} + f_2(\vec{x}, \Phi))}} \end{pmatrix}$$

$$= \begin{pmatrix} \exp\left(-\dfrac{i}{\hbar}\hat{\vec{p}}\cdot\vec{l}\right)\exp\left\{i\dfrac{e}{\hbar}\int_{\gamma\vec{x}}^{\vec{x}+\vec{l}} A(\vec{y},t)\cdot d\vec{y}\right\} \\ \exp\left(\dfrac{i}{\hbar}\hat{\vec{p}}\cdot\vec{l}\right)\exp\{-i\dfrac{e}{\hbar}\int_{\gamma\vec{x}}^{\vec{x}+\vec{l}} A(\vec{y},t)\cdot d\vec{y}\} \end{pmatrix}$$

(3.130)

(which is the Aharonov–Bohm length associated with the region of physical space generating a flux up on the screen), and

$$L_{Aharonov-Bohm\downarrow} = \begin{pmatrix} \dfrac{1}{\sqrt{(\nabla S_{Q1})^2 - \nabla^2 S_{Q1} - 2m(A_{\downarrow 1} + f_1(\vec{x}, \Phi))}} \\ \dfrac{1}{\sqrt{-(\nabla S_{Q2})^2 + \nabla^2 S_{Q2} - 2m(A_{\downarrow 2} + f_2(\vec{x}, \Phi))}} \end{pmatrix}$$

$$= \begin{pmatrix} \exp\left(-\dfrac{i}{\hbar}\hat{\vec{p}}\cdot\vec{l}\right)\exp\left\{i\dfrac{e}{\hbar}\int_{\gamma\vec{x}}^{\vec{x}+\vec{l}} A(\vec{y},t)\cdot d\vec{y}\right\} \\ \exp\left(\dfrac{i}{\hbar}\hat{\vec{p}}\cdot\vec{l}\right)\exp\left\{-i\dfrac{e}{\hbar}\int_{\gamma\vec{x}}^{\vec{x}+\vec{l}} A(\vec{y},t)\cdot d\vec{y}\right\} \end{pmatrix}$$

(3.131)

(which is the Aharonov–Bohm length associated with the region of physical space generating a flux down on the screen). Equations (3.130) and (3.131) indicate clearly that the features of the background space in the presence of the Aharonov–Bohm effect are determined by the symmetrized quantum entropy, by the symmetrized vector potential and by the symmetrized function linked with the phase difference. With the introduction of the quantum entropic lengths (3.130) and (3.131), it becomes so permissible the following reading of the formalism of the Aharonov–Bohm effect inside the symmetrized quantum potential approach based on the symmetrized quantum entropy at two components (the first regarding the forward-time process and the second regarding the time-reverse process): the distribution probability associated with the wavefunction under consideration determines a degree of order and chaos of the vacuum, a symmetrized quantum entropy corresponds to this degree of order and chaos of the vacuum given by Eq. (3.81), this symmetrized quantum entropy given by Eq. (3.81) determines, with the symmetrized vector potential and the symmetrized function linked with the phase difference, a change of the quantum geometry described by a quantum-entropic length given by Eqs. (3.130) and (3.131) (which provides thus a direct measure of the correlation degree, and thus of the degree of departure from the Euclidean geometry characteristic of classical physics, in the Aharonov–Bohm experiment). The Aharonov–Bohm effect is associated with the quantum entropic lengths (3.130) and (3.131) and thus is determined by the symmetrized quantum entropy, the symmetrized vector potential and the phase difference. The quantum-entropic lengths given by Eqs. (3.130) and (3.131) (and determined by the symmetrized quantum entropy, the symmetrized vector potential and the phase difference) can be considered as the ultimate source of quantum information as regards the Aharonov–Bohm effect.

On the basis of Eqs. (3.130) and (3.131) one can also say that — both in the region of physical space generating a flux up on the screen and in the region of physical space generating a flux down on the screen — there is an equivalence between the effect of the gauge field associated with $g_0 A_\mu^k T_k$, where T_k generates the gauge group and the symmetrized quantum entropy. On one hand, one can say that the effect of the gauge field associated with $g_0 A_\mu^k T_k$, where

T_k generates the gauge group, P is the path ordering and γ is a curve in space-time that is congruent to $\hat{\gamma}$ while $\bar{\gamma}$ is the curve γ traversed in the reversed order, is to determine a modification of the geometrical properties of the background space, with respect to the Euclidean geometry characteristic of classical physics, expressed by a symmetrized quantum entropy, a symmetrized vector potential and the symmetrized function linked with the phase difference. In other words, in this picture, Eqs. (3.130) and (3.131) express in what sense the effect of the gauge field associated with $g_0 A_\mu^k T_k$, where T_k generates the gauge group, produces the action of a 3D space as an immediate information medium as regards the Aharonov–Bohm effect. On the other hand, one can say that the action of the gauge field of determining a deformation of the geometry in the quantum regime derives just from the symmetrized quantum entropy, a symmetrized vector potential and the symmetrized function linked with the phase difference.

Although, in the Aharonov–Bohm setup, information is obtainable from the interference pattern associated with the deformation of physical space associated with the quantities (3.127) and (3.128), it is also important to mention that this information is not complete. In the classical Aharonov–Bohm effect, the flux Φ can only be determined modulo $2\pi \frac{\hbar c}{e}$. For $\Phi = 0$ modulo $2\pi \frac{\hbar c}{e}$, the interference pattern is the same as that for a positive and negative classical flux: there is no effect on the pattern. Superpositions of two magnetic fluxes (determined by the quantities (3.127) and (3.128) and thus associated with the geometries expressed by the quantum-entropic lengths (3.130) and (3.131)) which would individually be detectable via the Aharonov–Bohm effect can give rise to interference patterns which differ from that found in the classical (or nonsuperposed) case. Moreover, the information about superposition is contained from the full interference pattern.

One can also consider the case of fluxes which are characterized by unequal superpositions of "up" and "down" fluxes ($|\cos \frac{\vartheta}{2}|^2 \neq |\sin \frac{\vartheta}{2}|^2$). In this case, again, the information about the nature of the state can be obtained without any interaction which should cause the "collapse" of the state into one definite flux.

Finally, as regards our treatment of the Aharonov–Bohm effect in the symmetrized quantum potential approach, it is important to mention the role of the Berry phase. In his article *A new phase in quantum computation* Sjökvist remarked that a geometric quantum information can be obtained that employs one-dimensional geometric phase factors and that in this context the Berry phase, which occurs in situations like the Aharonov–Bohm setup may be used for quantum computation by encoding the logical states in nondegenerate energy levels, such as in the spin-up and spin-down states of a spin-1/2 particle in a magnetic field [236]. Inside the symmetrized quantum potential approach, the interesting perspective is opened that even the Berry phase characterizing the Aharonov–Bohm effect may be seen as a consequence of the symmetrized quantum potential and thus of the quantum entropy regarding the forward-time process and of the quantum entropy regarding the time-reverse process, which can be indeed considered as the ultimate source of quantum information (and thus of an opportune symmetrized quantum-entropic length providing a measure of the correlation degree and so a departure degree from the Euclidean geometry). In other words, the Berry phase characterizing the geometry of the Aharonov–Bohm setup can be seen as the result of the quantities (3.127) and (3.128) describing the deformation of the physical space in this process (and of the quantum-entropic lengths (3.130) and (3.131) associated with them).

3.8 About the Measurement Problem and the Collapse of the Wavefunction in the 3D Timeless non-Euclid Space

Consider an observer O who makes a measurement on a quantum system S. The measurement can be seen as a many-body interaction process that is special only insofar as the interaction is needed to leave the system under investigation in a particular state, an eigenfunction of an Hermitian operator (the "measured operator"), while the apparatus is left in a state such that its subsequent behavior in no way influences the system under consideration. The outcome of the measurement, as well as the location of the apparatus variable, allow us to infer the final wavefunction of the system, which is the initial wavefunction for all subsequent interactions, and hence the final actual value of the physical quantity that corresponds to the measured operator. The period of the measurement may be divided into two stages:

- a state preparation of a certain kind in which the wavefunction of the system under examination becomes correlated with the wavefunction of the observing apparatus and evolves into an eigenfunction of an Hermitian operator;
- an irreversible act of amplification which allows the observer to register the outcome and infer the value of the physical property of the particle corresponding to the operator.

If in a measurement the final wavefunction of the microsystem under investigation collapses into the state specified by the outcome of the measurement, the following questions become therefore natural. Is the collapse of the wavefunction instantaneous? Do properties of subatomic systems become manifest suddenly? When precisely does the final wavefunction corresponding with the outcome of a measurement emerge? When precisely can we say that a given event has happened?

In this regard, in his book *The landscape of theoretical physics: a global view*, Pavsic proposes the old idea (which appears however a little questionable) that the collapse of the wavefunction happens, and thus the final wavefunction actualizes, just at the moment when the information about the interaction between the microsystem and the macrosystem arrives in the observer's brain: according to this view, there would be no collapse until the signal reaches the observer's brain [237]. As regards the question of the timing of a quantum event, of when precisely properties of subatomic systems become manifest, it seems more interesting Rovelli's view which showed that quantum theory gives a precise answer to this question. Rovelli found that: (i) a precise (operational) sense can be given to the question of the timing of the measurement; (ii) one can compute the time at which the measurement happens using standard quantum techniques; (iii) the interpretation of the physical meaning of this time is no more problematic than the interpretation of any other quantum result [238]. Rovelli showed that the question "When does the measurement happen?" is quantum mechanical in nature, and not classical. Therefore, its answer must be probabilistic. For example, the sentence "half way through a measurement" would mean that the measurement is just "happened with probability 1/2", or "already realized in half of the repetitions of the experiment". The second idea is that the question "When does the measurement happen?" does not regard the measured quantum system S alone, but rather the coupled system formed by the observed system S and the observer system O. Therefore, the appropriate theoretical setting for answering this question is the quantum theory of the two coupled systems. More in detail, Rovelli introduced an operator measuring whether or not the measurement has happened. By considering the simple case of a physical system S (for example an electron) that interacts with another physical system O (an apparatus measuring the spin of the electron) and that the interaction between S and O qualifies as a quantum measurement of the variable q of the system S, if we suppose that q has only two eigenvalues a and b, during the interaction between S and O, the state of the combined system S–O is

$$\psi = c_a |a\rangle \otimes |Oa|\rangle + c_b |b\rangle \otimes |Ob\rangle, \qquad (3.132)$$

where $|a\rangle$ and $|b\rangle$ are the eigenstates of q corresponding to the eigenvalues a and b respectively, $|Oa\rangle$ and $|Ob\rangle$ are the states of O that can be identified as "the pointer of the apparatus indicates that q has value a" and "the pointer of the apparatus indicates that q has value b", respectively. When the combined system S–O is in the state (3.132), a definite correlation between the pointer variable, with eigenstates $|Oa\rangle$ and $|Ob\rangle$, and the system variable, with eigenstates $|a\rangle$ and $|b\rangle$, is established.

For some reason, at some point, we have to replace the pure state (3.132) with a mixed state. Equivalently, we have to replace (3.132) with either

$$\psi_a = |a\rangle \otimes |Oa\rangle \qquad (3.133)$$

or

$$\psi_b = |b\rangle \otimes |Ob\rangle, \qquad (3.134)$$

where, of course, the probability of having one or the other is $|c_a|^2$ and $|c_b|^2$ respectively. If the wavefunction has collapsed, and the state is either (3.133) or (3.134), the correlation between the pointer variable and the system variable is present as well.

Rovelli focused the attention on the question of what we can say about the precise time t at which the wavefunction changes from (3.132) to either (3.133) or (3.134) and the quantity q acquires correspondingly a definite value. In this regard, according to Rovelli's results, the operator M which measures the timing of the measurement is defined as the projection operator on the subspace spanned by the two states ψ_a and ψ_b:

$$M = |\psi_a\rangle\langle\psi_a| + |\psi_b\rangle\langle\psi_b|. \qquad (3.135)$$

M turns out to be a self-adjoint operator on the Hilbert space of the coupled system S–O. It may admit an interpretation as an observable property of the coupled system S–O. In all the eigenstates of M with eigenvalue 1 the pointer variable correctly indicates the value of q. In all the eigenstates of M with eigenvalue 0, it does not. Therefore, M has the following interpretation: $M = 1$ means that the pointer (correctly) measures q; $M = 0$ means that it does not. Therefore, we can say that $M = 1$ has the physical interpretation "the measurement has happened", and $M = 0$ has the physical interpretation "the measurement has not happened". By applying standard quantum mechanical rules to this operator, at every time t we can compute a precise (although probabilistic) answer to the question whether or not the measurement has happened: the probability that the measurement has happened at time t is given by relation $P(t) = \langle\psi(t)|M|\psi(t)\rangle$, where $\psi(t)$ is the state of the coupled system during the Schrödinger evolution. The probability density $p(t)$ that the measurement happens between time t and time $t+dt$ is

$$p(t) = \frac{d}{dt}\langle\psi(t)|M|\psi(t)\rangle = \langle\psi(t)|[M,H]|\psi(t)\rangle, \qquad (3.136)$$

where H is the total Hamiltonian. For a good measurement in which $P(t)$ grows smoothly and monotonically from zero to one, $p(t)$ will be a "bell shaped" curve, defining the time at which the measurement happens, and its quantum dispersion.

Now, in the symmetrized quantum potential approach developed by the author according to which the ultimate arena of quantum processes is a 3D space which, at the fundamental level of EPR-type experiments, acts as a direct information medium, Rovelli's results can be read in the following manner. As showed in the paper [214], the operator M can be interpreted as the operator which measures the numerical order of a measurement characterizing the interaction between a subatomic system and an apparatus; the probability density (3.136) defines the numerical order associated with the actualization of a measurement. In this approach, the fundamental essence, the fundamental arena of measurement processes is represented by the correlation, in the 3D space, between a physical system and an apparatus and thus by the entangled state (3.132) in the sense that it is just by starting from the geometry associated with this state that one can compute the numerical order corresponding to the actualization of the measured property of the physical system under consideration. The quantum superposition, the quantum entanglement between the measured physical system and the apparatus can be considered the fundamental reality of space in the quantum domain. If in Bohm's approach the quantum potential emerges as the ultimate entity which allows us to provide a causal description of measurement processes, inside the symmetrized quantum potential approach one can say that it is the geometry associated with the symmetrized quantum entropy (and thus with the symmetrized quantum-entropic length) the fundamental entity which allows us to explain and reproduce what physically happens in measurement processes. The symmetrized quantum potential deriving from the geometry associated with the symmetrized quantum entropy (and with the symmetrized quantum-entropic length) can provide a causal description both for forward-time process and for the time-reverse process of a measurement.

Finally, it is interesting to observe that, by introducing a 3D non-Euclid space which acts as a direct information medium (in the form of the symmetrized quantum potential deriving from a symmetrized quantum entropy) as fundamental arena of quantum processes, also Galapon's results regarding the QTOAP receive a new suggestive re-reading. In fact, one can say that the time of arrival operator linked with the appearance of a particle indicates only the numerical order associated with the appearance of the particle under consideration. In this picture, Eqs. (3.11)–(3.16) of Galapon's approach regard an underlying 3D background which acts as an immediate information medium where the quantum time of arrival associated with the appearance of a particle exists only in the sense of numerical order of material changes. In synthesis, according to the interpretation of Galapon's results inside the description of quantum processes in a 3D non-Euclid background proposed in this chapter, the appearance of a particle arises as a combination of the collapse of the initial wavefunction into one of the eigenfunctions of the time-of-arrival operator, followed by the unitary Schrödinger evolution of the eigenfunction, as a consequence of the fact that, at a fundamental level, the background of processes acts as an immediate information medium. In other words, the interesting perspective is opened that the collapse of the wavefunction on the appearance of particle is decomposable into a series of casually separated processes determined by a 3D non-Euclid space which acts as a direct medium of information.

The interpretation of the QTOAP here suggested according to which the underlying physics of quantum processes is determined by the fact that a 3D background space acts as a direct information medium and thus according to which time is only a mathematical parameter measuring the numerical order of the process of the appearance of particles, according to the author of this book, seems coherent from the physical point of view also because quantum mechanics is indeed inherently nonlocal in time [239–242]. The temporal nonlocality of quantum mechanics is for example illustrated by Wheeler's delayed-choice gedanken experiment. According to this experiment, the description of the past is not complete without regard to present actions. This experiment implies that the collapse right after the preparation (when arrival measurement is to be made) and the Schrödinger evolution right after the preparation (when some other measurement is to be made) are two mutually exclusive potentialities that are simultaneously true for the system, which one is realized depends on the decision what to do with the system at the moment. Temporal nonlocality replaces the spatial nonlocality inherent in the spontaneous localization of the wavefunction at the appearance of the particle in the standard interpretation. According to the approach of quantum processes considered in this chapter, the fact that in Wheeler's delayed-choice gedanken experiment the collapse of the wavefunction when arrival measurement is to be made and the Schrödinger evolution when some other measurement is to be made represent two potentialities that are simultaneously true for the system indicates clearly that the linear time past-present-future does not exist as a primary physical reality, but what exists is merely the numerical order of dynamical events, of material changes into a 3D fundamental background space which acts as a direct information medium.

Chapter 4

About Quantum Cosmology in a Background Space as an Immediate Information Medium

If at the fundamental level of quantum processes, nonlocal correlations are due to a 3D timeless space which acts as a direct information medium between the particles under consideration, the possibility is also opened that in the quantum gravity domain and in the quantum cosmology a timeless background space functions as a direct information medium. In this regard, in this chapter we want to focus the attention of the reader on a mathematical model of quantum gravity and cosmology (recently developed by the author) in which the idea of a background of processes as a direct, immediate information medium can be embedded. The crucial idea that is at the basis of this model lies in building a symmetrized version of the bohmian approach of Wheeler–DeWitt equation (WDW) and thus in introducing the considerations made in Chapter 3 inside WDW equation.

4.1 About Conceptual Problems in Quantum Cosmology

Quantum mechanics is a universal and fundamental theory, applicable to any physical system. The universe is, of course, a physical system which can be studied: there is a theory, standard cosmology, which allows us to describe it in physical terms, and to make predictions which can be confirmed or refuted by observations. In fact, the observations until now seem to confirm the standard cosmological scenario. Hence, supposing the universality of quantum mechanics, the universe itself must be described by quantum theory, from which standard cosmology could be recovered.

Quantum cosmology is the application of quantum theory to the universe as a whole. Conceptually, the construction of a quantum cosmology corresponds to the problem of formulating a quantum theory for a closed system from within, without reference to any external observers or measurement agencies. In its concrete formulation, it requires a quantum theory of gravity, since gravity is the dominating interaction at large scales. On the one hand, quantum cosmology may serve as a test bed for quantum gravity in a mathematically simpler setting. On the other hand, quantum cosmology may be directly relevant for an understanding of the real universe. As regards a general introduction to the conceptual issues of quantum cosmology, the reader can find details, for example, in the works of Coule, Halliwell, Kiefer, Sandhoefer, Wiltshire, Montani, Battisti, Benini and Imponente [243–248]. Supersymmetric quantum cosmology is analyzed at depth in Moniz's book *Quantum Cosmology — The Supersymmetric Perspective* [249]. An introduction to loop quantum cosmology has been provided recently by Bojowald [250], who together with Kiefer and Moniz made as well a comparison of standard quantum cosmology with loop quantum cosmology in the recent paper [251].

Quantum cosmology is usually discussed for homogeneous models (the models are then called minisuperspace models). The simplest case is to assume also isotropy. Then, the line element for the classical spacetime metric is given by

$$ds^2 = -N(t)^2 dt^2 + a(t)^2 d\Omega_3^2, \qquad (4.1)$$

where $d\Omega_3^2$ is the line element of a constant curvature space with curvature index $k = 0, \pm 1$. In order to consider a matter degree of freedom, a homogeneous scalar field ϕ with potential $V(\phi)$ is added. In this setting, WDW equation (1.1) becomes the following 2D partial differential equation for a wavefunction $\psi(a,\phi)$:

$$\left(\frac{\hbar^2\kappa^2}{12}a\frac{\partial}{\partial a}a\frac{\partial}{\partial a} - \frac{\hbar^2}{2}\frac{\partial^2}{\partial \phi^2} + a^6\left(V(\phi) + \frac{\Lambda}{\kappa^2}\right) - \frac{3ka^4}{\kappa^2}\right)\Psi(a,\phi) = 0, \quad (4.2)$$

where Λ is the cosmological constant and $\kappa^2 = 8\pi G$. Introducing $\alpha = \ln a$ (which has the advantage to have a range from $-\infty$ to $+\infty$), Eq. (4.2) may be expressed conveniently as:

$$\left(\frac{\hbar^2\kappa^2}{12}\frac{\partial}{\partial \alpha^2} - \frac{\hbar^2}{2}\frac{\partial^2}{\partial \phi^2} + e^{6\alpha}\left(V(\phi) + \frac{\Lambda}{\kappa^2}\right) - 3e^{4\alpha}\frac{k}{\kappa^2}\right)\Psi(a,\phi) = 0. \quad (4.3)$$

In many versions of quantum cosmology, Einstein's theory is the classical starting point. More general approaches include supersymmetric quantum cosmology [249], string quantum cosmology, non-commutative quantum cosmology [252], Hořava–Lifshitz quantum cosmology [253] and third-quantized cosmology [254].

In loop quantum cosmology, features from full loop quantum gravity are imposed on cosmological models [250]. Since in loop quantum gravity one of the main features is the discrete nature of geometric operators (as we will see in detail in Chapter 5), in this picture the WDW equation (4.2) is replaced by a difference equation. This difference equation becomes indistinguishable from the WDW equation at scales exceeding the Planck length, at least in certain models. The difference equation is difficult to be resolved in general, but one can use an effective theory [255].

Many features of quantum cosmology are discussed in the limit when the solution of the WDW equation assumes a semiclassical or WKB form [244]. This occurs, in particular, for concrete models in which one applies the no-boundary proposal or the tunneling proposal for the wavefunction. The no-boundary condition was originally formulated by Hawking in 1982 [256] and then elaborated by Hartle and Hawking in 1983 [257] and Hawking in 1984 [258]. The no-boundary condition, in which the wavefunction is expressed by an Euclidean path integral, states that — apart from the boundary where the three metric is specified — there is no other boundary on which initial conditions have to be specified. However, the no-boundary condition leads to many solutions, and here the integration has to be performed over complex metrics (see, for example the references [259, 260]). Moreover, the path integral defining the no-boundary condition can usually only be evaluated in a semiclassical limit (using the saddle-point approximation), so it is hard to make a general statement about singularity avoidance. In particular, in a Friedmann model with scale factor a and a scalar field ϕ with a potential $V(\phi)$, the no-boundary condition gives the semiclassical solution [258]

$$\psi_{NB} \propto (a^2 V(\phi) - 1)^{-1/4} \exp\left(\frac{1}{3V(\phi)}\right)$$

$$\times \cos\left(\frac{(a^2 V(\phi) - 1)^{3/2}}{3V(\phi)} - \frac{\pi}{4}\right), \quad (4.4)$$

which corresponds to the superposition of an expanding and a recollapsing universe (and implies that the no-boundary wavefunction is always real).

Another prominent boundary condition is the tunneling proposal, originally defined by the choice of taking "outgoing" solutions at singular boundaries of superspace (see, for example, the reference [261]). By applying this condition to the same Friedmann model here one obtains

$$\psi_T \propto (a^2 V(\phi) - 1)^{-1/4} \exp\left(-\frac{1}{3V(\phi)}\right)$$

$$\times \exp\left(-\frac{i}{3V(\phi)} \frac{(a^2 V(\phi) - 1)^{3/2}}{3V(\phi)} - \frac{\pi}{4}\right) \quad (4.5)$$

from which, considering the conserved Klein–Gordon current

$$j = \frac{i}{2}(\psi^* \nabla \psi - \psi \nabla \psi^*), \quad (4.6)$$

with $\nabla j = 0$, ∇ indicating the derivatives in minisuperspace, one obtains for a WKB solution of the form $\psi \approx C \exp(iS)$ the following expression

$$\psi \approx -|C|^2 \nabla S. \qquad (4.7)$$

Here, the tunneling proposal physically means that the current (4.6) should point outwards at large a and ϕ (if ψ is of WKB form). Although here a complex solution is chosen, again also in this proposal — in analogy to the no-boundary condition — the wavefunction is of semiclassical form. In this regard, it has even been suggested that the wavefunction of the universe may be interpreted only in the semiclassical WKB limit, because only in this case a time parameter and an approximate (functional) Schrödinger equation are available [262].

On the other hand, this can lead to a conceptual confusion. Implications for the meaning of the quantum cosmological wavefunctions should be derived as much as possible from exact solutions. This happens because the WKB approximation breaks down in many interesting situations, even for a universe of macroscopic size. One example is represented by a closed Friedmann universe with a massive scalar field [263]. The conceptual confusion regarding cosmological wavefunctions has been recently analyzed by Kiefer in his paper *Conceptual problems in quantum gravity and cosmology* [264]. In this paper, in his discussion of the present status of research on quantum gravity, Kiefer explains that the conceptual confusion regarding cosmological wavefunctions is linked to the fact that, for a classically recollapsing universe, one must impose the boundary condition that the wavefunction goes to zero for large scale factors, $\psi \to 0$ for large a. As a consequence, narrow wave packets do not remain narrow because of the ensuing scattering phase shifts of the partial waves (that occur in the expansion of the wavefunction into basis states) from the turning point. The correspondence to the classical model can only be understood if the quantum-to-classical transition (in the sense of decoherence) is invoked.

Another example is the case of classically chaotic cosmologies, as has been studied, for example, by Calzetta and Gonzalez [265] and by Cornish and Shellard [266]. Here, one can see that the WKB approximation breaks down in many situations. This is, of course, a situation well known from quantum mechanics. One of the moons of the planet Saturn, Hyperion, is characterized by chaotic rotational motion. Treating it quantum-mechanically, one recognizes that the semiclassical approximation breaks down and that Hyperion is expected to be in an extremely nonclassical state of rotation (in contrast to what is observed). This apparent conflict between theory and observation can be understood by invoking the influence of additional degrees of freedom in the sense of decoherence, as shown for example by Zurek and Paz in [267] and [268]. The same mechanism should rectify the situation for classically chaotic cosmologies [269].

4.2 The Quantum Potential in Quantum Cosmology

If one insists with the Copenhagen interpretation, at least in its present form, one has to assume that quantum theory is not universal, and therefore it derives that quantum cosmology should not make any sense at all. This is a perfect example of the adequacy of the following famous Albert Einstein sentence: "Contemporary quantum theory constitutes an optimum formulation of [certain] connections [but] offers no useful point of departure for future developments". Fortunately, as is well known, there are some alternative solutions to this quantum cosmological dilemma which can solve the measurement problem maintaining the universality of quantum theory. One can say that the Schrödinger evolution is an approximation of a more fundamental nonlinear theory which can accomplish the collapse, or that the collapse is effective but not real, in the sense that the other branches disappear from the observer but do not disappear from existence. In this category of approaches, one can mention Everett's Many-Worlds interpretation and the de Broglie–Bohm theory. Here, we focus our attention on the de Broglie–Bohm approach.

In the Bohm approach, the fundamental object of quantum gravity is the geometry of 3D space-like hypersurfaces, which is assumed to exist independently of any observation or measurement, and the same thing is valid for its canonical momentum, the extrinsic curvature of the space-like hypersurfaces. Its evolution, labeled by some time parameter, is dictated by a quantum evolution that is different from the classical one due to the presence of a quantum potential which appears naturally from the WDW equation. One of the important results obtained in the de Broglie–Bohm approach to quantum cosmology lies in the elimination of cosmological singularities, which has been proved at least for some particular but relevant cases. In the de Broglie–Bohm interpretation, the quantum potential becomes important near the singularity, yielding a repulsive quantum force counteracting the gravitational field, avoiding the singularity and yielding inflation.

As regards the active role of the quantum potential in redesigning the geometry of physical space in quantum cosmology, the discussion made here follows the references [270–275] which provide a Bohmian interpretation of quantum gravity and takes into consideration also the papers [14, 23, 276–298] for material on WDW equation and quantum gravity.

Before, all, starting from [272] one can write the Lagrangian density for general relativity in the form

$$L = \sqrt{-g}R = \sqrt{g}N(^{(3)}R + Tr(K^2)), \qquad (4.8)$$

where $^{(3)}R$ is the 3D Ricci scalar, K_{ij} is the extrinsic curvature, N is the lapse function, g_{ij} is the induced spatial metric, $g = \det g_{ij}$. The canonical momentum of the three-metric is given by

$$p^{ab} = \frac{\partial L}{\partial \dot{q}_{ab}} = \sqrt{q}(K^{ab} + q^{ab}Tr(K)). \qquad (4.9)$$

The classical Hamiltonian is

$$H = \int d^3x \sqrt{g}(NC + N_i C^i), \qquad (4.10)$$

where N_i is the shift function,

$$C = {}^{(3)}R + \frac{1}{q}\left(Tr(p^2) - \frac{1}{2}(Tr(p))^2\right) = -2G_{\mu\nu}n^\mu n^\nu, \qquad (4.11)$$

$$C^i = -2G_{\mu i}n^\mu, \qquad (4.12)$$

n^μ being the normal vector to the spatial hypersurfaces given by $n^\mu = (1/N, -\vec{N}/N)$. In a bohmian approach to cosmology, one has to add a "quantum potential for the gravitational field" (which emerges from WDW equation) to the classical Hamiltonian.

Here, following the references listed above we want to analyze the bohmian version of WDW equation

$$\left[(8\pi G)G_{abcd}(\hat{g}^3)p^{ab}p^{cd} + \frac{1}{16\pi G}\sqrt{g}(2\Lambda - {}^{(3)}R(\hat{g}^3))\right]\Psi(g^3) = 0, \quad (4.13)$$

and to see what results it allows us to obtain as regards the geometry of space. In a bohmian approach to WDW equation (4.6), by decomposing the wavefunctional Ψ in polar form $\Psi = Re^{iS/\hbar}$ one obtains a modified Hamilton–Jacobi equation

$$(8\pi G)G_{abcd}\frac{\delta S}{\delta g_{ab}}\frac{\delta S}{\delta g_{cd}} - \frac{1}{16\pi G}\sqrt{g}(2\Lambda - {}^{(3)}R) + Q_G = 0, \quad (4.14)$$

where

$$Q_G = \hbar^2 NqG_{abcd}\frac{1}{R}\frac{\delta^2 R}{\delta g_{ab}\delta g_{cd}}, \quad (4.15)$$

can be defined as "quantum potential for the gravitational field". Equation (4.14) suggests that the only difference between classical and quantum universes is the existence of the quantum potential in the latter. This means that, in order to obtain a quantum regime, one must modify the classical constraints via relation

$$C \to C + \frac{Q_G}{\sqrt{g}N}; C_i \to C_i. \quad (4.16)$$

As regards the constraint algebra one can use the integrated forms of the constraints defined as

$$C(N) = \int d^3x \sqrt{g}NC; \tilde{C}(\vec{N}) = \int d^3x \sqrt{g}N^iC_i, \quad (4.17)$$

which satisfy the following algebra

$$\{\tilde{C}(\vec{N}), \tilde{C}(\vec{N}')\} = \tilde{C}(\vec{N} \cdot \nabla \vec{N}' - \vec{N}' \cdot \nabla \vec{N});$$
$$\{\tilde{C}(\vec{N}), C(\vec{N})\} = C(\vec{N} \cdot \nabla \vec{N});$$
$$\{C(N), C(N')\} \approx 0. \quad (4.18)$$

On the basis of Eq. (4.11), the first 3-diffeomorphism subalgebra and the second 3-diffeomorphism subalgebra do not change with respect to the classical situation; instead, in the third the quantum potential changes the Hamiltonian constraint algebra according to relation

$$\frac{1}{N}\frac{\delta}{\delta q_{ab}}\frac{Q}{\sqrt{g}} = \frac{3}{4\sqrt{g}}g_{cd}p^{ab}p^{cd}\delta(x-z)$$

$$-\frac{\sqrt{g}}{2}g^{ab}(^{(3)}R - 2\Lambda)\delta(x-z) - \sqrt{g}\frac{\delta^{(3)}R}{\delta g_{ab}} \qquad (4.19)$$

giving for the Poisson bracket a result weakly equal to zero (namely zero when the equations of motion are satisfied).

The quantum potential for gravity (4.15) can be considered as a sort of generalization of the bohmian quantum potential to the universe as a whole. Just like in Bohm's interpretation of nonrelativistic quantum mechanics, the quantum potential guides the motion of subatomic particles in the regions where the wavefunction is more intense, in analogous way the quantum potential for gravity (4.15) can be considered the crucial element that guides the behavior of the universe. Moreover, as it was recently underlined in the reference [299], the quantum potential of the universe can be interpreted as the active information which characterizes the quantum creation after Planck's time and the evolution of the universe from the many-four geometries of the pre-planckian time. In this picture, the homogeneity and isotropy of cosmic microwave background radiation lies in the cosmic entanglement associated with the quantum potential of the universe: all the parts of the universe (and also photons of cosmic microwave background radiation) have a common origin and their entanglement emerges directly from the quantum potential of the universe (as regards the idea that photons of cosmic microwave background radiation cannot be considered as separable particles but parts of a unique fundamental field, the reader can find detail also in the reference [300]).

Here, one can remark that the presence of the quantum potential (4.15) means that the quantum algebra is the 3-diffeomorphism algebra times an Abelian subalgebra and that this resulting algebra is weakly closed. One can see that the algebra (4.18) is a clear projection of the general coordinate transformations to the spatial and temporal diffeomorphisms and in fact the equations of motion are invariant under such transformations. In particular, although the form of the quantum potential will depend on regularization and ordering, in the quantum constraint algebra the form of the quantum potential is not important; the algebra holds independently of the form of the quantum potential. Furthermore, the inclusion of matter terms does not seem to change anything.

The next step is to develop the quantum Einstein equations inside the geometry of space represented by the algebra mentioned here. In this regard, we follow the references [272, 301–304] of F. Shojai and A. Shojai of the Teheran school. Let us consider, before all, the quantum Einstein equations in absence of source of matter-energy. For the dynamical parts, by considering the Hamilton equations, one arrives to equation

$$G^{ab} = -\frac{1}{N}\frac{\delta \int d^3 x Q_G}{\delta g_{ab}}, \qquad (4.20)$$

which means that the quantum force modifies the dynamical parts of Einstein's equations (in Eq. (4.20), $G^{ab} = R^{ab} - \frac{1}{2}g^{ab}R$ is, of course, Einstein's tensor). For the nondynamical parts, by using the constraint relations (4.11) and (4.12), one obtains

$$G^{00} = \frac{Q_G}{2N^3\sqrt{g}}; \quad G^{0i} = -\frac{Q_G}{2N^3\sqrt{g}}N^i. \qquad (4.21)$$

Equations (4.21) can also be written as

$$G^{0\mu} = \frac{Q_G}{2\sqrt{-g}}g^{0\mu}, \qquad (4.22)$$

which shows that the nondynamical parts are modified by the quantum potential. On the basis of Eqs. (4.20)–(4.22), one can say that, in quantum cosmology, in absence of source of matter-energy, the quantum potential for the gravitational field determines a modification of the geometry of the physical space both for the dynamical parts and for the nondynamical parts. It is also interesting to observe that the modified Einstein's Eqs. (4.20)–(4.22) are covariant under spatial and temporal diffeomorphisms.

Then, by starting from Eqs. (4.20)–(4.22), inclusion of matter is straightforward. By inserting the matter quantum potential and by introducing the energy-momentum tensor in these equations, one obtains:

$$G^{ab} = -kT^{ab} - \frac{1}{N}\frac{\delta \int d^3x(Q_G + Q_m)}{\delta g_{ab}},$$

$$G^{0\mu} = -kT^{0\mu} - \frac{1}{N}\frac{(Q_G + Q_m)}{2\sqrt{-g}}g^{0\mu}, \qquad (4.23)$$

where

$$Q_m = \hbar^2 \frac{N\sqrt{H}}{2}\frac{\delta^2 R}{\delta \phi^2} \qquad (4.24)$$

(ϕ being the matter field) is the quantum potential for matter and, of course,

$$Q_G = \hbar^2 N g G_{abcd}\frac{1}{R}\frac{\delta^2 R}{\delta g_{ab}\delta g_{cd}} \qquad (4.15)$$

is the usual quantum potential for the gravitational field. Therefore, one can say that the bohmian approach to WDW equation based on Eq. (4.14) leads to general Bohm–Einstein equations of the form (4.23) which are in fact the quantum version of Einstein's equations. Here, since regularization only affects the quantum potential (cfr. also [272, 301–304]), for any regularization the Bohm–Einstein equations (4.23) are the same. Moreover, the Bohm–Einstein equations (4.23) are invariant under temporal ⊗ spatial diffeomorphisms and can be written also in the equivalent compact form

$$G^{\mu\nu} = -kT^{\mu\nu} + S^{\mu\nu}, \qquad (4.25)$$

where

$$S^{0\mu} = \frac{Q_G + Q_m}{2\sqrt{-g}}g^{0\mu}; \quad S^{ab} = -\frac{1}{N}\frac{\delta \int d^3x(Q_G + Q_m)}{\delta g_{ab}}. \qquad (4.26)$$

$S^{\mu\nu}$ is the quantum corrector tensor, under the temporal ⊗ spatial diffeomorphisms subgroup of the general coordinate transformations. One can say that Eq. (4.23) (and the equivalent Eq. (4.25)) describe the geometry of the background of processes which is derived from WDW equation and that the quantum corrector tensor $S^{\mu\nu}$ (defined by Eq. (4.26) and determined by the matter quantum potential and by the quantum potential for the gravitational field) expresses the deformation of the geometry of the background space produced by matter and gravity in WDW equation's regime.

In synthesis, according to F. Shojai's and A. Shojai's model developed in the papers [272, 301–304], in the bohmian approach to WDW equation the complete set of equations to be solved in order to describe and obtain the geometry of physical space in quantum cosmology is given by Eq. (4.25), the WDW equation and the appropriate equation of matter field given by matter lagrangian. It is also interesting to remark that in the papers [305, 306], the quantum Einstein's equations have also been derived for a special metric (Robertson–Walker metric), but there neither any attempt is done to write the equations for a general metric, nor the symmetries are investigated. Finally, in this bohmian model for WDW equation suggested by F. Shojai and A. Shojai, by getting the divergence of Eq. (4.25), one obtains:

$$\nabla_\mu T^{\mu\nu} = \frac{1}{k}\nabla_\mu S^{\mu\nu}, \qquad (4.27)$$

which can be interpreted as an energy conservation law for WDW equation.

In the bohmian approach to quantum cosmology, the fundamental object of quantum gravity, namely the geometry of 3D space-like hypersurfaces, is supposed to exist independently on any observation or measurement, as well as its canonical momentum, the extrinsic curvature of the space-like hypersurfaces. Its evolution, labeled by some time parameter, is dictated by a quantum evolution that is different from the classical one due to the presence of a quantum potential which emerges naturally from the WDW equation. The quantum potential, by introducing appropriate corrector terms in Einstein's equations, determines a deformation of the geometry of the background of processes with respect to the classical situation.

Bohm's approach to WDW equation has been applied to many minisuperspace models obtained by the imposition of homogeneity of the space-like hypersurfaces. In this regard, the reader can find details, for example, in the references [307–312]. Here, the classical limit, the singularity problem, the cosmological constant problem and the time issue have been investigated. For example, in some of these papers it was shown that in models involving scalar fields or radiation, which are good representatives of the matter content of the early universe, the singularity can be clearly avoided by quantum effects. In the de Broglie–Bohm interpretation of quantum cosmology, the quantum potential becomes important near the singularity, yielding a repulsive quantum force counteracting the gravitational field, avoiding the singularity and yielding inflation.

Other interesting results about the geometry of the background of processes determined by the quantum potential emerging from WDW equation have been obtained by Pinto-Neto and are summarized in his papers [275, 313–320]. As regards the bohmian approach to quantum cosmology in the case of homogeneous minisuperspace models, Pinto-Neto showed that there is no problem of time and that quantum effects can avoid the initial singularity, create inflation, and isotropize the universe. As regards the general case of superspace canonical quantum cosmology, Pinto-Neto proved that the bohmian evolution of the three-geometries, irrespective of any regularization and factor ordering of the WDW equation, can be obtained from a specific hamiltonian, which is of course different from the classical one.

By following Pinto-Neto's treatment, the de Broglie–Bohm interpretation of canonical quantum cosmology yields a quantum geometrodynamical picture where the bohmian quantum evolution of three geometries may give rise, depending on the wavefunctional, a consistent nondegenerate four-geometry (which can be Euclidean — for a very special local form of the quantum potential — or hyperbolic), and a consistent but degenerate four-geometry indicating the presence of special vector fields and the breaking of the space-time structure as a single entity (in a wider class of possibilities).

Pinto-Neto starts by writing the WDW equation in the following unregulated form in the coordinate representation

$$\left[-\hbar^2 \left(kG_{abcd} \frac{\delta}{\delta g_{ab}} \frac{\delta}{\delta g_{cd}} + \frac{1}{2\sqrt{g}} \frac{\delta^2}{\delta \phi^2} \right) + V \right] \Psi = 0, \qquad (4.28)$$

where V is the classical potential given by relation

$$V = \sqrt{g}\left[-\frac{1}{k}(^{(3)}R - 2\Lambda) + \frac{1}{2}q^{ab}\partial_a\phi\partial_b\phi + U(\phi)\right], \qquad (4.29)$$

and there is the constraint $-2g_{ab}\nabla_b\left(\frac{\delta\Psi}{\delta q_{ab}}\right) + \left(\frac{\delta\Psi}{\delta\phi}\right)\partial_a\phi = 0$. Writing the wavefunctional in polar form, Eq. (4.28) yields

$$kG_{abcd}\frac{\delta S}{\delta g_{ab}}\frac{\delta S}{\delta g_{cd}} + \frac{1}{2\sqrt{g}}\left(\frac{\delta S}{\delta\phi}\right)^2 + V + Q = 0, \qquad (4.30)$$

where the quantum potential is given by

$$Q = -\frac{\hbar^2}{R}\left(kG_{abcd}\frac{\delta^2 R}{\delta g_{ab}\delta g_{cd}} + \frac{1}{2\sqrt{g}}\frac{\delta^2 R}{\delta\phi^2}\right). \qquad (4.31)$$

Inside Pinto-Neto's model, a nondegenerate four-geometry can be attained if the quantum potential (4.31) assumes the specific form

$$Q = -\sqrt{g}\left[(\varepsilon + 1)\left(-\frac{1}{k}{}^{(3)}R + \frac{1}{2}g^{ab}\partial_a\phi\partial_b\phi\right)\right.$$
$$\left. + \frac{2}{k}(\varepsilon\bar{\Lambda} + \Lambda) + \varepsilon\bar{U}(\phi) + U(\phi)\right] \qquad (4.32)$$

(where ε is a constant which can be ± 1 depending if the four-geometry in which the three-geometries are embedded is Euclidean or hyperbolic, providing thus the conditions for the existence of spacetime). If $\varepsilon = -1$, condition in which the spacetime is hyperbolic, the quantum potential (4.32) becomes

$$Q = -\sqrt{g}\left[\frac{2}{k}(\varepsilon\bar{\Lambda} + \Lambda) - \varepsilon\bar{U}(\phi) + U(\phi)\right], \qquad (4.33)$$

and thus is like a classical potential. The effect of the quantum potential (4.33) is to renormalize the cosmological constant and the classical scalar field potential, nothing more. In this domain, the quantum geometrodynamics turns out to be indistinguishable from the classical one. If $\varepsilon = +1$, condition in which the space-time is

Euclidean, the quantum potential (4.32) becomes

$$Q = -\sqrt{g}\left[2\left(-\frac{1}{k}{}^{(3)}R + \frac{1}{2}g^{ab}\partial_a\phi\partial_b\phi\right)\right.$$
$$\left.+ \frac{2}{k}(\varepsilon\bar{\Lambda} + \Lambda) + \varepsilon\bar{U}(\phi) + U(\phi)\right]. \qquad (4.34)$$

In this case, the quantum potential not only renormalizes the cosmological constant and the classical scalar field potential but also changes the signature of space-time. The total potential $V + Q$ may describe some era of the early universe when it had Euclidean signature, but not the present era, when it is hyperbolic. The transition between these two phases must happen in a hypersurface where $Q = 0$, which is the classical limit. On the basis of Pinto-Neto's treatment, one can conclude that if a quantum space-time exists with different features from the classical observed one, then it must be Euclidean. In other words, the sole relevant quantum effect which maintains the nondegenerate nature of the four-geometry of space-time is its change of signature to a Euclidean one. The other quantum effects are either irrelevant or break the space-time structure completely.

However, as shown by Pinto-Neto, the three geometries evolved under the influence of a quantum potential do not in general stick together to form a nondegenerate four-geometry, a single space-time with the causal structure of relativity. Among the consistent bohmian evolutions, the most general structures that are formed are degenerate four-geometries with alternative causal structures. In the case of consistent quantum geometrodynamical evolution but with degenerate four-geometry, Pinto-Neto showed that any real solution of the WDW equation leads to a structure which is the idealization of the strong gravity limit of general relativity. This type of geometry, which was already investigated also in the reference [321], might well be the correct quantum geometrodynamical description of the young universe. It would also be interesting to investigate if these structures have a classical limit yielding the usual four-geometry of classical cosmology.

Moreover, for nonlocal quantum potentials, Pinto-Neto showed that apparently inconsistent quantum evolutions are in fact consistent if restricted to the bohmian trajectories satisfying the guidance relations corresponding to the Bohm–Einstein equations (4.25). If one wants to be strict and impose that quantum geometrodynamics does not break space-time by determining a nondegenerate four-geometry, then one will have stringent boundary conditions, like the form (4.27) for the quantum potential (which implies a severe restriction on the solutions to the WDW equation).

4.3 A Background Space as an Immediate Information Medium in the Quantum Gravity and Cosmology Domain: The Symmetrized Extension of WDW Equation and the Symmetrized Quantum Potential for Gravity

Just like the quantum potential of Bohm's interpretation of nonrelativistic quantum mechanics, also the quantum potential for gravity (4.15) has a like-space, instantaneous action. As a consequence, if the original Bohm's quantum potential can be associated — in order to characterize the nonlocal geometry of the quantum world — with the idea of space as an immediate information medium between subatomic particles, in analogous way the quantum potential for gravity (4.15) — just in virtue of its nonlocal, instantaneous action — can be associated with the idea of a background space in the quantum gravity and cosmology domain as an immediate information medium. The quantum potential for gravity (4.15) can be considered an appropriate mathematical candidate in order to represent the special state of space in the quantum gravity and cosmology domain as an immediate, direct information medium. In this paragraph, by following the treatment made in the papers [210, 212] we want to analyze the perspectives introduced in this regard by the research of the author of this book.

In analogy to the quantum domain, the background space which acts as a direct information medium in the quantum gravity and cosmology domain can be characterized by defining a "quantum entropy for the gravitational field" of the form

$$S_Q = g \ln \rho. \quad (4.35)$$

The quantum entropy (4.35) indicates a degree of order and chaos that can be associated with the gravitational field characterized by the three-metric q_{ab}. More precisely, we can say that the density of particles $\rho = R^2$ associated with the wavefunctional Ψ of the universe determines a deformation of the gravitational space described by the quantum entropy for the gravitational field (4.35). With the introduction of the quantum entropy (4.35), the quantum potential for the gravitational field (4.15) can be expressed in the following entropic way:

$$Q_G = \hbar^2 N g G_{abcd} \frac{1}{R} \left(\frac{\delta^2}{\delta g_{ab} \delta g_{cd}} \right)_g S_Q. \quad (4.36)$$

On the basis of Eq. (4.36), the quantum entropy for the gravitational field can be considered as the fundamental entity, as the ultimate visiting card which determines the action of the quantum potential for the gravitational field. The following interesting perspective is therefore opened: just like in nonrelativistic de Broglie–Bohm theory the quantum entropy represents the fundamental entity from which the behavior of subatomic particles is derived, in analogous way as regards the bohmian approach to WDW equation the quantum entropy for the gravitational field can be considered as the ultimate visiting card which produces the behavior of the universe in the presence of a gravitational field. And, in particular, one can say that the nonlocal, instantaneous action of the quantum potential for the gravitational field derives just from the quantum entropy (4.35).

Moreover, just like it happens in the original Bohm's pilot wave theory, also the original Bohm's approach to quantum gravity cannot be considered completely convincing from the point of view of the mathematical symmetry in time. The standard quantum laws regarding WDW equation (and thus also the Bohm's approach to WDW equation which derives from it) are not time-symmetric and therefore if one inverts the sign of time, the filming of a process in the quantum gravity and cosmology domain could not correspond to what physically happens. On the basis of these considerations, in the papers [210, 212] the author and Sorli introduced a time-symmetric extension of WDW equation which leads to a symmetrized quantum potential for the gravitational field. According to this approach, the symmetrized version of the quantum potential for gravity can be considered as the mathematical state of the background space in the quantum gravity and cosmology domain which can explain and reproduce the processes of quantum gravity and cosmology both from the forward-time perspective and from the time-reverse perspective.

By following the treatment in the papers [210, 212], the author and Sorli undertook pioneering effort to build a time-symmetric version of WDW equation in analogous manner to the considerations made by Wharton [322] in his attempt to develop a time-symmetric formulation of standard quantum mechanics, namely by considering a time-symmetric extension of WDW equation of the form

$$\begin{pmatrix} H & 0 \\ 0 & -H \end{pmatrix} C = 0, \qquad (4.37)$$

where

$$H = \left[(8\pi G) G_{abcd} p^{ab} p^{cd} + \frac{1}{16\pi G} \sqrt{g} \left(2\Lambda - {}^{(3)}R \right) \right], \qquad (4.38)$$

and $C = \begin{pmatrix} \Psi \\ \Phi \end{pmatrix}$, Ψ is the solution of the standard WDW equation, Φ is the solution of the time-reversed WDW equation.

Then, a time-symmetric reformulation of the bohmian approach to WDW equation in the light of the symmetrized WDW equation (4.37) was made in analogous way to the program followed in Sec. 3.5 as regards the time-symmetric extension of Bohm's nonrelativistic quantum mechanics. The crucial point is to decompose the time-symmetric WDW equation (4.37) into two real equations, by expressing the wavefunctionals Ψ and Φ in polar form:

$$\Psi = R_1 e^{iS_1}, \qquad (4.39)$$

$$\Phi = R_2 e^{iS_2}, \qquad (4.40)$$

where R_1 and R_2 are real amplitude functionals and S_1 and S_2 are real phase functionals. Inserting (4.39) and (4.40) into Eq. (4.37) and separating into real and imaginary parts one obtains the following quantum Hamilton–Jacobi equation for quantum general relativity

$$(8\pi G)G_{abcd} \begin{pmatrix} \dfrac{\delta S_1}{\delta g_{ab}} \dfrac{\delta S_1}{\delta g_{cd}} \\ \dfrac{\delta S_2}{\delta g_{ab}} \dfrac{\delta S_2}{\delta g_{cd}} \end{pmatrix} - \dfrac{1}{16\pi G}\sqrt{g} \begin{pmatrix} 2\Lambda - {}^{(3)}R \\ -2\Lambda + {}^{(3)}R \end{pmatrix}$$

$$\times (2\Lambda - {}^{(3)}R) + \begin{pmatrix} Q_{G1} \\ Q_{G2} \end{pmatrix} = 0, \qquad (4.41)$$

where

$$Q_{G1} = \hbar^2 N g G_{abcd} \dfrac{1}{R_1} \dfrac{\delta^2 R_1}{\delta g_{ab} \delta g_{cd}}, \qquad (4.42)$$

$$Q_{G2} = -\hbar^2 N g G_{abcd} \dfrac{1}{R_2} \dfrac{\delta^2 R_2}{\delta g_{ab} \delta g_{cd}}. \qquad (4.43)$$

In this way, a symmetrized extension of bohmian version of WDW equation emerges which is characterized by a symmetrized quantum potential for gravity at two components having the following form

$$Q_G = \begin{pmatrix} \hbar^2 N g G_{abcd} \dfrac{1}{R_1} \dfrac{\delta^2 R_1}{\delta g_{ab} \delta g_{cd}} \\ -\hbar^2 N g G_{abcd} \dfrac{1}{R_2} \dfrac{\delta^2 R_2}{\delta g_{ab} \delta g_{cd}} \end{pmatrix}. \qquad (4.44)$$

The first component (4.42) coincides with the original "quantum potential for gravity" (4.15): it can explain the forward-time process of the space-like, instantaneous action of quantum gravity in cosmology. The second component (4.43) must be introduced in order to reproduce also the time-reverse of a process in the quantum gravity and cosmology domain through the instantaneous action (this second component usually assures that one could see what really happens if one would imagine to film the process by going backwards in time). In this way, the symmetrized quantum potential for gravity can be considered a good mathematical candidate for the state of space in the quantum gravity and cosmology domain that expresses a direct, immediate information medium (just like the symmetrized quantum potential — (3.65) for one-body systems and (3.71) for many-body systems — can be considered the starting point to develop mathematically the idea of space as an immediate information medium between elementary particles in nonrelativistic quantum mechanics). The symmetrized quantum potential for gravity implies that also in the quantum gravity and cosmology domain a fundamental stage of physical processes exists which acts as an immediate information medium.

Moreover, in analogy to nonrelativistic quantum mechanics, also in the quantum gravity in the context of WDW equation both the components (4.42) and (4.43) of the symmetrized quantum potential for gravity can be considered as physical quantities deriving from the quantum entropy for the gravitational field (4.35). The first component can be expressed as

$$Q_{G1} = \hbar^2 N g G_{abcd} \frac{1}{R_1} \left(\frac{\delta^2}{\delta g_{ab} \delta g_{cd}} \right)_g S_{Q1}, \qquad (4.45)$$

while the second component can be expressed as

$$Q_{G2} = -\hbar^2 N g G_{abcd} \frac{1}{R_1} \left(\frac{\delta^2}{\delta g_{ab} \delta g_{cd}} \right)_g S_{Q2}, \qquad (4.46)$$

where $S_{Q1} = g \ln \rho_1$ is the quantum entropy expressing the change of the geometry of the gravitational space for the forward-time processes (where here $\rho_1 = |\Psi|^2$, $\Psi = R_1 e^{iS_1}$ being the forward-time wavefunctional, solution to the standard WDW equation) and $S_{Q2} = g \ln \rho_2$ is the quantum entropy indicating the change of the geometry of the gravitational space for the time-reverse processes (where here $\rho_2 = |\Phi|^2$, $\Phi = R_2 e^{iS_2}$ being the time-reverse wavefunctional, solution to the time-reversed WDW equation). Therefore, the symmetrized quantum potential for gravity (4.44) reads as

$$Q_G = \begin{pmatrix} \hbar^2 N g G_{abcd} \frac{1}{R_1} \left(\frac{\delta^2}{\delta g_{ab} \delta g_{cd}} \right)_g S_{Q1} \\ -\hbar^2 N g G_{abcd} \frac{1}{R_2} \left(\frac{\delta^2}{\delta g_{ab} \delta g_{cd}} \right)_g S_{Q2} \end{pmatrix}. \qquad (4.47)$$

250   *The Timeless Approach*

Thus the nonlocal geometry of the quantum gravity and cosmology domain in the form of a background which acts as a direct information medium can be defined by a symmetrized quantum entropy for gravity at two components of the form

$$S_Q = \begin{pmatrix} g \ln \rho_1 \\ g \ln \rho_2 \end{pmatrix}. \qquad (4.48)$$

The symmetrized quantum potential for gravity can be considered as the ultimate entity which expresses the deformation of the background of processes of the quantum gravity and cosmology caused by the gravitational field. The action of the background of quantum gravity and cosmology as an immediate information medium in its special mathematical state represented by the symmetrized quantum potential for gravity derives from the symmetrized quantum entropy for gravity (4.48).

Moreover, in the symmetrized quantum potential for gravity approach based on the symmetrized quantum entropy for gravity, the Bohm–Einstein equations of the form (4.23) may be generalized into a symmetrized form:

$$\begin{pmatrix} G_1{}^{ab} = -kT^{ab} - \dfrac{1}{N} \dfrac{\delta \int d^3 x (Q_{G1} + Q_{m1})}{\delta g_{ab}} \\ G_2{}^{ab} = -kT^{ab} - \dfrac{1}{N} \dfrac{\delta \int d^3 x (Q_{G2} + Q_{m2})}{\delta g_{ab}} \end{pmatrix};$$

$$\begin{pmatrix} G_1{}^{0\mu} = -kT^{0\mu} - \dfrac{1}{N} \dfrac{(Q_{G1} + Q_{m1})}{2\sqrt{-g}} g^{0\mu} \\ G_2{}^{0\mu} = -kT^{0\mu} - \dfrac{1}{N} \dfrac{(Q_{G2} + Q_{m2})}{2\sqrt{-g}} g^{0\mu} \end{pmatrix}, \qquad (4.49)$$

where

$$Q_{m1} = \hbar^2 \dfrac{N\sqrt{H}}{2} \dfrac{\delta^2 S_{Q1}}{\delta \phi^2}, \qquad (4.50)$$

$$Q_{m2} = \hbar^2 \dfrac{N\sqrt{H}}{2} \dfrac{\delta^2 S_{Q1}}{\delta \phi^2}, \qquad (4.51)$$

are the forward-time and the time-reverse components of the quantum potential for matter, which may be written in the compact form

$$Q_m = \begin{pmatrix} \hbar^2 \dfrac{N\sqrt{H}}{2} \dfrac{\delta^2 S_{Q1}}{\delta \phi^2} \\ \hbar^2 \dfrac{N\sqrt{H}}{2} \dfrac{\delta^2 S_{Q2}}{\delta \phi^2} \end{pmatrix}. \tag{4.52}$$

On the basis of Eq. (4.49), one can say that the geometry of the background of processes which is derived from WDW equation is linked with the behavior of the symmetrized quantum entropy for gravity with respect to the spatial metric and the matter field. One can also express the deformation of the geometry of the background produced by the behavior of the symmetrized quantum entropy for gravity with respect to matter and gravity in WDW equation's regime by the following symmetrized quantum corrector tensor

$$S^{0\mu} = \begin{pmatrix} \dfrac{Q_{G1}+Q_{m1}}{2\sqrt{-g}} g^{0\mu} \\ \dfrac{Q_{G2}+Q_{m2}}{2\sqrt{-g}} g^{0\mu} \end{pmatrix}; \quad S^{ab} = \begin{pmatrix} -\dfrac{1}{N}\dfrac{\delta \int d^3x \,(Q_{G1}+Q_{m1})}{\delta g_{ab}} \\ -\dfrac{1}{N}\dfrac{\delta \int d^3x \,(Q_{G1}+Q_{m1})}{\delta g_{ab}} \end{pmatrix}.$$

$$\tag{4.53}$$

Finally, also Pinto-Neto's approach to WDW equation based on Eqs. (4.28)–(4.34) can be generalized into a symmetrized picture where the symmetrized quantum entropy for gravity (4.48) emerges as the ultimate entity which describes the geometrical properties of the background. Here, the quantum Hamilton–Jacobi equation for quantum general relativity assumes the form

$$kG_{abcd} \begin{pmatrix} \dfrac{\delta S_{Q1}}{\delta g_{ab}} \dfrac{\delta S_{Q1}}{\delta g_{cd}} \\ \dfrac{\delta S_{Q2}}{\delta g_{ab}} \dfrac{\delta S_{Q2}}{\delta g_{cd}} \end{pmatrix} + \dfrac{1}{2\sqrt{g}} \begin{pmatrix} \dfrac{\delta S_{Q1}}{\delta \phi} \\ \dfrac{\delta S_{Q2}}{\delta \phi} \end{pmatrix} + \begin{pmatrix} V \\ -V \end{pmatrix} + \begin{pmatrix} Q_1 \\ Q_2 \end{pmatrix} = 0,$$

$$\tag{4.54}$$

where the quantum potential is given by relation

$$Q = \begin{pmatrix} -\dfrac{\hbar^2}{R}\left(kG_{abcd}\dfrac{\delta^2 S_{Q1}}{\delta g_{ab}\delta g_{cd}} + \dfrac{1}{2\sqrt{g}}\dfrac{\delta^2 S_{Q1}}{\delta\phi^2}\right) \\ \dfrac{\hbar^2}{R}\left(kG_{abcd}\dfrac{\delta^2 S_{Q2}}{\delta g_{ab}\delta g_{cd}} + \dfrac{1}{2\sqrt{g}}\dfrac{\delta^2 S_{Q2}}{\delta\phi^2}\right) \end{pmatrix}. \qquad (4.55)$$

Inside the symmetrized extension of Pinto-Neto's model, the results regarding the evolution of the three geometries which may form a consistent nondegenerate four-geometry (which can be Euclidean — for a very special local form of the quantum potential — or hyperbolic), or a consistent but degenerate four-geometry indicating the presence of special vector fields and the breaking of the space-time structure as a single entity can be seen as a consequence of the peculiar behavior of the symmetrized quantum potential and thus of the symmetrized quantum entropy, with respect to the induced spatial metric and the matter field.

Chapter 5

The Gravitational Space in an A-Temporal Quantum-Gravity Space Theory

In the effort to unify quantum mechanics with gravity, by following the covariant Hamiltonian philosophy that is the basis for loop quantum gravity, in the papers [323–327] the author of this book developed an a-temporal quantum-gravity space theory, which starts from the idea that the gravitational space is characterized by a wave structure. This model has the following interesting perspectives at the basis of its foundations: the characterization of the gravitational space through different levels of description and the possibility that, at a fundamental level, the universal space is a nonempty, granular, wave, timeless and nonlinear structure. It suggests the possibility to unify the gravitational and quantum aspects of the universe at the fundamental level of physical reality represented by the quantum-gravity space, therefore introducing a modification in the 20th century field-geometry paradigm towards a real granular-wave-dynamic picture and deeply holistic view of the universe. In this chapter, after a brief analysis of the main results of loop quantum gravity regarding the granularity of the spatial geometry and of the perspectives about the holographic and nonlocal features of the ultimate texture of space-time foam in the light of some relevant current research, by following the treatment provided in the papers [323–327] we want to analyze the fundamental mathematical and conceptual features of this approach.

254 The Timeless Approach

5.1 About the Discreteness of the Spatial Geometry in Loop Quantum Gravity

Twenty-six years ago, Abhay Ashtekar, building on earlier work by Sen, laid the foundations of Loop Quantum Gravity by reformulating general relativity in terms of canonical connection and triad variables. The completion of the kinematics — the quantum theory of spatial geometry — led to the prediction of a granular structure of space, described by specific discrete spectra of geometric operators, area [328], volume [328–330], length [331–333] and angle [334] operators.

The discreteness of quantum geometry predicted by loop quantum gravity is understood to be a genuine property of space, independent of the strength of the actual gravitational field at any given location. Thus, there opens an interesting perspective to observe quantum-gravity effects even without strong gravitational field, in the flat space limit. In the paper [335] Amelino-Camelia and other authors proposed that granularity of space influences the propagation of particles, when their energy is comparable with the quantum-gravity energy scale. Further, the assumed invariance of this energy scale, or the length scale, respectively, is in apparent contradiction with special relativity. So it is expected that the energy-momentum dispersion relation could be modified to include dependence on the ratio of the particle's energy and the quantum-gravity energy.

In the years following this work, the nascent field of quantum gravity phenomenology developed from ad hoc effective theories, like isolated isles lying between the developing quantum gravity theories and reality, linked to the former ones loosely by plausibility arguments [336]. Today the main efforts of quantum gravity phenomenology go in two directions: to establish a bridge between the intermediate effective theories and the fundamental quantum gravity theory and the refinement of observational methods, through new effective theories and experiments that could shed new light on quantum-gravity effects. These are exceptionally healthy developments for the field. The development of physical theory relies on the link between theory and experiment. Now these links between current observation and quantum gravity theory are possible and under active development.

Loop quantum gravity hews close to the classical theory of general relativity, taking the notion of background independence and apparent four dimensionality of space-time seriously. Loop quantum gravity takes the novel view of the world provided by general relativity and quantum theory, by incorporating the notions of space and time from general relativity directly into quantum field theory. The quantization is performed in stages: kinematics, quantization of spatial geometry, dynamics and finally the full description of space-time.

The kinematics of the theory is given by a Hilbert space carrying an algebra of operators that have a physical interpretation in terms of observables quantities of the system considered. In loop quantum gravity the operators representing area, angle, length, and volume have discrete spectra, so discreteness is naturally incorporated into loop quantum gravity.

The Hilbert space H on which the theory is defined is given by relation

$$\tilde{H} = \underset{\Gamma}{\oplus} H_\Gamma, \qquad (5.1)$$

where the Γ are abstract graphs defined by sets of links l, sets of nodes n. Spin network states represent a convenient basis in H. Spin network states are eigenstates of the area and volume operators. A spin network state can be given a simple geometrical interpretation. It represents a "granular" space where each node n represents a "grain" or "chunk" of space. Two grains n and n' are adjacent if there is a link l connecting the two, and in this case the area of the elementary surface separating the two grains is $8\pi\gamma\hbar G\sqrt{j_l(j_l+1)}$, where γ is a parameter called the Barbero-Immirzi parameter and j_l are quantum numbers analogous to the spin angular momentum numbers of quantum mechanics. In the recent Rovelli's paper *A new look at loop quantum gravity*, the states in $\tilde{H}$ are interpreted as "boundary states", in the sense that they describe the quantum space surrounding a given finite four-dimensional (4D) region of space-time [337].

The area operator depends on a surface Σ cutting the links $l_1, \ldots, l_S$. The area operator is defined as

$$A_\Sigma = \sum_{l \in \Sigma} \sqrt{L_l^i L_l^i}, \qquad (5.2)$$

$$\vec{L}_l = \{L_l^i\}, \quad i = 1, 2, 3, \qquad (5.3)$$

being the gravitational field operator corresponding with the flux of Ashtekar's electric field, or the flux of the inverse triad, across "an elementary surface cut by the link l". The action of the area operator (5.2) on a spin network function ψ_Γ is

$$A_\Sigma |\psi_\Gamma\rangle = \frac{8\pi\gamma}{\kappa^2} \sum_{p \in \Gamma \cap \Sigma} \sqrt{j_p(j_p + 1)} |\psi_\Gamma\rangle, \qquad (5.4)$$

where j_p is the spin, or color, of the link that intersects Σ at p and γ is the Barbero-Immirzi parameter. The area operator acts only on the intersection points of the surface with the spin network graph, $\gamma \cap \Sigma$ and so gives a finite number of contributions. Spin network functions are eigenfunctions. The eigenvalues are obviously discrete. The expression (5.4) gives the "spectrum of the area" of loop quantum gravity. The quanta of area lie on the edges of the graph and are the simplest elements of quantum geometry in the picture of loop quantum gravity. There is a minimal eigenvalue, the so-called area gap, which is the area when a single edge with $j = \frac{1}{2}$ intersects Σ,

$$\Delta A = 4\sqrt{3}\pi\hbar G c^{-3} \approx 10^{-70} m^2. \qquad (5.5)$$

This is the minimal quantum of area, which can be carried by a link.

The eigenvalues (5.4) form only the main sequence of the spectrum of the area operator. When nodes of the spin network lie on Σ and some links are tangent to it the relation is modified (the reader can find details about this topic in the papers [23, 338]). As regards the area operator, the important fact is that discreteness of area with the spin network links, carrying its quanta, emerges in a natural way.

The interpretation of discrete geometric eigenvalues as observable quantities goes back to early work in [339]. This discreteness made the calculation of black hole entropy possible by counting the number of microstates of the gravitational field that lead to a given area of the horizon within some small interval. Moreover, area operators acting on surfaces that intersect in a line fail to commute, when spin network nodes line in that intersection [340]. In the recent paper [341] additional insight into this noncommutativity comes from the formulation of discrete classical phase space of loop gravity, in which the flux operators also depend on the connection.

As regards the volume there are two definitions of the operator, one due to Rovelli and Smolin [334] and the other due to Ashtekar and Lewandowski [342]. In Rovelli's paper [337], the operator volume of a region R (which is a collection of nodes) is given by

$$V_R = \sum_{n \in R} V_n, \tag{5.6}$$

where

$$V_n^2 = \frac{2}{9} |\varepsilon_{ijk} L_{la}^i L_{lb}^j L_{lc}^k|, \tag{5.7}$$

where l_a, l_b, l_c are any three distinct of the four links of n. In Ashtekar's and Lewandowski's definition, for a given spin network function based on a graph Γ, the operator $\hat{V}_{R,\Gamma}$ is

$$\hat{V}_{R,\Gamma} = \left(\frac{l_p}{2}\right)^3 \sum_{v \in R} \sqrt{\left|\frac{i}{3! \cdot 8} \sum_{I,J,K} s(e_I, e_J, e_K) \varepsilon_{ijk} \hat{X}_{v,e_I}^i \hat{X}_{v,e_J}^j \hat{X}_{v,e_K}^k\right|}, \tag{5.8}$$

where l_p is Planck length, the three derivative operators $\hat{X}_{v,e_I}^i$ act at every node or vertex v on each triple of adjacent edges e_I and the dependence on the tangent space structure of the embedding is manifest in $s(e_I, e_J, e_K)$. This is $+1$ or -1 whether e_I, e_J and e_K are positive or negative oriented, and is zero when the edges are coplanar. The action of the operators $\hat{X}_{v,e_I}^i$ on a spin network function $\psi_\Gamma = \psi(h_{e_1}(A), \ldots, h_{e_n}(A))$ based on the graph Γ is

$$\hat{X}_{v,e_I}^i \psi_\Gamma = i \, tr\left(h_{e_I}(A) \tau^i \frac{\partial \psi}{\partial h_{e_I}(A)}\right), \tag{5.9}$$

when e_I is outgoing at v. Equation (5.9) expresses the action of the left-invariant vector field on $SU(2)$ in the direction of τ^i; for ingoing edges it would be the right-invariant vector field. In virtue of the "triple-product" action of the operator (5.8), vertices carry discrete quanta of volume. The volume operator of a small region containing a node does not change the graph, nor the colors of the adjacent edges, it acts in the form of a linear transformation in the space of intertwiners at the vertex for given colors of the adjacent edges. This space of intertwiners forms the "atoms of quantum geometry".

The spectrum of the volume operator (5.8) has been investigated in the papers [343–348]. In the thorough analysis of [343, 344], Brunnemann and Rideout showed that the volume gap, i.e. the lower boundary for the smallest nonzero eigenvalue, depends on the geometry of the graph and does not in general exist. In the simplest nontrivial case, for a four-valent vertex, the existence of a volume gap is demonstrated analytically.

The two volume operators (5.6) and (5.8) are inequivalent, yielding different spectra. While the details of the spectra of the Rovelli–Smolin and the Ashtekar–Lewandowski definitions of the volume operator differ, they do share the property that the volume operator vanishes on all gauge invariant trivalent vertices [330]. According to an analysis performed in the papers [349, 350] the Ashtekar–Lewandowski operator is compatible with the flux operators, on which it is based, and the Rovelli–Smolin operator is not. On the other hand, thanks to its topological structure the Rovelli–Smolin volume does not depend on tangent space structure; the operator is 'topological' in the sense that is invariant under spatial homeomorphisms. It is also covariant under "extended diffeomorphisms", which are everywhere continuous mappings that are invertible everywhere except at a finite number of isolated points; the Ashtekar–Lewandowski operator is invariant under diffeomorphisms.

Physically, the distinction between the two operators is the role of the tangent space structure at spin network nodes. There is some tension in the community about the role of this structure. Recent developments in twisted discrete geometries [351] and the polyhedral point of view [352] may help resolve these issues. In [353] Bianchi and Haggard, show that the volume spectrum of the four-valent node may be obtained by direct Bohr–Sommerfeld quantization of geometry. The description of the geometry goes all the way back to Minkowski, who showed that the shapes of convex polyhedra are determined from the areas and unit normals of the faces. Kapovich and Millson showed that this space of shapes is a phase space, and it is this phase space — the same as the phase space of intertwiners — that Bianchi and Haggard used for the Bohr–Sommerfeld quantization. The agreement between the spectra of the Bohr–Sommerfeld and loop quantum gravity volume turns out to be quite good [353].

A fundamental result of the geometric operators described above (as well as of length and angle) is their discrete spectra. In other words, the geometry associated with spin network states is characterized by a discrete quantized three-dimensional (3D) metric. It is natural to ask whether this discreteness of loop quantum gravity operators and thus the granularity of spatial geometry is physical. Can it be used as a basis for the phenomenology of quantum geometry? Using examples, Dittrich and Thiemann [354] argue that the discreteness of the geometric operators, being gauge noninvariant, may not survive implementation in the full dynamics of loop quantum gravity. On the contrary, Rovelli [355] argues in favor of the reasonableness of physical geometric discreteness, showing in one case that the preservation of discreteness in the generally covariant context is immediate. In phenomenology, this discreteness has been a source of inspiration for models. Nonetheless as the discussion of these operators makes clear, there are subtleties that wait to be resolved, either through further completion of the theory or, perhaps, through observational constraints on phenomenological models. As regards the phenomenology of loop quantum gravity and its physical consequences and applications, in the recent paper *Loop quantum gravity phenomenology: linking loops to observational physics*, Girelli, Hinterleitner and Major [356] described ways in which loop quantum gravity, mainly by means of discreteness of the spatial geometry, may lead to experimentally viable predictions and furthermore explored the possibility of a large variety of modifications of special relativity, particle physics and field theory in the weak field limit on the basis of the discreteness of the spatial geometry.

5.2 About the Holographic and Nonlocal Features of the Ultimate Texture of Space-Time Foam

What is the connection between the discrete and quantized geometry of the fundamental foam predicted by loop quantum gravity and the smooth physical geometry that we perceive in our everyday life? In this chapter, we want to review some relevant results as regards the texture of spin network weaves of loop quantum gravity and of other similar models of space-time foams characterized by an holographic mapping.

By following Rovelli's treatment in his book *Quantum Gravity* [23], let us start by considering a classical macroscopic 3D gravitational field which determines a 3D metric $g_{\mu\nu}$ and in this metric let us fix a region R of area S with a size much larger than $l \gg l_P$ and slowly varying at this scale. A spin network state $|S\rangle$, if it is an eigenstate of the volume operator $V(R)$ (and of the area operator $A(S)$), with eigenvalues equal to the volume of R (and of the area of S) determined by the metric $g_{\mu\nu}$, up to small corrections in l/l_P, namely if it satisfies the following equations

$$V(R)|S\rangle = \left(V[g_{\mu\nu}, R] + O\left(\frac{l}{l_P}\right)\right)|S\rangle, \qquad (5.10)$$

$$A(S)|S\rangle = \left(A[g_{\mu\nu}, S] + O\left(\frac{l}{l_P}\right)\right)|S\rangle, \qquad (5.11)$$

is called a weave state of the metric g. The texture of reality which emerges from loop quantum gravity is a weave of spin network states satisfying Eqs. (5.10) and (5.11).

At large scale, the state $|S\rangle$ determines precisely the same volumes and areas as the metric g. Equations (5.10) and (5.11) are valid also to the diffeomorphism invariant level: the s-knot state $|s\rangle = P_{diff}|S\rangle$ is called the weave state of the 3D geometry $[g]$, namely the equivalence class of three-metric to which the metric g belongs.

Several weave states were constructed and analysed in the early days of loop quantum gravity, for various 3D metrics, including those for flat space, Schwarzschild and gravitational waves. They satisfied Eqs. (5.10) and (5.11) or equations similar to these. Weave states have played an important role as regards the historical development of loop quantum gravity. In particular, they provided an explanation of the emergence of the Planck-scale discreteness. The intuition was that a macroscopic geometry could be built by taking the limit of an infinitely dense lattice of loops, as the lattice size goes to 0. With increasing density of loops, the eigenvalue of the operator turned out to increase.

In order to define a weave on a 3D manifold with coordinates $\vec{x}$ that approximates the flat 3D metric $g^{(0)}_{\mu\nu}(\mathbf{x}) = \eta_{\mu\nu}$, one can build a spatially uniform weave state $|S_{\mu_0}\rangle$ constituted by a set of loops of coordinate density $\rho = \mu_0^{-2}$. The loops are then at an average distance μ_0 from each other. By decreasing the "lattice spacing" μ_0, namely by increasing the coordinate density of the loops, one obtains

$$A(S)|S_\mu\rangle \approx \frac{\mu_0^2}{\mu^2}\left(A[g^{(0)}, S] + O\left(\frac{l}{l_P}\right)\right)S\rangle, \qquad (5.12)$$

which indicates an increasing of the area. Since

$$\frac{\mu_0^2}{\mu^2}A[g^{(0)}, S] = \left(A\left[\frac{\mu_0}{\mu}g^{(0)}, S\right] + O\left(\frac{l}{l_P}\right)\right) = A[g^{(\mu)}, S], \qquad (5.13)$$

the weave with increased loop density approximates the metric

$$g^{(\mu)}{}_{\mu\nu}(\vec{x}) = \frac{\mu_0^2}{\mu^2}\eta_{\mu\nu}. \qquad (5.14)$$

At the same time, however, the physical density ρ_μ, defined as the ratio between the total length of the loops and the total volume, determined by the metric $g^{(\mu)}$, remains μ_0, irrespective of the density of the loops μ chosen:

$$\rho_\mu = \frac{L_\mu}{V_\mu} = \frac{(\mu_0/\mu)L}{(\mu_0/\mu)^3 V} = \frac{\mu^2}{\mu_0^2}\rho = \frac{\mu^2}{\mu_0^2}\mu^{-2} = \mu_0^{-2}. \qquad (5.15)$$

Equation (5.15) implies that, if μ_0 is not determined by the density of the loops, it must be given by the only dimensional constant of the theory, namely the Planck length:

$$\mu_0 \approx l_P. \quad (5.16)$$

By substituting Eq. (5.16) into Eq. (5.14), the metric approximated by the increased loop density becomes

$$g^{(\mu)}{}_{\mu\nu}(\vec{x}) = \frac{l_P^2}{\mu^2}\eta_{\mu\nu}. \quad (5.17)$$

The physical meaning of the approach based on Eqs. (5.12)–(5.17) is that, in loop quantum gravity, a smooth geometry cannot be approximated at a physical scale lower than the Planck length. Each loop carries a quantum geometry of the Planck scale: more loops give more size, not a better approximation to a given geometry.

Another significant feature of the texture of the weaves of the fundamental quantum geometry of the Planck scale is its holographic nature. In this regard, in the paper [357] Gambini and Pullin showed that from the framework of loop quantum gravity in spherical symmetry an holography emerges in the form of an uncertainty in the determination of volumes that grows radially. Gambini and Pullin found that, in the kinematical structure of spherical loop quantum gravity, the minimal increment in the volume, which derives from the Hamiltonian constraint due to the fact that one cannot take the continuum limit in the loop representation, is given by the relation

$$\Delta V = 8\pi\gamma\rho l_P(x + 2M), \quad (5.18)$$

where $x + 2M$ is the radial coordinate in Schwarzschild coordinate, ρ is the coordinate density of the loops (whose minimum possible value leads to interpret (5.18) as the elementary volume). As a consequence, the number ΔN of elementary volumes in a shell with width Δx is

$$\Delta N = \frac{x\Delta x}{2\gamma\rho l_P^2}, \quad (5.19)$$

and its entropy is

$$\Delta S = \nu_V \frac{x\Delta x}{2\gamma\rho l_P^2}, \quad (5.20)$$

where ν_V is the mean entropy per unit volume. According to Eqs. (5.18)–(5.20) of Gambini's and Pullin's approach, the kinematical structure of loop quantum gravity in spherical symmetry implies holography. This is a very general result which stems from the fact that the elementary volume which any dynamical operator may involve behaves as xl_P^2.

An interesting model of fundamental space-time foam of holographic nature has been recently proposed by Jack Ng in the papers [358–361], in which the concept of entropy plays a crucial role. By following Ng's treatment in the papers [358–361], let us review the fundamental features and results of this model. In Ng's model, space-time undergoes quantum fluctuations which appear when we measure a distance l, in the form of uncertainties in the measurement. The quantum fluctuations of space-time manifest themselves in the form of uncertainties in the geometry of space-time and thus the structure of space-time foam can be inferred from the accuracy with which we can measure its geometry. Let us consider mapping out the geometry of space-time for a spherical volume of radius l over the amount of time $T = 2l/c$ it takes light to cross the volume. The total number of operations, including the ticks of the clocks and the measurements of signals, is determined by the Margolus–Levitin theorem in quantum computation, which establishes that the rate of operations for any computer cannot exceed the amount of energy E that is available for computation divided by $\pi\hbar/2$ [362]. A total mass M of clocks then yields, via the Margolus–Levitin theorem, the bound on the total number of operations given by $(2Mc^2/\pi\hbar) \cdot 2l/c$. But to prevent black hole formation, the mass M must be less than $lc^2/2G$. Together, these two limits imply that the total number of operations that can occur in a spatial volume of radius l for a time period $2l/c$ is no greater than $2(l/l_p)^2/\pi$. To maximize spatial resolution, each clock must tick only once during the entire time period. The operations can be regarded as partitioning the space-time volume into "cells", yielding an average separation between neighboring cells no less than $(2\pi^2/3)^{1/3} l^{1/3} l_P^{2/3}$. This spatial separation can be interpreted as the average minimum uncertainty [363], and thus the accuracy in the measurement of a distance l, namely

$$\delta l \geq (2\pi^2/3)^{1/3} l^{1/3} l_P^{2/3}. \qquad (5.21)$$

One can easily understand in what sense this quantum foam model turns out to be a holographic model. Dropping the multiplicative factor of order 1, since on the average each cell occupies a spatial volume of ll_P^2, a spatial region of size l can contain no more than $l^3/(ll_P^2) = (l/l_P)^2$ cells. Thus, this model corresponds to the case of maximum number of bits of information $(l/l_P)^2$ in a spatial region of size l, that is allowed by the holographic principle [364–369] which implies that, although the world around us appears to have three spatial dimensions, its contents can actually be encoded on a two-dimensional (2D) surface, like a hologram. In other words, the maximum entropy, i.e. the maximum number of degrees of freedom, of a region of space is given by its surface area in Planck units. In order to see explicitly in what sense the holographic principle has its origin in the quantum fluctuations of space-time, namely how space-time foam manifests itself holographically, let us consider a cubic region of space with linear dimension l. In this case, from conventional wisdom the number of degrees of freedom of the region should be bounded by $(l/l_P)^3$, namely the volume of the region in Planck units. But conventional wisdom is wrong, because, on the basis of Eq. (5.21), the smallest cubes into which one can partition the region cannot have a linear dimension smaller than $(ll_P^2)^{1/3}$. Therefore, the number of degrees of freedom of the region is bounded by $[l/(ll_P^2)^{1/3}]^3$, i.e. the area of the region in Planck units, in agreement with the holographic principle [364–369].

Assuming that there is unity of physics connecting the Planck scale to the cosmic scale, Ng also applied his holographic space-time foam model to cosmology [358, 370]. In this regard, the consistency of the holographic space-time foam model is assured if the average minimum uncertainty (5.21) corresponds to a maximum energy density

$$\rho = \frac{3}{8\pi}(ll_P)^{-2}, \qquad (5.22)$$

for a sphere of radius l that does not collapse into a black hole. Hence, according to the holographic space-time foam cosmology proposed by Ng, one obtains that the cosmic energy density is given by

$$\rho = \frac{3}{8\pi}(R_H l_P)^{-2}, \qquad (5.23)$$

where R_H is the Hubble radius. The energy density (5.23) is the critical cosmic energy density as observed.

Moreover, if one divides a large distance l into l/λ equal parts each of which has length λ (that can be taken as small as l_P), the cumulative factor C characterizing the cumulative effects of the space-time fluctuations over this large distance defined by $C = \frac{\delta l}{\delta \lambda}$ turns out to be $C = (l/\lambda)^{1/3}$. Because of its holographic features and quantum-gravity effects, the individual fluctuations cannot be completely random: successive fluctuations appear to be entangled and somewhat anti-correlated in such a way that one obtains a cube root dependence in the number l/l_P of fluctuations for the total fluctuation of l.

In Ng's model, as a consequence of the holographic principle, the physical degrees of freedom of the space-time foam, at the Planck scale, must be considered as infinitely correlated, with the result that the space-time location of an event may lose its invariant significance. In other words, in virtue of its holographic nature, the space-time foam gives rise to nonlocality. This argument is also supported by the observation that the long-wavelength (hence "nonlocal") "particles" constituting dark energy in the holographic space-time foam cosmology obey an exotic statistics which has attributes of nonlocality [358].

In this regard, let us consider a perfect gas of N particles obeying Boltzmann statistics at temperature T in a volume V. At the lowest-order approximation, since one can neglect the contributions from matter and radiation to the cosmic energy density for the recent and present eras, the Friedmann equations (for the relativistic case) yield the partition function

$$Z_N = (N!)^{-1}(V/\lambda^3)^N, \qquad (5.24)$$

where $\lambda = (\pi)^{2/3}/T$, and the entropy

$$S = N\left[\ln\frac{V}{N\lambda^3} + \frac{5}{2}\right].\qquad(5.25)$$

Here, since $V \approx \lambda^3$, the entropy S becomes nonsensically negative unless $N \approx 1$ which is equally nonsensical because $N \approx (R_H/l_P)^2 \gg 1$. The solution comes with the observation that the N inside the log term for S somehow must be absent. Then $S \approx N \approx (R_H/l_P)^2 \gg 1$ without N being small (of order 1) and S is nonnegative as physically required. In this case, the Gibbs $1/N!$ factor is absent from the partition function (5.24) and thus the entropy (5.25) becomes

$$S = N\left[\ln\frac{V}{\lambda^3} + \frac{3}{2}\right].\qquad(5.26)$$

Taking account of Eq. (5.26), the only consistent statistics in greater than two space dimensions without the Gibbs factor is the infinite statistics, called also "quantum Boltzmann statistics", which is characterized by a q deformation of the commutation relations of the oscillators:

$$a_k a_l^+ - q a_l^+ a_k = \delta_{kl},\qquad(5.27)$$

with q between -1 and 1. Infinite statistics can be thought of as corresponding to the statistics of identical particles with an infinite number of internal degrees of freedom, which is equivalent to the statistics of nonidentical particles since they are distinguishable by their internal states. As shown in [371–374], a theory of particles obeying infinite statistics turns out to be explicitly nonlocal. In Ng's model, since the holographic principle is believed to be an important ingredient in the formulation of quantum gravity, it is just the nonlocal features of the space-time foam that make it easier to incorporate gravitational interactions in the theory. In this approach, quantum gravity and infinite statistics appear to fit together nicely, and nonlocality seems to be a common feature of both of them [358].

On the other hand, other recent researches seem to support the idea of nonlocality as a fundamental feature of the quantum gravity domain. Using the Matrix theory approach, Jejjala, Kavic and Minic showed that dark energy quanta obey infinite statistics and also concluded that the nonlocality present in systems obeying infinite statistics and the nonlocality present in holographic theories may be related [375–377]. Giddings remarked that the nonperturbative dynamics of gravity is nonlocal [378]. His argument is based on several reasons: lack of a precise definition in quantum gravity, connected with the apparent absence of local observables; indications from high-energy gravitational scattering; hints from string theory, particularly the AdS/CFT correspondence; conundrums of quantum cosmology; and the black hole information paradox. Horowitz remarked that quantum gravity may need some violation of locality, in particular if one reconstructs the string theory from the gauge theory (in the AdS/CFT correspondence), physics may not be local on all length scales [379].

Violation of locality in the form of nonlocal links between the fundamental weaves of the space-time foam appear also in quantum graphity models, which are spin system toy models for emergent geometry and gravity. They are based on quantum, dynamical graphs whose adjacency is dynamical: their edges can be on (connected), off (disconnected), or in a superposition of on and off. In these approaches, one can interpret the graph as a pregeometry (and the connectivity of the graph tells us who is neighbouring whom). A particular graphity model is given by such graph states evolving under a local Ising-type Hamiltonian. For example, in the reference [380] a graphity toy model was proposed by Hamma, Markopoulou, Lloyd, Caravelli, Severini and Markstrom for interacting matter and geometry, characterized by a Bose Hubbard model where the interactions are quantum variables.

In [381], Caravelli, Hamma, Markopoulou and Riera solved the model of [380] in the limit of no back-reaction of the matter on the lattice, and for states with certain symmetries called rotationally invariant graphs. In this case, the problem reduces to an one-dimensional (1D) Hubbard model on a lattice with variable vertex degree and multiple edges between the same two vertices. The probability density for the matter obeys a (discrete) differential equation closed in the classical regime. This is a wave equation in which the vertex degree is related to the local speed of propagation of probability. This approach allows thus an interpretation of the probability density of particles similar to what is usually considered in similar gravity systems in the sense that here matter sees a curved space-time.

In the recent paper *Disordered locality and Lorentz dispersion relations: an explicit model of quantum foam* [382], Caravelli and Markopoulou, using the framework of Quantum Graphity, suggested an explicit model of a quantum foam, a quantum space-time with spatial nonlocal links. The quantum states describing this space-time foam depend on two parameters: the minimal size of the link and their density with respect to this length. In particular, Caravelli and Markopoulou considered the case in which the quantum state of the space-time background is a superposition of many graph states. In this picture, the minimum scale of the space-time foam is provided by the intrinsic discreteness of the graph picture. Caravelli and Markopoulou also studied the effect of the discreteness of the graph picture on the dispersion relations and showed that these states with nonlocal links violate macrolocality and give corrections to Lorentz invariance.

Finally, as regards the violation of locality and of Lorentz invariance characterizing the fundamental foam of physical processes, it is also interesting to remark that Mignani, Cardone and Petrucci recently suggested that the local violation of Lorentz invariance can be interpreted in terms of a energy-dependent deformation of the Minkowski geometry, whose corresponding metrics provide an effective dynamical description of the fundamental interactions (at the energy scale and in the energy range considered). In Mignani's, Cardone's and Petrucci's approach, the deformation of space-time affects the external fields that deform the geometry of space and, differently from multi-dimensional theories (whose prototype is the well known Kaluza–Klein formalism), gauge fields do not need to be added from the outside but emerge in a natural way as a direct consequence of the metric deformation [383, 384].

In synthesis, in the light of the current research, in particular of Ng, of Caravelli and Markopoulou, and of Mignani, Cardone and Petrucci the suggestive perspective is opened that the fundamental space-time background of processes at the Planck scale is a nonlocal, holographic quantum foam characterized by a Lorentz invariance and a deformation of the geometry.

5.3 The Postulates of the A-Temporal Quantum-Gravity Space Theory

The a-temporal quantum-gravity space theory — developed by the author in the papers [323–327] — provides a geometrical description of gravitation and quantum effects which explores the perspective introduced by loop quantum gravity about the discreteness of the spatial geometry. This model is based on the following foundational ideas (which can be considered as the postulates of this model):

1. The gravitational space is characterized by two levels of description: a universal "cosmic space", which is a primordial timeless pre-quantum pre-space, and a quantum-gravity space, which emerges from the cosmic space and exhibits a wave nature.
2. The cosmic space is defined in terms of two fundamental quantities: the density of cosmic space (which is linked with the amount of matter present in the region under consideration) and a quantum number indicating a sort of "rotation-orientation" of each point of the gravitational space. The quantum-gravity space is characterized by a wavefunction (which depends on the density of cosmic space and on the quantum number of "rotation-orientation") satisfying appropriate nonlinear generalized Klein–Gordon and Dirac equations.
3. In the quantum-gravity space, the duration of material motions does not have a primary physical existence: time does not flow on its own as an independent variable of evolution but exists only as a measuring system of the numerical order of material changes.

In the a-temporal quantum gravity-space theory, the density of cosmic space is the fundamental physical entity which corresponds to the appearance of a material object in a given region of space. The density of cosmic space associated with a material object of mass m in the points situated at distance r from the centre of this object is defined by the relation

$$D(r) = \frac{Gm}{r^2}, \quad (5.28)$$

where G is gravitational constant. More in general, if a region of space is endowed with n densities of cosmic space $D_1(r_1)$, $D_2(r_2)$, ..., $D_n(r_n)$ and in the generic point P there is a gravitational field given by equation

$$\begin{aligned}\vec{g}_{\text{tot}}(P) &= \vec{g}_1(P) + \vec{g}_2(P) + \cdots + \vec{g}_n(P) \\ &= D_1(r_1)\hat{r}_1 + D_2(r_2)\hat{r}_2 + \cdots + D_n(r_n)\hat{r}_n \\ &= D_{tot}(P)\frac{\hat{r}_1 + \hat{r}_2 + \cdots + \hat{r}_n}{\sqrt{n}}, \end{aligned} \quad (5.29)$$

one can say that in this region there are n material objects.

The rotation-orientation of space is defined by the quantum vector

$$\vec{j} = \frac{G\vec{s}}{r^3}, \quad (5.28a)$$

where $\vec{s}$ is the spin of the particle associated with the density of cosmic space (5.28). The quantum vector $\vec{j} = \frac{G\vec{s}}{r^3}$ provides a sort of rotational-orientational degree of freedom to the gravitational space. As regards its physical meaning, this quantum vector is analogous to the rotational degrees-of-freedom characterizing the Cosserat continuum of solid mechanics, which were introduced by the Cosserat brothers at the turn of the last century, and which motivated E. Cartan's development of the torsion tensor of space-time in general relativity [385]. As regards this quantum vector **j**, it is assumed that its quantum number j multiplied by r^3/G can assume integer or half-integer multiple values of the Planck reduced constant in order to assure consistency with the results of the standard quantum mechanics. We will say thus that each point of the gravitational space is characterized by a determinate value of the density of cosmic space and is endowed with a particular rotation-orientation linked with a particular value of the quantum number j which can assume integer of half integer multiple values of $\frac{\hbar G}{r^3}$. The density of cosmic space and the quantum vector indicating the rotation-orientation are interpreted as the mass and spin of a quasi-particle which may be termed the "Planck–Fiscaletti granule". In the a-temporal quantum-gravity space theory developed by the author in the papers [323–327], the "Planck–Fiscaletti granules" may be considered as the fundamental quanta of the gravitational space.

The primordial cosmic space is a timeless structure as a consequence of the a-temporal character of the interaction of Planck–Fiscaletti granules. Gravitational interaction between Planck–Fiscaletti granules is a-temporal in the sense that no movement of particle-wave is needed for its acting: gravity is transmitted directly by the density of cosmic space characterizing each Planck–Fiscaletti granule. Two Planck–Fiscaletti granules characterized by a different density of cosmic space attract each other on the basis of the relation

$$\vec{F}_g = \frac{D_1(r_1) \cdot D_2(r_2) \cdot r_1^2 \cdot r_2^2}{Gr^2} \hat{r}, \qquad (5.30)$$

which represents the general law of interaction. Equation (5.30) can be considered as a new way to express the famous Newton's law

$$\vec{F}_g = G \frac{m_1 \cdot m_2}{r^2} \hat{r}. \qquad (5.31)$$

According to Eq. (5.30)) the following interpretation of masses' interaction emerges: material objects move in the direction where the density of cosmic space is increasing. Areas of lower density have a tendency to move towards areas of higher density. On the basis of this treatment of the gravitational interaction, objects with lower mass have a tendency to move towards objects with bigger mass just as a consequence of the behavior of the density of cosmic space, the most universal property of the primordial cosmic space.

According to the a-temporal quantum-gravity space theory, the fundamental level of description of the gravitational space is a quantum-gravity space, which is a manifold which exhibits a wave and granular structure. In the case $j = 0$, the quantum-gravity space is defined by the wavefunction of space

$$\psi_S = A \exp\left[2\pi i \left(\frac{D(r)}{c^2} \hat{r} \cdot \vec{r}_0 - \frac{D(r) \cdot v \cdot sen\vartheta}{c^2} t + \varphi_0\right)\right], \qquad (5.32)$$

where **v** is the speed of the particle associated with the density of cosmic space $D(r)$, ϑ is the angle between **r** and **v** and the amplitude

A is a function of the point (x, y, z) of space. The wavefunction of space (5.32) is solution of the generalized nonlinear Klein–Gordon equation for the density of cosmic space

$$\nabla^2 \psi_S - \frac{1}{c^2} \frac{\partial^2 \psi_S}{\partial t^2} = \frac{[D(r)]^2}{c^4} \psi_S. \tag{5.33}$$

In the case $j = \frac{G\hbar}{2r^3}$, the wavefunction of space can be expressed as

$$\psi_S(x) = \psi_S^{(P)}(x) + \psi_S^{(A)}(x), \tag{5.34}$$

where $\psi_S^{(P)}(x)$ represents the wavefunction of quantum-gravity space associated with the appearance of particles of spin $\frac{1}{2}$ and $\psi_S^{(A)}(x)$ represents the wavefunction of quantum-gravity space associated with the appearance of the corresponding antiparticles. In analogy to Nikolic's model of relativistic fermionic quantum field theory developed in the papers [386, 387], the wavefunctions of quantum-gravity space associated with the appearance of particles of spin $\frac{1}{2}$ and the wavefunction of quantum-gravity space associated with the appearance of the corresponding antiparticles can be expanded as

$$\psi_S^{(P)}(x) = \sum_k b_k u_k(x), \tag{5.35}$$

$$\psi_S^{(A)}(x) = \sum_k d_k^* v_k(x), \tag{5.36}$$

respectively. Here u_k are positive-frequency 4-spinors of quantum-gravity space while v_k are negative frequency 4-spinors of quantum-gravity space; they together form a complete set of orthonormal solutions to the nonlinear generalized Dirac equation for the density of cosmic space

$$\left(i\gamma^\mu \partial_\mu - \frac{D(r)}{c^2} \right) \psi_S = 0, \tag{5.37}$$

where $x = (x^0, x^1, x^2, x^3) = (t, \mathbf{x})$ and γ_μ are the well-known relativistic matrices $\gamma^0 = 1 \otimes \begin{pmatrix} 1 & 0 \\ 0 & -1 \end{pmatrix}$, $\gamma^i = \sigma_i \otimes \begin{pmatrix} 0 & 1 \\ -1 & 0 \end{pmatrix}$ (and $\boldsymbol{\sigma}$ are the Pauli matrices which are linked with $\mathbf{j}$ through the relation $\mathbf{j} = \frac{1}{2} G\hbar \frac{\boldsymbol{\sigma}}{r^3}$).

In Eqs. (5.35)) and (5.36), the label k is an abbreviation for the set $(\mathbf{k}, j)$ where $\mathbf{k}$ is the 3-momentum $\mathbf{k} = (p_1, p_2, p_3)$ and $j = \frac{Gs}{r^3}$

Gravitational Space in A-Temporal Quantum-Gravity Space Theory   273

with $s = \frac{1}{2}\hbar$, is the label of the rotation-orientation of the generic Planck–Fiscaletti granule of cosmic space. As regards the expressions of u_k and v_k, Eq. (5.37) leads to the following results:

$$u = W^z(\mathbf{k}) \exp\left[-\frac{i}{\hbar} p_\mu x^\mu\right], \qquad (5.38)$$

with $z = 1, 2$ and

$$v = W^z(\mathbf{k}) \exp\left[\frac{i}{\hbar} p_\mu x^\mu\right], \qquad (5.39)$$

with $z = 3, 4$, where $p_0 = \frac{\hbar D(r)}{c}$ and

$$W^1 = \sqrt{\frac{cE + \hbar D(r)}{2\hbar D(r)}} \begin{bmatrix} 1 \\ 0 \\ \dfrac{p_3 c^2}{cE + \hbar D(r)} \\ \dfrac{(p_1 + ip_2)c^2}{cE + \hbar D(r)} \end{bmatrix}, \qquad (5.40)$$

$$W^2 = \sqrt{\frac{cE + \hbar D(r)}{2\hbar D(r)}} \begin{bmatrix} 0 \\ 1 \\ \dfrac{(p_1 - ip_2)c^2}{cE + \hbar D(r)} \\ \dfrac{-p_3 c^2}{cE + \hbar D(r)} \end{bmatrix}, \qquad (5.41)$$

$$W^3 = \sqrt{\frac{cE + \hbar D(r)}{2\hbar D(r)}} \begin{bmatrix} \dfrac{p_3 c^2}{cE + \hbar D(r)} \\ \dfrac{(p_1 + ip_2)c^2}{cE + \hbar D(r)} \\ 1 \\ 0 \end{bmatrix}, \qquad (5.42)$$

$$W^4 = \sqrt{\frac{cE + \hbar D(r)}{2\hbar D(r)}} \begin{bmatrix} \frac{(p_1 - ip_2)c^2}{cE + \hbar D(r)} \\ \frac{-p_3 c^2}{cE + \hbar D(r)} \\ 0 \\ 1 \end{bmatrix}, \quad (5.43)$$

where $E = \left[k^2 c^2 + \frac{\hbar^2 (D(r))^2}{c^2}\right]^{1/2}$ is the energy of the quantum-gravity space (which depends on the density of cosmic space $D(r)$ and on the modulus of the momentum of the particle associated with the density of cosmic space $D(r)$).

5.4 About the Mathematical Formalism of the Wavefunction of Quantum-Gravity Space

As the author showed in the papers [323–327], the a-temporal quantum-gravity space theory allows us to obtain some interesting and relevant results.

Before all, in the quantum-gravity space a particular and significant link emerges between the standard Klein–Gordon and Dirac equations for subatomic particles and the corresponding generalized Klein–Gordon and generalized Dirac equations for the density of cosmic space. In fact, the standard Klein–Gordon equation

$$\nabla^2 \psi - \frac{1}{c^2} \frac{\partial^2 \psi}{\partial t^2} = \frac{m^2 c^2}{\hbar^2} \psi \quad (5.44)$$

and the standard Dirac equation

$$\left(i\gamma^\mu \partial_\mu - \frac{mc}{\hbar}\right)\psi = 0 \quad (5.45)$$

describing the behavior of relativistic particles can be considered particular cases of the generalized Klein–Gordon equation (5.33) and the generalized Dirac equation (5.37) respectively in the sense that they can be obtained from these equations in the particular case $r = l_p$. This suggests that the density of cosmic space associated with a given material particle is a quantity which cannot be defined for $r < l_p$. The maximum, and physically sensed, value that it can assume, which corresponds with the centre of the appearance of the material particle, concern $r = l_p$, namely is $D(l_p) = \frac{Gm}{l_p^2}$. The fact that Eqs. (5.44) and (5.45) can be seen as special cases of the correspondent generalized Eqs. (5.33) and (5.37) means that the density of cosmic space cannot assume all values, but has a discrete structure, and that its maximum (and physically consistent) value occurs for $r = l_p$ because this is the condition which corresponds to the appearance of the material particle into consideration. The granularity of the density of cosmic space implies therefore that the gravitational space is endowed with a granular structure at the Planck scale. In sum, an important result that a-temporal quantum-gravity space theory allows us to obtain is that space turns out to have a granular structure and in particular is endowed with elementary grains, quanta of space, having the size of Planck length, as a consequence of the link between, on one hand, the standard Klein–Gordon equation (5.44) and the standard Dirac equation (5.45) for subatomic particles and, on the other hand, the corresponding generalized Klein–Gordon equation (5.33) and generalized Dirac equation (5.37) for the density of cosmic space. The a-temporal quantum-gravity space theory shows in a clear and elegant way that the idea of a wave gravity-space and the idea of a granular (not infinitely divisible) space are not in contradiction, are not incompatible [388].

Moreover, as the author clearly showed in the paper [325], the fact that the Klein–Gordon and Dirac equations of standard quantum theory can be seen as special cases of the generalized Klein–Gordon and generalized Dirac equations for the density of cosmic space suggests a new way to face the matter regarding the emergence of quantum mechanics. On the basis of this approach, standard quantum mechanics (in the form of the standard Klein–Gordon and Dirac equation) is not considered as a fundamental theory but rather as a theory which emerges from a more fundamental descriptive level of reality, the quantum-gravity space characterized by Planck–Fiscaletti granules described by nonlinear generalized wave equations for the density of cosmic space.

Another significant result of the a-temporal quantum-gravity space theory is to provide a geometrization of gravity, a sort of unification of the gravitational and quantum aspects of the density of cosmic space in a geometric picture (that shows some analogy with the results of the approach to gravity and quantum obtained by A. Shojai and F. Shojai, for example in [389, 390]). By following the treatment provided in the papers [323–326], let us show in what sense, in the a-temporal quantum-gravity space approach, in the particular case $j = 0$ the wavefunction of quantum-gravity space (5.32) can be considered the element that determines a curved a-temporal space. In this regard, before all, by changing the ordinary partial derivative ∂_μ with the covariant derivative ∇_μ and by changing the Lorentz metric $\eta_{\mu\nu}$ with the curved metric $g_{\mu\nu}$ inside the quantum Hamilton–Jacobi equation for the wavefunction of quantum-gravity space

$$\partial_\mu S \partial^\mu S = \frac{\hbar^2 [D(r)]^2}{c^4} \exp(1 + Q) \tag{5.46}$$

(where

$$S = \frac{2\pi}{\hbar}\left(\frac{D(r)}{c^2}\hat{r}\cdot \mathbf{r}_0 - \frac{D(r)\cdot v \cdot sen\vartheta}{c^2}t + \varphi_0\right) \tag{5.47}$$

is the phase of the wavefunction (5.32) of quantum-gravity space), we obtain the equation valid in the general relativistic domain

$$g^{\mu\nu}\nabla_\mu S \nabla_\nu S = \frac{\hbar^2 [D(r)]^2}{c^4}\exp Q, \tag{5.48}$$

where

$$Q = \frac{c^4}{[D(r)]^2}\frac{\left(\nabla^2 - \frac{1}{c^2}\frac{\partial^2}{\partial t^2}\right)_g A}{A} \tag{5.49}$$

is the quantum potential. Here, in analogy with an interesting observation of de Broglie [391], Eq. (5.48) can also be written in the equivalent form

$$\frac{1}{\exp Q}g^{\mu\nu}\nabla_\mu S \nabla_\nu S = \frac{\hbar^2 [D(r)]^2}{c^4}. \tag{5.50}$$

From this relation it can be concluded that the quantum effects are identical with the change of the a-temporal space metric (in which time exists as a numerical order of material change) from $g_{\mu\nu}$ to $g_{\mu\nu} \to g_{\mu\nu}^I = \frac{g_{\mu\nu}}{\exp Q}$ which is a conformal transformation. Therefore, Eq. (5.50) can also be written in the form

$$g^{\mu\nu I} \nabla_\mu^I S \nabla_\nu^I S = \frac{\hbar^2 [D(r)]^2}{c^4}, \qquad (5.51)$$

where ∇_μ^I represents the covariant differentiation with respect to the metric $g_{\mu\nu}^I$. Moreover, in this new curved a-temporal space metric, the continuity equation assumes the form

$$g_{\mu\nu}^I \nabla_\mu^I J^\mu = 0, \qquad (5.52)$$

where

$$J^\mu = -\frac{c^2 A^2}{D(r)} \partial^\mu S \qquad (5.53)$$

is the current associated with the wavefunction of space (5.32) (that defines a congruence of an ensemble of particles connected with the wavefunction of quantum-gravity space (5.32), their density being given by J^0).

From the mathematical treatment of the case $j = 0$, the following important conclusion can be drawn: the presence of the quantum potential (5.49) associated with the wavefunction (5.32) of quantum-gravity space is equivalent to a curved a-temporal space with its metric being given by $g_{\mu\nu}^I = \frac{g_{\mu\nu}}{\exp Q}$. In other words, the curving of the a-temporal space can be considered an effect, in a certain sense, of the quantum potential associated with the wavefunction (5.32) of quantum-gravity space. Thus, a significant, relevant link between the wavefunction (5.32) of quantum-gravity space and the idea of the curved a-temporal space emerges. One can therefore say that in the a-temporal quantum-gravity space approach it is possible to provide a geometrization of the quantum aspects of a density of cosmic space $D(r)$ (in the case in which it determines the presence of a particle of spin zero) and thus an important unification, in a geometric picture, between the gravitational and quantum aspects of matter. This means that, on the basis of this approach to the gravitational space, it seems that there is a dual aspect in the role of geometry in physics. The a-temporal space geometry sometimes looks like what we call gravity and sometimes look like what we understand as quantum behavior, and the wavefunction of space represents a sort of link between these two elements, which indeed can be seen as two aspects of the same coin.

As regards the link between the wavefunction of space and the curvature of space, one can also say that the a-temporal quantum-gravity space approach implies an interesting as well as strange feedback loop in nature, which can be expressed as follows: "The wavefunction of quantum-gravity space tells the a-temporal space medium what it is and in turn the a-temporal space tells matter that derives from the wavefunction of quantum–gravity space how it must move". The reader might disbelieve this strange result. But there is no doubt that, as regards the feedback loop under consideration, the a-temporal quantum-gravity space approach has a certain analogy with general relativity. On the other hand, if in general relativity a varying density of matter and energy is referred to as curvature of space and determines the paths of moving particles which follow the curvature, in analogous way in the a-temporal quantum-gravity space approach the wavefunction associated with a density of cosmic space determines the curvature of space and thus the motion of particles. General relativity and the a-temporal quantum-gravity space approach contain an analogous feedback loop. This can be considered a relevant result of the a-temporal quantum-gravity space theory proposed by the author in the papers [323–327].

5.5 A Scalar-Tensor Description of the Gravitational Space in the A-Temporal Quantum-Gravity Space Theory

The following step of the a-temporal quantum-gravity space theory is the construction of a scalar-tensor description of the gravitational space. To review the fundamental features of this approach, we base our discussion on the papers [323, 325, 326].

A general relativistic system consisting of gravity and classical matter (namely without quantum effects) which emerges from a density of cosmic space $D(r)$ is associated with the action for the gravitational space given by relation

$$A_{no\text{-}quantum} = \frac{1}{2k}\int d^4x\sqrt{-g}R$$
$$+ \int d^4x\sqrt{-g}\frac{\hbar}{D(r)}\left(\frac{\rho}{\hbar^2}\partial_\mu S\partial^\mu S - [D(r)]^2\rho\right), \quad (5.54)$$

where $\rho = J^0$ is the ensemble density of the particles, $k = 8\pi G$ and hereafter we chose the units in which $c = 1$. Since the equivalence between the wavefunction of quantum-gravity space and the change of the a-temporal space metric is determined by the conformal transformation $g_{\mu\nu} \to g^I_{\mu\nu} = \frac{g_{\mu\nu}}{\exp Q}$, in order to characterize the quantum effects of the gravitational space, one can make this conformal transformation [392] by writing our action with quantum effects as:

$$A[\bar{g}_{\mu\nu}, \Omega, S, \rho, \lambda] = \frac{1}{2k}\int d^4x\sqrt{-\bar{g}}(\bar{R}\Omega^2 - 6\bar{\nabla}_\mu\Omega\bar{\nabla}^\mu\Omega)$$
$$+ \int d^4x\sqrt{-\bar{g}}\left(\frac{\rho}{\hbar D(r)}\Omega^2\bar{\nabla}_\mu S\bar{\nabla}^\mu S - \hbar D(r)\rho\Omega^4\right)$$
$$+ \int d^4x\sqrt{-\bar{g}}\lambda\left(\Omega^2 - \left(1 + \frac{\left(\nabla^2 - \frac{\partial^2}{\partial t^2}\right)\sqrt{\rho}}{[D(r)]^6\sqrt{\rho}}\right)\right)$$
$$(5.55)$$

where $\Omega^2 = \exp Q$, a bar over any quantity means that it corresponds to no-quantum regime and λ is a Lagrange multiplier introduced in order to identify the conformal factor with its Bohmian value.

Starting from the action (5.55), by the variation of the above action with respect to $\bar{g}_{\mu\nu}$, Ω, ρ, S and λ one arrives at the following relations as our equations of motion:

1. The equation of motion for Ω:

$$\bar{R}\Omega + 6\left(\bar{\nabla}^2 - \frac{\bar{\partial}^2}{\partial t^2}\right)\Omega$$
$$+ 2\frac{k}{\hbar D(r)}\rho\Omega(\bar{\nabla}_\mu S\bar{\nabla}^\mu S - \hbar^2[D(r)]^2\Omega^2) + 2k\lambda\Omega = 0. \quad (5.56)$$

2. The continuity equation for the particles associated with the density of cosmic space $D(r)$:

$$\bar{\nabla}_\mu(\rho\Omega^2\bar{\nabla}^\mu S) = 0. \tag{5.57}$$

3. The equation of motion for the particles associated with the density of cosmic space $D(r)$:

$$(\bar{\nabla}_\mu S\bar{\nabla}^\mu S - \hbar^2[D(r)]^2\Omega^2)\Omega^2\sqrt{\rho}$$

$$+ \frac{\hbar}{2D(r)}\left[\left(\bar{\nabla}^2 - \frac{\bar{\partial}^2}{\partial t^2}\right)\left(\frac{\lambda}{\sqrt{\rho}}\right) - \lambda\frac{\left(\bar{\nabla}^2 - \frac{\bar{\partial}^2}{\partial t^2}\right)\sqrt{\rho}}{\rho}\right] = 0.$$

$$\tag{5.58}$$

4. The modified Einstein equations for $\bar{g}_{\mu\nu}$:

$$\Omega^2\left[\bar{R}_{\mu\nu} - \frac{1}{2}\bar{g}_{\mu\nu}\bar{R}\right] - \left[\bar{g}_{\mu\nu}\left(\bar{\nabla}^2 - \frac{\bar{\partial}^2}{\partial t^2}\right) - \bar{\nabla}_\mu\bar{\nabla}_\nu\right]\Omega^2$$

$$- 6\bar{\nabla}_\mu\Omega\bar{\nabla}_\nu\Omega + 3\bar{g}_{\mu\nu}\bar{\nabla}_\alpha\Omega\bar{\nabla}^\alpha\Omega + \frac{2k}{\hbar D(r)}\rho\Omega^2\bar{\nabla}_\mu S\bar{\nabla}_\nu S$$

$$- \frac{k}{\hbar D(r)}\rho\Omega^2\bar{g}_{\mu\nu}\bar{\nabla}_\alpha S\bar{\nabla}^\alpha S + k\hbar D(r)\rho\Omega^4\bar{g}_{\mu\nu}$$

$$+ \frac{k}{[D(r)]^2}\left[\bar{\nabla}_\mu\sqrt{\rho}\bar{\nabla}_\nu\left(\frac{\lambda}{\sqrt{\rho}}\right) + \bar{\nabla}_\nu\sqrt{\rho}\bar{\nabla}_\mu\left(\frac{\lambda}{\sqrt{\rho}}\right)\right]$$

$$- \frac{k}{[D(r)]^2}\bar{g}_{\mu\nu}\bar{\nabla}_\alpha\left[\lambda\frac{\bar{\nabla}^\alpha\sqrt{\rho}}{\sqrt{\rho}}\right] = 0. \tag{5.59}$$

5. The constraint equation:

$$\Omega^2 = 1 + \frac{1}{[D(r)]^2}\frac{\left(\bar{\nabla}^2 - \frac{\bar{\partial}^2}{\partial t^2}\right)\sqrt{\rho}}{\sqrt{\rho}}. \tag{5.60}$$

The back-reaction effects of the quantum factor on the background are contained in the highly coupled Eqs. (5.56)–(5.60). By combining Eqs. (5.56) and (5.58) and by using the trace of (5.59), after some mathematical manipulations one arrives at equation

$$\lambda = \frac{1}{[D(r)]^2}\bar{\nabla}_\mu\left[\lambda\frac{\bar{\nabla}^\mu\sqrt{\rho}}{\sqrt{\rho}}\right]. \quad (5.61)$$

If one resolves Eq. (5.61) in a perturbative way in terms of the parameter $\alpha = \frac{1}{[D(r)]^2}$ by writing $\lambda = \lambda^{(0)} + \alpha\lambda^{(1)} + \alpha^2\lambda^{(2)} + \cdots$ and $\sqrt{\rho} = \sqrt{\rho}^{(0)} + \alpha\sqrt{\rho}^{(1)} + \alpha^2\sqrt{\rho}^{(2)} + \cdots$ the perturbative solution is $\lambda = 0$ which is its trivial solution. In this way the equations of the quantum-gravity space dynamics may be written as:

$$\bar{\nabla}_\mu(\rho\Omega^2\bar{\nabla}^\mu S) = 0, \quad (5.62)$$

$$\bar{\nabla}_\mu S \bar{\nabla}^\mu S = \hbar^2[D(r)]^2\Omega^2, \quad (5.63)$$

$$G_{\mu\nu} = -kT^{(m)}_{\mu\nu} - kT^{(\Omega)}_{\mu\nu}, \quad (5.64)$$

where $T^{(m)}_{\mu\nu}$ is the matter energy-momentum tensor and

$$kT^{(\Omega)}_{\mu\nu} = \frac{\left[g_{\mu\nu}\left(\nabla^2 - \frac{\partial^2}{\partial t^2}\right) - \nabla_\mu\nabla_\nu\right]\Omega^2}{\Omega^2} + 6\frac{\nabla_\mu\Omega\nabla_\nu\Omega}{\omega^2}$$

$$- 3g_{\mu\nu}\frac{\nabla_\alpha\Omega\nabla^\alpha\Omega}{\Omega^2} \quad (5.65)$$

and

$$\Omega^2 = 1 + \alpha\frac{\overline{\left(\nabla^2 - \frac{\partial^2}{\partial t^2}\right)\sqrt{\rho}}}{\sqrt{\rho}}. \quad (5.66)$$

Here, (5.63) is a Bohmian-type equation of motion, and if we write it in terms of the physical metric $g_{\mu\nu}$, it reads as

$$\nabla_\mu S \nabla^\mu S = \hbar^2[D(r)]^2. \quad (5.67)$$

The next step is to make dynamical the conformal factor and the quantum potential of the quantum-gravity space. In this regard, one can start from the most general scalar-tensor action

$$A = \int d^4x \left\{ \phi R - \frac{\omega}{\phi} \nabla^\mu \phi \nabla_\mu \phi + 2\Lambda\phi + L_m \right\} \quad (5.68)$$

in which ω is a constant independent of the scalar field, and Λ is the cosmological constant. The equations of motion are

$$R + \frac{2\omega}{\phi}\left(\nabla^2 - \frac{1}{c^2}\frac{\partial^2}{\partial t^2}\right)\phi - \frac{\omega}{\phi^2}\nabla^\mu\phi\nabla_\mu\phi + 2\Lambda + \frac{\partial L_m}{\partial \phi} = 0, \quad (5.69)$$

$$G^{\mu\nu} - \Lambda g^{\mu\nu} = -\frac{1}{\phi}T^{\mu\nu} - \frac{1}{\phi}\left[\nabla^\mu\nabla^\nu - g^{\mu\nu}\left(\nabla^2 - \frac{1}{c^2}\frac{\partial^2}{\partial t^2}\right)\right]\phi$$
$$+ \frac{\omega}{\phi^2}\nabla^\mu\phi\nabla^\nu\phi - \frac{1}{2}\frac{\omega}{\phi^2}g^{\mu\nu}\nabla^\alpha\phi\nabla_\alpha\phi. \quad (5.70)$$

The scalar curvature can be evaluated from the contracted form of (5.70), and it can be substituted in the relation (5.69). Then, we have:

$$\frac{2\omega - 3}{\phi}\left(\nabla^2 - \frac{1}{c^2}\frac{\partial^2}{\partial t^2}\right)\phi = -\frac{T}{\phi} + 2\Lambda - \frac{\partial L_m}{\partial \phi}. \quad (5.71)$$

The density of cosmic space lagrangian without any quantum contribution is

$$L_{m(no\text{-}quantum)} = \frac{\rho c^3}{\hbar D(r)}\nabla_\mu S \nabla^\mu S - \frac{\rho \hbar D(r)}{c^3}. \quad (5.72)$$

This lagrangian can be generalized if one assumes that there is some interaction between the scalar field and the matter field. In the scalar-tensor description of the gravitational space provided by the author in the papers [323, 325, 326], for simplicity, one assumes that the interaction between the scalar field and the matter field is in the form of powers of ϕ. The quantum effects are introduced by adding terms containing the quantum potential. If one assumes that, on the basis of physical intuition, some interaction between the cosmological constant and the quantum potential of the quantum-gravity space exists, the density of cosmic space lagrangian may be written as:

$$L_m = \frac{\rho c^3}{\hbar D(r)}\phi^a \nabla_\mu S \nabla^\mu S - \frac{\rho \hbar D(r)\phi^b}{c^3} - \Lambda(1+Q)^f \quad (5.73)$$

in which a, b, f are constant that can be fixed with the following precedure. Using the density of cosmic space lagrangian (5.73) and

Gravitational Space in A-Temporal Quantum-Gravity Space Theory   283

the energy-momentum tensor

$$\begin{aligned}T^{\mu\nu} &= -\frac{1}{\sqrt{-g}}\frac{\delta}{\delta g_{\mu\nu}}\int d^4x\sqrt{-g}L_m\\ &= -\frac{1}{2}g^{\mu\nu}L_m + \frac{\rho c^3}{\hbar D(r)}\phi^a\nabla^\mu S\nabla^\nu S - \frac{1}{2}\Lambda fQ(1+Q)^{f-1}g^{\mu\nu}\\ &\quad -\frac{1}{2}\alpha\Lambda f\nabla_\alpha\sqrt{\rho}\nabla_\beta\left(\frac{(1+Q)^{f-1}}{\sqrt{\rho}}\right)[g^{\mu\nu}g^{\alpha\beta}-g^{\alpha\mu}g^{\beta\nu}-g^{\beta\mu}g^{\alpha\nu}],\end{aligned}$$
(5.74)

one can calculate the first and third terms in the relation (5.68). The other equations, namely the continuity equation and the quantum Hamilton–Jacobi equation, are expressed respectively as:

$$\nabla^\mu(\rho\phi^a\nabla_\mu S) = 0,\qquad(5.75)$$

$$\begin{aligned}\nabla^\mu S\nabla_\mu S &= \frac{\hbar^2[D(r)]^2}{c^6}\phi^{b-a} - \frac{\Lambda\hbar D(r)Q}{2c^2\rho\phi^a}(1+Q)^{f-1}\\ &\quad + \frac{1}{2}\frac{\Lambda\hbar D(r)}{c^2\sqrt{\rho}\phi^a}\left(\nabla^2 - \frac{1}{c^2}\frac{\partial^2}{\partial t^2}\right)\left(\frac{(1+Q)^{f-1}}{\sqrt{\rho}}\right).\end{aligned}$$
(5.76)

In order to simplify the calculations, one can choose ω to be $3/2$. Then a perturbative expansion for the scalar field and matter distribution density can be used as:

$$\begin{aligned}\phi &= \phi_0 + \alpha\phi_1 + \cdots\\ \sqrt{\rho} &= \sqrt{\rho_0} + \alpha\sqrt{\rho_1} + \cdots\end{aligned}$$
(5.77)

In the zeroth order approximation, the scalar field equation gives: $b = a + 1$; $\phi_0 = 1$. In the first order approximation one obtains:

$$\alpha\phi_1 = \frac{f}{2}(1-a)Q + \frac{a}{2}f\tilde{Q},\qquad(5.78)$$

where

$$\tilde{Q} = \alpha\frac{\nabla_\mu\sqrt{\rho}\nabla^\mu\sqrt{\rho}}{\rho}.\qquad(5.79)$$

Since the scalar field is the conformal factor on the a-temporal space metric and some arguments show that this field is a function of the quantum potential associated with the density of cosmic space, one might choose the constant a equal to zero. Then the scalar field is independent of $\tilde{Q}$ and we have:

$$\alpha \phi_1 = \frac{f}{2} Q. \tag{5.80}$$

Also the Bohmian equations of motion give:

$$\nabla^\mu S \nabla_\mu S = \frac{\hbar^2 [D(r)]^2}{c^4}(1 + fQ/2) - \frac{\Lambda \hbar D(r) f(Q - \tilde{Q})}{c^2 \rho_0}. \tag{5.81}$$

It is necessary to choose $f = 2$ in order that the first term on the right hand be the same as the quantum term $\frac{\hbar D(r)}{c^3} \sqrt{\exp Q}$. These choices for the parameters a, b and f lead to the nonperturbative quantum gravity equations as follows:

$$\phi = 1 + Q - \frac{\alpha}{2} \left(\nabla^2 - \frac{1}{c^2} \frac{\partial^2}{\partial t^2} \right) Q, \tag{5.82}$$

$$\nabla^\mu S \nabla_\mu S = \frac{\hbar^2 [D(r)]^2}{c^4} \phi - \frac{2\Lambda \hbar D(r)}{c^2 \rho}(1 + Q)(Q - \tilde{Q})$$

$$+ \frac{\alpha \Lambda \hbar D(r)}{c^2 \rho} \left[\left(\nabla^2 - \frac{1}{c^2} \frac{\partial^2}{\partial t^2} \right) Q - 2 \nabla_\mu Q \frac{\nabla^\mu \sqrt{\rho}}{\sqrt{\rho}} \right], \tag{5.83}$$

$$\nabla^\mu (\rho \nabla_\mu S) = 0, \tag{5.84}$$

$$G^{\mu\nu} - \Lambda g^{\mu\nu} = -\frac{1}{\phi} T^{\mu\nu} - \frac{1}{\phi} \left[\nabla^\mu \nabla^\nu - g^{\mu\nu} \left(\nabla^2 - \frac{1}{c^2} \frac{\partial^2}{\partial t^2} \right) \right] \phi$$

$$+ \frac{\omega}{\phi^2} \nabla^\mu \phi \nabla^\nu \phi - \frac{1}{2} \frac{\omega}{\phi^2} g^{\mu\nu} \nabla^\alpha \phi \nabla_\alpha \phi. \tag{5.85}$$

We can call these equations as the generalized quantum gravity equations for the density of cosmic space which provide a scalar-tensor description of the gravitational space in the a-temporal quantum-gravity space theory.

The scalar-tensor version of a-temporal quantum-gravity space (synthesized in Eqs. (5.82)–(5.85)) allows us to draw some important conclusions:

— On the basis of Eq. (5.85), the causal structure of the a-temporal space metric $g^{\mu\nu}$ is determined by the density of cosmic space.

Gravitational Space in A-Temporal Quantum-Gravity Space Theory   285

Equation (5.82) implies that quantum effects determine directly the scale factor of the a-temporal space metric.
— The density of cosmic space field given by the right-hand side of Eq. (5.83) consists of two parts. One part, which is proportional to α, is a purely quantum effect, while the other part, which is proportional to $\alpha\Lambda$, is a mixture of the quantum effects and the large scale structure introduced via the cosmological constant.
— In this model, the scalar field produces the quantum force which appears on the right hand and violates the equivalence principle (just like, in Kaluza–Klein theory, the scalar field — dilaton — produces a fifth force which leads to the violation of the equivalence principle [393]).

5.6 The Geometrization of the Quantum Effects and the Generalized Equivalence Principle in the A-Temporal Quantum-Gravity Space Theory

One of the fundamental aspects of the a-temporal quantum-gravity space theory provided by the author in the papers [323–327] lies in the geometrization of the quantum effects. On the basis of this approach, one can say that there is a dual role of the geometry in physics. The gravitational effects associated with the density of cosmic space determine the causal structure of a-temporal space as long as quantum effects give its conformal structure. Quantum effects can act on the causal structure through back-reaction terms appearing in the metric field equations (in this regard it is important to mention that a similar situation happens in the model of toy quantum gravity in curved space-time developed by Shojai F., Shojai A. and Golshani [394–396]). The dominant role in the causal structure belongs to the gravitational effects and thus to the density of cosmic space. The same is true for the conformal factor. The conformal factor of the metric is a function of the quantum potential of the quantum-gravity space and the mass of a relativistic particle is a field produced by quantum corrections to the classical mass. In the toy model developed by the author the presence of the quantum potential is equivalent to a conformal mapping of the metric of the a-temporal space. Thus in conformally related frames one measures different quantum masses and different curvatures. One can say that different conformal frames are equivalent pictures of the gravitational and quantum phenomena.

Considering the quantum force, the conformally related frames are not distinguishable just like, as regards the treatment of gravity in general relativity, different coordinate systems are equivalent. Since the conformal transformation changes the length scale locally, one measures different quantum forces in different conformal frames. This is analogous to what happens in general relativity in which general coordinate transformation changes the gravitational force at any arbitrary point. One can say that according to this approach the geometrization of quantum effects implies conformal invariance just as gravitational effects imply the general coordinate invariance. If in the development of general relativity, the general covariance principle leads to the identification of gravitational effects of matter with the geometry of space-time and the crucial element which supports this identification is the equivalence principle (according to which one can always remove the gravitational field at some point by a suitable coordinate transformation), in a similar way, in the a-temporal quantum gravity space developed by the author in the papers [323–327], at any point (or even globally) the quantum effects of a density of cosmic space can be removed by a suitable conformal transformation. Thus in that point(s) the matter determined by this density of cosmic space behaves classically. In this way inside this approach one can introduce a new quantum equivalence principle, similar to the standard equivalence principle, the conformal equivalence principle. According to the conformal equivalence principle gravitational effects can be removed by going to a freely falling frame while quantum effects can be eliminated by choosing an appropriate scale. Einstein's equivalence principle interconnects gravity and general covariance while the conformal equivalence principle has the same role about quantum and conformal covariance. Both these principles state that there is no preferred frame, either coordinate or conformal.

Moreover, according to the model suggested by the author, the quantum potential associated with the density of cosmic space justifies Mach's principle leading to the existence of interrelation between the global properties of the universe (space-time structure, the large scale structure of the universe) and its local properties (local curvature, motion in a local frame, etc.). In the a-temporal quantum-gravity space approach, it can be easily seen that the space geometry is determined by the density of cosmic space. A local variation of the density of cosmic space changes the quantum potential acting on the geometry. Thus the geometry is altered globally (in conformity with Mach's principle). In this sense the quantum gravity-space approach is highly nonlocal as it is forced by the nature of the quantum potential. What we call geometry is only the gravitational and quantum effects of a density of cosmic space. Without density of cosmic space the geometry of the gravitational space would be meaningless.

5.7 The Generalized Dirac Equation with Electromagnetic Interaction and the Curvature of the A-Temporal Space

In the papers [324–327], the author provided a treatment of the a-temporal quantum-gravity space characterized by a density of cosmic space which determines the presence of a particle of spin $\frac{1}{2}$ which is subjected to an electromagnetic interaction. This treatment is based on the following generalized Dirac equation

$$\left(i\gamma^\mu \partial_\mu - \frac{D_e(r) l_p^3 G}{K c^4} A_\mu - \frac{D(r)}{c^2} \right) \psi_S = 0, \qquad (5.86)$$

where

$$D_e(r) = \frac{Kq}{l_p^3 r^2} \qquad (5.87)$$

is the electric density of space (physical quantity which indicates the amount of charge characterizing the region of space into consideration and therefore which indicates the electric properties of this region), $K = \frac{1}{4\pi\varepsilon_0}$ is the electrostatic constant, indicating the strength of the electric force (with ε_0 being the dielectric constant of the vacuum), r is the distance from the centre of the charge q, A_μ is the electromagnetic potential. Equation (5.86) can be called as the generalized Dirac equation for the density of cosmic space coupled with electromagnetic interaction. In this approach, the well known Dirac equation for relativistic particles of spin $\frac{1}{2}$ with electromagnetic interaction

$$\left(i\gamma^\mu\partial_\mu - \frac{q}{c\hbar}A_\mu - \frac{mc}{\hbar}\right)\psi = 0 \qquad (5.88)$$

can be seen as a particular case of the more general Eq. (5.86) in the sense that it can be directly obtained from (5.86) in the conditions $\frac{q}{c\hbar} = \frac{D_e(r)l_p^3 G}{Kc^4}$ (namely $D_e(r) = \frac{Kqc^3}{\hbar G l_p^3}$) and $\frac{mc}{\hbar} = \frac{D(r)}{c^2}$ (namely $D(r) = \frac{mc^3}{\hbar}$) which together correspond to the condition $r^2 = l_p^2$. As a consequence, Eq. (5.88) can be interpreted as the equation of the density of cosmic space coupled with electric density of space and electromagnetic interaction for the particular points of space situated at distance $r = l_p$ from the centre of the density of cosmic space and the electric density of space into consideration. Moreover, in this picture, the standard wavefunctions of Dirac particles with electromagnetic interaction can be seen as particular cases of more general wavefunctions of quantum-gravity space coupled with electromagnetic interaction for the particular points situated at distance $r = l_p$ from the centre of the density of cosmic space and the electric density of space into consideration. The generalized Dirac equation with electromagnetic interaction (5.86) can be resolved, analogously to the corresponding standard Dirac equation of quantum theory, by introducing a Feynman propagator inside a perturbative method with respect to the electric density of space (5.87). In absence of the electric density of space we have the wavefunctions of quantum-gravity space without electromagnetic interaction, solutions to the generalized Dirac equation (5.37) (these solutions, on the basis of this perturbative treatment, can be interpreted as approximations at the zeroth order of the solutions to the generalized Dirac equation with electromagnetic interaction (5.86)). When the electric density of space is present and is lead to its physical value, we have the solutions to the generalized Dirac equation for the density of cosmic space with electromagnetic interaction (5.86) at the first order.

As regards the behavior of a fermion of rest mass m and charge e in a curved space-time in the presence of electromagnetic interaction, in 1929 Fock [397] found that the wavefunction ψ of the fermion obeys the following equation

$$\left(i\gamma^k\left(\partial_k - \Gamma_k - \frac{ieA_k}{c\hbar}\right) - \frac{mc}{\hbar}\right)\psi = 0, \qquad (5.89)$$

where γ_i are the generalized gamma matrices defining the covariant Clifford algebra [398, 399], $\gamma_i\gamma_j + \gamma_j\gamma_i = 2g_{ij}$, g_{ij} is the space-time metric, Γ_i is the spinorial affine connection and A_i is the electromagnetic four-vector potential. According to the a-temporal quantum-gravity space theory, by following the treatment provided in the papers [324, 327], Fock's equation (5.89) can be seen as a special case of a more general equation of the form

$$\left(i\gamma^k\left(\partial_k - \Gamma_k - \frac{iD_e(r)l_p^3 GA_k}{Kc^4}\right) - \frac{D(r)}{c^2}\right)\psi_S = 0, \qquad (5.90)$$

which can be defined as the generalized Dirac equation for the wavefunction of quantum-gravity space in the curved a-temporal space (characterized by a density of cosmic space and an electric density of space which determine the appearance of a particle of spin $\frac{1}{2}$). In the a-temporal space, the electromagnetic gauge contribution to the generalized Dirac equation (5.90) for a charged spinor corresponds to the separate entity $e\gamma^k A_k$ which can be automatically incorporated into the geometrical connection Γ_i, for an arbitrary background four-vector potential A_i. That is, by changing the definition of Γ_i, one can express equation (5.90) in the following form

$$\left(i\gamma^k\left(\partial_k - \Gamma_k^I\right) - \frac{D(r)}{c^2}\right)\psi_S = 0, \qquad (5.91)$$

where

$$\Gamma_k^I = \Gamma_k + \frac{iD_e(r)l_p^3 GA_k}{Kc^4}. \qquad (5.92)$$

As a consequence, the gauge symmetry of the standard Dirac equation (enunciated by Weyl [400] in Minkowski space-time) can be generalized, in a geometric perspective, in a curved a-temporal space described by the generalized affine connection (5.92) where it appears as an invariance under the simultaneous transformations

$$\partial_k \to \partial_k - i\Gamma_k^{II}, \psi(x^i) \to \psi^I(x^i) = e^{i\int \Gamma_k^{II} dx^k} \psi(x^i), \qquad (5.93)$$

where

$$\Gamma_k^{II} = \frac{D_e(r) l_p^3 G A_k}{K c^4} \mathbf{1}, \qquad (5.94)$$

(here **1** is the usual unit matrix), implying $\Gamma_k^{II+} = \Gamma_k^{II}$ and hence invariance of the probability current. An interesting result obtained inside the approach of the wavefunction of quantum-gravity space in the curved a-temporal space coupled with electromagnetic interaction based on Eqs. (5.90)–(5.94) lies in the possibility to obtain the geometrical connection of the standard quantum theory $\Gamma_k^{II} = ieA_k \mathbf{1}$ proposed recently by Pollock [401] for the standard Dirac equation as a special case of the generalized affine connection (5.94). The geometrical connection of standard quantum theory suggested by Pollock emerges from the generalized affine connection (5.94) just in the particular case $r = l_p$. Planck's constant emerges thus as the fundamental constant which allows us to explore the link between the domain of standard quantum theory and the domain of the a-temporal quantum-gravity space theory proposed by the author, also as regards the treatment of motion in a curved a-temporal space in the presence of electromagnetic interaction.

It is also possible to extend the treatment based on the generalized Dirac equation with electromagnetic interaction in such a way to include the Kaluza–Klein mechanism [402, 403] in the picture of a five-dimensional (5D) curved space. In this regard, before all, in analogy to a result found by Schrödinger [404], one can introduce a modified Klein–Gordon equation by squaring the operator in equation (5.90):

$$\left(\frac{1}{\sqrt{-g}} D_k^I \left(\sqrt{-g} g^{kl} D_l^I \right) - \frac{1}{4} R - \frac{i D_e(r) l_p^3 G}{2 K c^4} F_{ij} j^{ij} + \frac{D(r)}{c^2} \right) \psi = 0, \qquad (5.95)$$

where $D_k^I = \partial_k - \Gamma_k^I$, R is the Ricci scalar and $F_{jk} = \partial_j A_k - \partial_k A_j$ is the electromagnetic field tensor. The presence of the term

$\frac{iD_e(r)l_p^3 G}{2Kc^4} F_{ij} j^{ij}$ derives from the fact that in the presence of an electron the rotation-orientation of space is quantized and is given by odd multiples of $j = \frac{G\hbar}{2r^3}$. On the basis of Eq. (5.95) one can say that the appearance of an electron derives just from the term $\frac{iD_e(r)l_p^3 G}{2Kc^4} F_{ij} j^{ij}$. Here, the term $\frac{iD_e(r)l_p^3 G}{2Kc^4} F_{ij} j^{ij}$ can reproduce and explain a set of significant experimental results in ferromagnetic materials, especially the phenomenon of diamagnetism, and also some X-ray crystallographic experiments [405, 406], the rotation of the plane of polarization by optically active substances [407] and the helical motion revealed by particle tracks in the cloud-chamber observations by Wilson.

The interaction term $F_{ij} j^{ij}$ has a relevant role also as regards the generalized Dirac equation for the density of cosmic space with electromagnetic interaction. In fact, in this picture the generalized Dirac equation with electromagnetic interaction can be expressed as follows

$$\left(i\gamma^k \left(D_k - \frac{iD_e(r)l_p^3 G A_k}{Kc^4} \right) - \frac{il_0}{2} F_{ij} j^{ij} - \frac{D(r)}{c^2} + \cdots \right) \psi_S = 0,$$
(5.96)

where $l_0 = \sqrt{\alpha_4}\beta/4$, $D_k = \partial_k - \Gamma_k$ and it has been made a reduction from a 5D Kaluza–Klein manifold without density of cosmic space and electric density of space to a 4D space via the ansatz

$$\hat{g}_{ij} = g_{ij} + \beta^2 A_i A_j \hat{g}_{44},$$

$$\hat{g}^{ij} = g^{ij}, \frac{\hat{g}_{i4}}{\hat{g}_{44}} = \beta A_i,$$
(5.97)

$$\hat{g}^{i4} = -\beta A^i, \quad |\hat{g}_{44}| = \alpha_4,$$

in which $\alpha_4 \beta^2 = 2\kappa^2$ and $\kappa^2 = 8\pi(m_p)^{-2}$ is the 4D gravitational coupling and m_p is the Planck mass.

The extra term $-\frac{il_0}{2} F_{ij} j^{ij}$ can be defined as an "anomalous electromagnetic interaction energy" with the quantum-gravity space. In standard quantum theory already Pauli [408, 409] studied an anomalous interaction term of this type by underlining that the requirements of relativistic invariance, gauge invariance and correspondence do not determine the Dirac equation uniquely in the Minkowski space limit. In the treatment of the wavefunction of quantum-gravity space in a curved a-temporal space coupled with electromagnetic interaction, the "anomalous electromagnetic interaction energy" can be understood in a simple way by assuming that in the presence of an electromagnetic interaction the effective density of cosmic space becomes

$$D_{eff}(r) = D(r) + \frac{icl_0}{2\hbar}F_{ij}j^{ij} = D(r) + \frac{c}{\hbar}\mu_{\text{anomalous}} \cdot \mathbf{B} - \frac{c}{\hbar}il_0\gamma^0\boldsymbol{\gamma} \cdot \mathbf{E}, \quad (5.98)$$

where the electric and the magnetic three-vector fields are defined by

$$E_\alpha = -F_{0\alpha} \quad \text{and} \quad B_\alpha = \frac{1}{2}\frac{\delta^{\alpha\beta\gamma}}{\sqrt{-g/g_{00}}}F_{\beta\gamma} \quad (5.99)$$

respectively, while the spatial component of the rotation-orientation operator can be written as

$$j_{\alpha\beta} = -i\frac{G\hbar}{r^3}\frac{\delta^{\alpha\beta\gamma}}{\sqrt{-g/g_{00}}}\sigma^\gamma, \quad (5.100)$$

which follows from the commutator of the Pauli matrices σ_α.

Ignoring the electric field, on the basis of Eq. (5.98) we can say that a region of space characterized by a density of cosmic space (but without an electric density of space) is endowed with an anomalous magnetic moment defined by the relation

$$\boldsymbol{\mu}_{\text{anomalous}} = l_0\boldsymbol{\sigma}, \quad (5.101)$$

and thus that this region is describable by means of an effective density of cosmic space which turns out to be increased by the amount $\boldsymbol{\mu}_{\text{anomalous}} \cdot \mathbf{B}$ in an external magnetic field.

In analogous way, also starting from the modified Klein–Gordon equation (5.95) one can define an effective density of cosmic space given by relation

$$D_{eff}(r) = D(r) - \frac{1}{2}\frac{c}{\hbar}\frac{D_e(r)l_p^3 G}{K}F_{ij}j^{ij} \approx D(r) - \frac{c}{\hbar}\frac{D_e(r)l_p^3 G}{K}\sigma_\alpha B^\alpha. \quad (5.102)$$

In the nonrelativistic limit $\left|\frac{D_e(r)l_p^3 G}{Kc^4}\sigma_\alpha B^\alpha\right| \ll \frac{\hbar^2(D(r))^2}{c^2}$, Eq. (5.102) implies a change in the effective mass of the electron which is compatible with an intrinsic magnetic moment $\mu = \frac{e\hbar}{2mc}$.

Gravitational Space in A-Temporal Quantum-Gravity Space Theory   293

By introducing the anomalous electromagnetic interaction energy term in the generalized Dirac equation with electromagnetic interaction, we obtain the following equation

$$\left(i\gamma^k \left(\partial_k - \frac{iD_e(r)l_p^3 G A_k}{Kc^4}\right) - \frac{D_{eff}(r)}{c^2}\right)\psi_S = 0, \quad (5.103)$$

where

$$D_{eff}(r) = D(r) + \frac{icl_0}{2\hbar}F_{ij}j^{ij}. \quad (5.104)$$

Here, the net magnetic moment can be obtained by squaring the operator in Eq. (5.102):

$$\left(\partial'_k \partial'^k - \frac{1}{2}l_0\gamma^k \partial_k(F_{ij}j^{ij}) + \frac{D'_{eff}(r)}{c^2}\right)\psi_S = 0, \quad (5.105)$$

where $\partial'_k = \partial_k - \frac{iD_e(r)l_p^3 G A_k}{Kc^4}$ and

$$D'_{eff}(r) = D_{eff}(r) - \frac{D_e(r)l_p^3 G}{2\hbar K c^3}F_{ij}j^{ij}$$

$$= D(r) - \frac{iD_e(r)l_p^3 G}{2\hbar K c^3}\left(1 - \sqrt{\frac{1}{\alpha}\frac{D(r)r^2}{Gm_p}}\right)F_{ij}j^{ij}$$

$$- \frac{c}{4\hbar}l_0^2(F_{ij}j^{ij})^2, \quad (5.106)$$

where $\alpha = \frac{e^2}{4\pi\hbar c\varepsilon_0}$ is the fine structure constant, ε_0 is the permittivity of free space.

Equation (5.106) implies that the anomalous electromagnetic interaction produces a correction to the intrinsic magnetic moment of quantum-gravity space by the factor $1 + \delta$ where

$$\delta = -\sqrt{\frac{1}{\alpha}\frac{D(r)r^2}{Gm_p}}, \quad (5.107)$$

which, in the particular case $r = l_p$, allows us to obtain the following correction to the intrinsic magnetic moment of the electron:

$$\delta \approx -4,90 \cdot 10^{-22}. \quad (5.108)$$

Quantum electrodynamics, on the other hand, leads to an anomalous magnetic moment of the electron which, according to Schwinger's calculations, is

$$\delta_{\text{QED}} = \frac{\alpha}{2\pi} \approx -1,16 \cdot 10^{-3}. \qquad (5.109)$$

The result (5.108) which derives from (5.107) of the a-temporal quantum-gravity space approach, turns out to be smaller with respect to Schwinger's result (5.109) by a factor $-4,2 \cdot 10^{-19}$. Although the quantity (5.108) is beyond our current technology and not susceptible to experimental tests, it is however relevant that the 5D Kaluza–Klein theory differs in its predictions from the 4D theory by the presence of this additional anomalous electromagnetic energy term of the quantum-gravity space. The a-temporal quantum-gravity space suggested by the author suggests the possibility to interpret the anomalous magnetic moment of the electron as the result, at a more "macroscopic" level, of the wavefunction of quantum-gravity space in a curved space in the presence of electromagnetic interaction. The derivation, on the basis of a generalized Dirac equation with electromagnetic interaction (and with the presence of the anomalous electromagnetic interaction energy term) of the intrinsic magnetic moment of the electron (as a consequence of a more fundamental intrinsic magnetic moment of quantum-gravity space) can be considered a very interesting perspective of the a-temporal quantum-gravity space theory.

In synthesis, the approach based on a generalized Dirac equation for the density of cosmic space with electromagnetic interaction leads to a treatment of the curvature of the a-temporal space which allows us to interpret and re-read some important results obtained as regards the Dirac equation (Fock's formalism regarding the wavefunction of a fermion in a curved space-time, the anomalous electromagnetic interaction energy which emerges by applying a Kaluza–Klein mechanism and the resulting intrinsic magnetic moment of the electron) in a new holistic picture.

5.8 Towards an Unification of the Generalized Klein–Gordon and Generalized Dirac Equations

In the recent article *Dirac and Klein–Gordon equations in curved space*, A. D. Alhaidari and A. Jellal provided a new systematic formulation of the Dirac and Klein–Gordon equations in a curved space by introducing a matrix operator algebra involving the Dirac gamma matrices with a universal length scale constant as a measure of the curvature of space. These two authors found that spin connections or vierbeins are no longer required in order to derive the Dirac equation and that the Klein–Gordon equation emerges in its canonical form without first order derivatives [410]. By following the philosophy that is on the basis of Alhaidari's and Jellal's approach, in the paper [327] the author of this book developed a suggestive unification of the generalized Klein–Gordon and generalized Dirac equations for the density of cosmic space in a scheme based on a matrix operator algebra of the density of cosmic space describing the curvature of the a-temporal space.

The unifying picture of the generalized Klein–Gordon and generalized Dirac equations starts by introducing the following version of the generalized Dirac equation for the wavefunction of quantum-gravity space in the curved a-temporal space:

$$(i\gamma^\mu \partial_\mu + \lambda\beta)\psi_S = \frac{D(r)}{c^2}\psi_S, \tag{5.110}$$

where β is a space dependent matrix, λ is a universal real constant of inverse length dimension that gives a measure of the curvature of space (e.g., the inverse of the "effective radius of curvature" of space). The generalized Klein–Gordon equation can be derived directly from equation

$$[-\gamma^\mu\gamma^\nu \partial_\mu \partial_\nu + (-\gamma^\mu \partial_\mu \gamma^\nu + i\lambda\{\beta,\gamma^\nu\})\partial_\nu + i\lambda\gamma^\mu \partial_\mu \beta + \lambda^2 \beta^2]\psi_S$$

$$= \left(\frac{D(r)}{c^2}\right)^2 \psi_S \tag{5.111}$$

(which is obtained by squaring Eq. (5.110)) under appropriate conditions for the matrix β. In fact, by defining $\hat{D} = \gamma^\mu \partial_\mu$ and by requiring that

$$-\frac{i}{\lambda}\hat{D}\gamma^\mu = \{\beta, \gamma^\mu\}, \tag{5.112}$$

$$-\frac{i}{\lambda}\hat{D}\beta = \{\beta, \beta\} = \beta^2, \tag{5.113}$$

Eq. (5.111) becomes

$$\left(g^{\mu\nu}\partial_\mu\partial_\nu + \lambda^2\beta^2 - \frac{[D(r)]^2}{c^4}\right)\psi_S = 0, \tag{5.114}$$

which is the generalized Klein–Gordon equation for the density of cosmic space in a curved a-temporal space in its simple canonical form (namely with no first order derivatives). Equation (5.110) with the $n+2$ matrices $\{\beta, \gamma^\mu\}_{\mu=0}^n$ satisfying the matrix algebra conditions (5.112) and (5.113) can be considered as the generalized Dirac equation for the density of cosmic space in a curved a-temporal space whose metric is $g^{\mu\nu}$ and whose curvature parameter is λ. The corresponding generalized Klein–Gordon equation for the density of cosmic space in a curved a-temporal space (which determines the appearance of a spinless particle) is Eq. (5.114). One of the advantages of the algebra defined by Eqs. (5.112) and (5.113) is the absence of first order derivatives in the resulting generalized Klein–Gordon equation. One can also remark that in the flat Lorentz space limit ($\lambda \to 0$) the generalized Dirac and Klein–Gordon equations for the density of cosmic space in the domain of special relativity emerge provided that $\lim_{\lambda \to 0} g^{\mu\nu} = \eta^{\mu\nu}$ where $\eta_{\mu\nu}$ is the flat a-temporal space metric.

The extension of the unified approach of generalized Klein–Gordon and generalized Dirac equations for the density of cosmic space based on a matrix operator algebra to the a-temporal space characterized by an electromagnetic interaction is straightforward. In the presence of electromagnetic interaction the unifying picture of the generalized Klein–Gordon and Dirac equations emerges directly from the following version of the generalized Dirac equation with electromagnetic interaction in the curved a-temporal space:

$$(i\gamma^\mu \partial_\mu + \lambda\beta)\psi_S = \left(\frac{D_e(r)l_p^3 G}{Kc^4} A_\mu + \frac{D(r)}{c^2}\right)\psi_S. \quad (5.115)$$

In this case, squaring Eq. (5.115) yields

$$[-\gamma^\mu\gamma^\nu \partial_\mu \partial_\nu + (-\gamma^\mu \partial_\mu \gamma^\nu + i\lambda\{\beta, \gamma^\nu\})\partial_\nu + i\lambda\gamma^\mu \partial_\mu \beta + \lambda^2\beta^2]\psi_S$$

$$= \left(\frac{(D_e(r))^2 l_p^6 G^2}{K^2 c^8}(A_\mu)^2 + \frac{(D(r))^2}{c^4} + \frac{2D_e(r)D(r)l_p^3 G}{Kc^6} A_\mu\right)\psi_S.$$
$$(5.116)$$

Just like in the case of the density of cosmic space without electromagnetic interaction, also in the presence of electromagnetic interaction a generalized Klein–Gordon equation with electromagnetic interaction can be derived under the conditions (5.112) and (5.113) for the matrix β. Under those conditions Eq. (5.116) becomes:

$$[g^{\mu\nu}\partial_\mu \partial_\nu + \lambda^2\beta^2]\psi_S$$

$$= \left(\frac{(D_e(r))^2 l_p^6 G^2}{K^2 c^8}(A_\mu)^2 + \frac{(D(r))^2}{c^4} + \frac{2D_e(r)D(r)l_p^3 G}{Kc^6} A_\mu\right)\psi_S$$
$$(5.117)$$

which is the generalized Klein–Gordon equation for the density of cosmic space in a curved a-temporal space in its simple canonical form (namely with no first order derivatives). Equation (5.117) with the $n+2$ matrices $\{\beta, \gamma^\mu\}_{\mu=0}^n$ satisfying the matrix algebra conditions (5.112) and (5.113) can be considered as the generalized Dirac equation for the density of cosmic space with electromagnetic interaction in a curved a-temporal space whose metric is $g^{\mu\nu}$ and whose curvature parameter is λ.

Moreover, inside the unified approach based on Eqs. (5.110)–(5.117), the matrix operator algebra introduced by Alhaidari and Jellal for the standard Dirac and Klein–Gordon equations can be seen as a special case of the more general matrix operator algebra describing the curvature of the a-temporal space associated with the generalized Klein–Gordon and Dirac equations for the density of cosmic space.

Alhaidari's and Jellal's approach can be derived from the generalized approach based on Eqs. (5.110)–(5.117) in the particular case $r = l_p$: this result provides another important perspective about the role of Planck's constant as the fundamental entity which links the standard approach and the a-temporal quantum gravity-space theory.

5.9 Perspectives about the Geometry of the Gravitational Space in the A-Temporal Quantum-Gravity Space Theory

As we have shown in chapter 3, in the quantum domain the nonlocal geometry associated with the 3D timeless space which acts as an immediate information medium can be characterized by introducing opportune notions of quantum distances deriving from a quantum entropy indicating the deformation of the geometrical properties of space with respect to the Euclidean geometry characteristic of classical physics. In analogy to the quantum domain, one can think to describe also the a-temporal quantum-gravity space analysed in this chapter through an opportune quantum distance deriving from an opportune quantum entropy. In this section, we want to make some considerations on how one could perform a programme of this type, by considering the simple case $j = 0$ in which the wavefunction of the quantum-gravity space satisfies the generalized Klein–Gordon equation for the density of cosmic space (5.33).

In analogy with chapter 3, the quantity

$$S_Q = -\frac{1}{2} \ln \rho \qquad (5.118)$$

can be defined as the quantum entropy associated with the wavefunction (5.32) of the quantum-gravity space. The quantum entropy (5.118) can be interpreted as the physical entity that, in the quantum-gravity domain, describes the degree of order and chaos of the vacuum supporting the density ρ describing the distribution of the ensemble of quasi-particles associated with the wavefunction of the quantum-gravity space (in the case $j = 0$). In this entropic approach, the equations of motion for Planck–Fiscaletti granules with $j = 0$ assume the following form:

$$g^{\mu\nu I}\frac{1}{c}\frac{\partial S_Q}{\partial t} = g^{\mu\nu I}\left[-(p_\mu \nabla^{I\mu} S_Q) + \frac{1}{2}\nabla^I_\mu p^\mu\right] \tag{5.119}$$

and

$$g^{\mu\nu I}\nabla^I_\mu S \nabla^I_\nu S = \frac{\hbar^2 [D(r)]^2}{c^4}, \tag{5.120}$$

where

$$g^I_{\mu\nu} = \frac{M^2 c^6}{\hbar^2 [D(r)]^2} g_{\mu\nu} \tag{5.121}$$

is a conformal transformation that indeed is determined by the quantum entropy (5.118),

$$M^2 = \frac{\hbar^2 [D(r)]^2}{c^6}$$

$$\times \exp\left[-\frac{c^4}{[D(r)]^2}(\nabla_\mu S_Q)^2 + \frac{c^4}{[D(r)]^2}\left(\left(\nabla^2 - \frac{1}{c^2}\frac{\partial^2}{\partial t^2}\right)_g S_Q\right)\right], \tag{5.122}$$

and the quantum potential is

$$Q = -\frac{c^4}{[D(r)]^2}(\nabla_\mu S_Q)^2 + \frac{c^4}{[D(r)]^2}\left(\left(\nabla^2 - \frac{1}{c^2}\frac{\partial^2}{\partial t^2}\right)_g S_Q\right). \tag{5.123}$$

In the entropic approach, the presence of the nonlocal quantum potential is equivalent to a curved a-temporal space with its metric being given by (5.121) determined by the quantum entropy on the basis of Eq. (5.122). In this way, by introducing the quantum entropy (5.118), a geometrization of the quantum aspects of Planck–Fiscaletti granules with $j = 0$ emerges in a picture based on the idea that the density of particles associated with the wavefunction of quantum-gravity space determines a nonlocal modification of the geometry.

In the entropic approach suggested in this paragraph, the real key of reading of the link between gravity and quantum behavior of a density of cosmic space lies just in the quantum entropy: the effects

300 *The Timeless Approach*

of gravity on geometry and the quantum effects on the geometry of the a-temporal space are highly coupled because they are both determined by the nonlocal geometry of the background described by the quantum entropy. The quantum entropy appears indeed as a real nonlocal intermediary between gravitational and quantum effects of the density of cosmic space.

Now, the geometrical properties of the background can be characterized by introducing the following quantum length associated with the metric of the a-temporal space (5.121):

$$L_{\text{quantum}} = \frac{1}{\sqrt{(\nabla_\mu S_Q)^2 - \left(\nabla^2 - \frac{1}{c^2}\frac{\partial^2}{\partial t^2}\right)_g S_Q}}. \qquad (5.124)$$

The quantum length (5.124) can be used to evaluate the strength of quantum effects in the a-temporal quantum-gravity space domain. Once the quantum length (5.124) becomes nonnegligible a Planck–Fiscaletti granule with $j = 0$ goes into a regime where the quantum and gravitational effects are highly related.

The nonlocal quantum geometry of the a-temporal quantum-gravity space theory described by Eq. (5.124) can also be put in correspondence with the discreteness at the Planck scale of loop quantum gravity and with the holographic foam of Ng's model, providing thus a new suggestive geometrodynamic entropic of some fundamental equations of loop quantum gravity and of Ng's model. In fact, by substituting Eqs. (5.121) and (5.122) into Eq. (5.17) expressing the metric which characterizes loop quantum gravity, one obtains

$$g^I_{\mu\nu} = \frac{l_P^2}{\mu^2}\eta_{\mu\nu}$$

$$= \exp\left[-\frac{c^4}{[D(r)]^2}(\nabla_\mu S_Q)^2 + \frac{c^4}{[D(r)]^2}\left(\left(\nabla^2 - \frac{1}{c^2}\frac{\partial^2}{\partial t^2}\right)_g S_Q\right)\right]\eta_{\mu\nu}$$

(5.125)

namely

$$\frac{l_P^2}{\mu^2} = \exp\left[-\frac{c^4}{[D(r)]^2}(\nabla_\mu S_Q)^2 + \frac{c^4}{[D(r)]^2}\left(\left(\nabla^2 - \frac{1}{c^2}\frac{\partial^2}{\partial t^2}\right)_g S_Q\right)\right].$$
(5.126)

This means that, in an entropic bohmian approach, the lattice size of the discrete foam corresponding to the metric (5.17) of loop quantum gravity may be written as

$$\mu = \left(\frac{l_P^2}{\exp\left[-\frac{c^4}{[D(r)]^2}(\nabla_\mu S_Q)^2 + \frac{c^4}{[D(r)]^2}\left(\left(\nabla^2 - \frac{1}{c^2}\frac{\partial^2}{\partial t^2}\right)_g S_Q\right)\right]}\right)^{1/2}.$$
(5.127)

Therefore, Eq. (5.85) showing the link between the causal structure of the a-temporal space metric and the gravitational effects of the density of cosmic space (and thus the quantum entropy), in the quantum gravity domain characterized by a discrete nature of the space-time foam, becomes

$$G^{\mu\nu} - \Lambda\frac{l_P^2}{\mu^2}\eta^{\mu\nu} = -\frac{1}{\phi}T^{\mu\nu} - \frac{1}{\phi}\left[\nabla^\mu\nabla^\nu - \frac{l_P^2}{\mu^2}\eta^{\mu\nu}\left(\nabla^2 - \frac{\partial^2}{\partial t^2}\right)\right]\phi$$
$$+ \frac{\omega}{\phi^2}\nabla^\mu\phi\nabla^\nu\phi - \frac{1}{2}\frac{\omega}{\phi^2}\frac{l_P^2}{\mu^2}\eta^{\mu\nu}\nabla^\alpha\phi\nabla_\alpha\phi, \quad (5.128)$$

where the energy-momentum tensor is

$$T^{\mu\nu} = -\frac{1}{\sqrt{-g}}\frac{\delta}{\delta\frac{l_P^2}{\mu^2}\eta_{\mu\nu}}\int d^4x\sqrt{-\eta}L_m.$$
(5.129)

Moreover, substituting the quantum length (5.124) into Eq. (5.18) one obtains

$$8\pi\gamma\rho l_P(x+2M) = \frac{1}{\left(\sqrt{(\nabla_\mu S_Q)^2 - \left(\nabla^2 - \frac{1}{c^2}\frac{\partial^2}{\partial t^2}\right)_g S_Q}\right)^3},$$
(5.130)

namely

$$\rho = \frac{1}{8\pi\gamma l_P(x+2M)\left(\sqrt{(\nabla_\mu S_Q)^2 - \left(\nabla^2 - \frac{1}{c^2}\frac{\partial^2}{\partial t^2}\right)_g S_Q}\right)^3}, \quad (5.131)$$

which provides the condition which is satisfied by the quantum entropy associated to the wavefunction of the quantum-gravity space (of the a-temporal quantum-gravity space theory) in order to give origin to the holographic nature of the kinematical structure of loop quantum gravity in spherical symmetry of Gambini's and Pullin's approach.

In analogous way, the nonlocal geometry of the a-temporal quantum-gravity space theory described by Eq. (5.124) can be put in correspondence with Ng's holographic space-time foam characterized by quantum fluctuations in the form of the average minimum uncertainty (5.21). In this regard, by substituting Eq. (5.124) into Eq. (5.21), the fundamental equation of Ng's holographic space-time foam model becomes

$$(2\pi^2/3)^{1/3} l^{1/3} l_P^{2/3} = \frac{1}{\sqrt{(\nabla_\mu S_Q)^2 - \left(\nabla^2 - \frac{1}{c^2}\frac{\partial^2}{\partial t^2}\right)_g S_Q}}, \quad (5.132)$$

namely

$$l^{1/3} = \frac{1}{(2\pi^2/3)^{1/3} l_P^{2/3} \sqrt{(\nabla_\mu S_Q)^2 - \left(\nabla^2 - \frac{1}{c^2}\frac{\partial^2}{\partial t^2}\right)_g S_Q}}. \quad (5.133)$$

Equation (5.132) provides an a-temporal description of Ng's holographic model by stating that the average separation between neighbouring cells of Ng's quantized space-time foam — and thus the accuracy in the measurement of the geometry of the space-time foam — derives from the quantum entropy of the a-temporal quantum-gravity space. Conversely, according to Eq. (5.133), one can say that the measure of a distance l in the space-time foam emerges from Planck's length as well as from the quantum entropy of the a-temporal quantum-gravity space.

As a consequence, Ng's cumulative factor $C = (l/\lambda)^{1/3}$ characterizing the cumulative effects of the space-time fluctuations over the distance l becomes

$$C = \frac{1}{(2\pi^2\lambda/3)^{1/3} l_P^{2/3} \sqrt{(\nabla_\mu S_Q)^2 - \left(\nabla^2 - \frac{1}{c^2}\frac{\partial^2}{\partial t^2}\right)_g S_Q}}, \quad (5.134)$$

while the maximum energy density (5.22) for a sphere of radius l that does not collapse into a black hole becomes

$$\rho = \frac{3}{8\pi}(8\pi^6/27)^2 l_P^2 \left((\nabla_\mu S_Q)^2 - \left(\nabla^2 - \frac{1}{c^2}\frac{\partial^2}{\partial t^2}\right)_g S_Q\right)^3, \quad (5.135)$$

namely

$$\rho = \frac{8\pi^{11}}{3^5} l_P^2 \left((\nabla_\mu S_Q)^2 - \left(\nabla^2 - \frac{1}{c^2}\frac{\partial^2}{\partial t^2}\right)_g S_Q\right)^3, \quad (5.136)$$

where thus the Hubble radius may be written as

$$R_H = \frac{1}{(2\pi^2/3) l_P^2 \left(\sqrt{(\nabla_\mu S_Q)^2 - \left(\nabla^2 - \frac{1}{c^2}\frac{\partial^2}{\partial t^2}\right)_g S_Q}\right)^3}. \quad (5.137)$$

According to the entropic approach of the background of gravitational space suggested in this paragraph, one can say that the holographic, nonlocal features of the fundamental foam at the Planck scale is determined by the quantum entropy associated with the wavefunction of the quantum-gravity space. One can conclude that the quantum length (5.124) associated with the quantum entropy (5.118) may be considered as the ultimate physical entity which describes, in an a-temporal picture, the geometry of quantum gravity in the sense that the nonlocal holographic foam at the Planck scale, both of loop quantum gravity and of Ng's model emerge directly from it.

Chapter 6

A Three-Dimensional Timeless Quantum Vacuum as the Fundamental Bridge Between Gravitation and the Quantum Behavior

The 20th century theoretical physics brought the idea of a unified quantum vacuum as a fundamental medium subtending the observable forms of matter, energy and space-time. In the quantum field theories which underlie modern particle physics, the notion of empty space has been replaced with that of a quantum vacuum state, defined to be the ground (lowest energy density) state of a collection of quantum fields, the medium that carries the zero-point field, where energies turn out to be present even when all classical forms of energy vanish. A peculiar and truly quantum mechanical feature of the quantum fields of the vacuum is that they exhibit zero-point fluctuations everywhere in space, even in regions which are otherwise 'empty' (namely devoid of matter and radiation). These zero-point fluctuations of the quantum fields, as well as other 'vacuum phenomena' of quantum field theory, give rise to an enormous vacuum energy density. On the basis of the current research exploring the behavior of subatomic particles and their interactions, one can say that the unified quantum vacuum must be considered not only as a theoretical artefact, a requirement of the mathematics of the field theory, but really as a part of a physical reality, as a fundamental medium in the universe.

The realistic concept of the vacuum completes and complements Einstein's theory of relativity. Relativity theory views space-time as a relative and dynamic manifold, interacting with matter and energy. It is the "background" against which the events of the manifest world unfold. But the origins of this background are not accounted for in relativity theory: space-time is simply "given," together with matter and energy. In general relativity, the standard interpretation of phenomena in gravitational fields is in terms of a fundamentally curved space-time. However, this approach leads to well known problems if one aims to find a unifying picture which takes into account some basic aspects of the quantum theory. In order to escape this situation of impasse, several authors advocated alternative ways in order to treat gravitational interaction, in which the space-time manifold can be considered as an emergence of the deepest processes situated at the fundamental level of quantum gravity. In this regard, for example, one can mention the germinal proposal of Sacharov for deducing gravity as "metric elasticity" of space [411] (the reader can also see the reference [412] for a review of this concept), Haisch's and Rueda's model [413] regarding the interpretation of inertial mass and gravitational mass as effects of the electromagnetic quantum vacuum, Puthoff's polarizable vacuum model of gravitation [414] and, more recently, a model developed by Consoli based on ultra-weak excitations in a condensed manifold in order to describe gravitation and Higgs mechanism [18, 415–417].

Under the construction of all of these models there is probably one underlying fundamental observation: as light in Euclid space deviates from a straight line in a medium with variable density, an "effective" curvature might originate from the same physical flat-space vacuum.

In reference to the treatment of gravity in terms of density fluctuations of a physical vacuum, the author of this book and Sorli have recently introduced a model of a three-dimensional (3D) timeless quantum vacuum as a fundamental arena of gravity and quantum phenomena, where time exists only an emergent quantity providing the numerical order of changes. According to this model, 3D quantum vacuum's most fundamental physical properties are its energy density and state-reduction (**RS**) processes (of creation–annihilation of quantum particles). Despite at the first stages of development, this model allows us to re-read in a unitary picture the idea of gravity as a phenomenon determined by the density fluctuations of the same physical flat-space vacuum (which is at the basis of Consoli's model as well as of Sacharov's model, Haisch's and Rueda's model and Puthoff's model) and, at the same time, introduces interesting perspectives in the interpretation of quantum mechanics. According to this model, the change of the quantum vacuum energy density can be considered as the fundamental medium which determines a bridge between gravity and the quantum behavior of matter, leading thus to new interesting perspectives about the problem of unifying gravity with quantum theory. In this chapter, after reviewing some relevant theoretical results about the quantum vacuum, our aim is to analyse the fundamental features of this approach.

6.1 About the Quantum Vacuum: An Overview

The notions of physical vacuum as a fundamental arena of processes originated after the birth of quantum mechanics, in connection with the development of the idea of spontaneous emission of an isolated excited atom [418]. In particular, the laws of quantum mechanics applied to electromagnetic radiation imply the existence of a fundamental level, a background 'sea' of electromagnetic zero-point energy that can be called as the electromagnetic component of the physical vacuum or briefly electromagnetic quantum vacuum. This stems from the Planck's black body radiation law (1900) and was explained by Einstein and Stern (1913). It was later found that the electromagnetic quantum vacuum could manifest itself in the spatial "smearing" of the electron and a change, as a result, of the potential energy of its interaction with the nucleus, thus providing conditions for the removal of the degeneracy of the energies of the $2S\frac{1}{2}$ and $2P\frac{1}{2}$ states in the hydrogen atom — the Lamb shift [419]. It was also demonstrated that the quantum fluctuations of the electromagnetic quantum vacuum in regions contiguous with material objects could alter the relativistic quantum relationships in the near-surface regions of the objects and thus give rise to its macroscopic manifestations — the Casimir ponderomotive effect [420, 421], Josephson contact noise [422]. All these effects mentioned here are of electromagnetic nature: they vanish if the fine structure constant $\alpha = \frac{e^2}{\hbar c}$ (where e is the electron charge, c is the velocity of light in a vacuum, and $\hbar$ is Planck's reduced constant) tends to zero [420, 421].

The quantization of the electromagnetic field in terms of quantum-mechanical operators is found in standard textbooks. As stated by Loudon (1984): "The electromagnetic field is quantized by the association of a quantum-mechanical harmonic oscillator with each mode of the radiation field" [423]. The same $\frac{h\nu}{2}$ zero-point energy expression exists for each mode of the electromagnetic field as for a mechanical oscillator. Summing up the energy over the modes for all frequencies, directions and polarization states, one finds that the background space of physics can be characterized by a zero-point energy density of the electromagnetic quantum vacuum. In this picture, space can be described in terms of fluctuations which are just the consequence of the energy density of the electromagnetic quantum vacuum (and of Heisenberg's uncertainty principle). An energy of $\frac{h\nu}{2}$ per mode of the field characterizes the fluctuations of the quantized radiation field in quantum field theory. In the semi-classical representation of stochastic electrodynamics (see for example the monographs by de la Peña and Cetto, 1996 and Milonni, 1994 [424, 425]) the electromagnetic quantum vacuum is represented by propagating classical electromagnetic plane waves of random phase having this average energy, $\frac{h\nu}{2}$, in each mode. The volumetric density of modes between frequencies ν and $\nu + d\nu$ is given by the density of states function

$$N(\nu)d\nu = \frac{8\pi\nu^2}{c^3}d\nu. \tag{6.1}$$

Each state has a minimum $\frac{h\nu}{2}$ of energy, and using this density of states function and this minimum zero-point energy per state one gets the spectral energy density of the electromagnetic quantum vacuum:

$$\rho(\nu)d\nu = \frac{8\pi\nu^2}{c^3}\frac{h\nu}{2}d\nu, \quad (6.2)$$

which is similar to the spectral energy density of the blackbody radiation if one takes away all the thermal energy.

It was shown that a Planck like component of the electromagnetic quantum vacuum will arise in a uniformly accelerated coordinate system having constant proper acceleration a with what amounts to an effective temperature $T_a = \frac{\hbar a}{2\pi c k}$ [426]. More precisely an observer who accelerates in the conventional quantum vacuum of Minkowski space will perceive a bath of radiation, while an inertial observer perceives nothing. This is a quantum phenomenon and the temperature is negligible for most accelerations and become significant only in extremely large gravitational fields. Thus, for the case of no true external thermal radiation ($T = 0$), but including this acceleration effect T_a, equation (6.2) becomes

$$\rho(\nu, T_a)d\nu = \frac{8\pi\nu^2}{c^3}\left[1 + \left(\frac{a}{2\pi c\nu}\right)^2\right]\left[\frac{h\nu}{2} + \frac{h\nu}{e^{h\nu/kT_a} - 1}\right]d\nu, \quad (6.3)$$

where the pseudo-Planckian component at the end is in general very small.

Then, the development of quantum chromodynamics [427] brought in a fresh knowledge of the nature of the physical vacuum on high-energy scales. It turned out in particular that at an energy density ("effective temperature") of $E_{\text{QED}} \approx 200\,\text{MeV}$ a confinement–deconfinement transition was realized inside the nucleus: quarks were no longer bound in nucleons and formed a quark-gluon plasma or quark soup. In this context, the strong interaction constant α_S proved to be dependent on the excitation energy: its magnitude changed from $\alpha_S \approx 1$ at low energies to $\alpha_S \approx 0,3$ at energies of a few gigaelectron-volts, depending but weakly on energy thereafter [427].

In the last decade, the notions of physical vacuum have come into wide use in cosmology [428–431] in connection with the standard model of the dynamics of the universe, at whose root lie the Friedmann equations of the general theory of relativity, with "dark energy" that accounts for 73% of the entire energy of the universe. It is believed that "dark energy" is uniformly "spilled" in the universe, its unalterable density being $\varepsilon_V = \lambda c^4/8\pi G$, where λ and G are the cosmological and the gravitational constant, respectively. The standard model also considers another physically hard-to-imagine substance — dark matter — whose energy content amounts to 23%, which is introduced into the Friedmann equations in order to remove contradictions between the magnitudes of the apparent masses of gravitationally bound objects, as well as systems of such objects, and their apparent parameters, including the structural stability of galaxies and galactic clusters in the expanding universe. Apart from the introduction of the physically obscure entities here mentioned — dark energy and dark matter — there are relevant problems in the construction of the standard model as a consequence of the unsuccessful attempts to tie in the apparent value $\varepsilon_V \approx 0,66 \cdot 10^{-8} \text{erg/cm}^3$ [432] with the parameters of the physical vacuum introduced in elementary particle physics, the quantum chromodynamics vacuum (QCD vacuum). The above discrepancies come to more than 40 orders of magnitude if the characteristic energy scale of the quantum chromodynamics vacuum is taken to be $E_{\text{QCD}} \approx 200 \text{ MeV}$ [428, 433, 434], with its energy density being $\varepsilon_{\text{QCD}} = E_{\text{QCD}}^4/(2\pi\hbar c)^3$, and over 120 orders of magnitude if one is orientated towards the vacuum of physical fields, wherein quantum effects and gravitational effects would manifest themselves simultaneously, with the Planck energy density

$$\rho_{pE} = \sqrt{\frac{c^{14}}{\hbar^2 G^4}} \approx 4,641266 \cdot 10^{113} \, J/m^3 \cong 10^{97} \, \text{Kg/m}^3 \quad (6.4)$$

playing the part of the characteristic energy scale.

The Planck energy density (6.4) is usually considered as the origin of the dark energy and thus of a cosmological constant, if the dark energy is supposed to be owed to an interplay between quantum mechanics and gravity. However, the observations are compatible with a dark energy

$$\rho_{\text{DE}} \cong 10^{-26} \, \text{Kg/m}^3. \quad (6.5)$$

Thus, Eqs. (6.4) and (6.5) give rise to the so-called "cosmological constant problem" because the dark energy (6.5) is 123 orders of magnitude larger than (6.4). In order to solve this problem, Santos proposed an explanation for the actual value (6.5) of the dark energy which invokes the fluctuations of the quantum vacuum [435, 436]. In these papers, Santos showed that quantum vacuum fluctuations determine a curvature of space and made a calculation, involving plausible hypotheses within quantized gravity, which establishes a relation between the two-point correlation of the vacuum fluctuations and the space curvature. In Santos' approach ρ_{DE} is the effect of the quantum vacuum fluctuations on the curvature of space-time according to equations

$$\rho_{\text{DE}} \cong 70\, G \int_0^\infty C(s) s\, ds, \qquad (6.6)$$

where $C(s)$ is a two-point correlation function of vacuum density fluctuations.

By following Santos' treatment in [435, 436], one finds that, in a semiclassical Newtonian gravity, the existence of vacuum fluctuations of energy necessarily implies a gravitational energy associated with the vacuum, with density

$$\rho_{grav} c^2 = -4\pi G \int_0^\infty C(r_{12}) r_{12}\, dr_{12}, \qquad (6.7)$$

where $C(|\mathbf{r_1} - \mathbf{r_2}|)$ is the two-point correlation function of the quantum vacuum fluctuations which, for equal times, depends only on the distance between the points, namely

$$C(|\mathbf{r_1} - \mathbf{r_2}|) = \frac{1}{2}\langle \text{vac}|\hat{\rho}(\mathbf{r}_1, t)\hat{\rho}(\mathbf{r}_2, t) + \hat{\rho}(\mathbf{r}_2, t)\hat{\rho}(\mathbf{r}_1, t)|\text{vac}\rangle. \qquad (6.8)$$

The relation (6.7) suggests that quantized general relativity may provide a connection between the two-point correlation function $C(|\mathbf{r_1} - \mathbf{r_2}|)$ and the curvature of space-time induced by the quantum vacuum fluctuations. The main hypothesis of Santos' model is that the constant ρ_{DE} is not an actual density, but the effect of the quantum vacuum fluctuations on space-time.

In order to provide a measure of the correlation function, in Santos' approach two different scales are considered: a cosmic scale (with typical distances of megaparsecs) and the atomic scale of the correlations between vacuum fluctuations (which involves distances smaller than nanometers). In the latter scale quantization is essential, but in the former we may consider everything as classical. For any two quantum observables, $\hat{a}(x)$ and $\hat{b}(y)$ at the space points x and y respectively, the correlation is defined as

$$C_{ab}(x,y) \equiv \langle \phi|\hat{a}(x)\hat{b}(y)|\phi\rangle - \langle \phi|\hat{a}(x)|\phi\rangle\langle|\hat{b}(y)|\phi\rangle, \quad (6.9)$$

where ϕ is the usual gravitational potential function. The correlation (6.9), although relevant at the atomic scale, can be assumed to go to zero when the distance increases toward the cosmic scale. As a consequence, in the cosmic scale the expectations of quantum observables may be treated as classical variables, and the expectations of products of observables can be seen as products of the corresponding classical variables.

Here, one can assume that the expectation of the stress-energy tensor operator of the quantum fields at any point gives the matter-energy, without any additional contribution of the vacuum. In order to have the correct Friedmann–Robertson–Walker metric, this assumption means that

$$\langle\phi|\hat{T}_4^4|\phi\rangle = \rho, \quad \langle\phi|\hat{T}_\mu^\nu|\phi\rangle \cong 0 \quad \text{for } \mu\nu \neq 00. \quad (6.10)$$

This leads to a definition of a vacuum stress-energy tensor operator as

$$\hat{T}_{\mu\nu}^{vacuum} \equiv \hat{T}_{\mu\nu} - \langle\phi|\hat{T}_{\mu\nu}|\phi\rangle\hat{I}, \quad (6.11)$$

where $\hat{I}$ is the identity operator. The existence of vacuum fluctuations means that, although the expectation of $\hat{T}_{\mu\nu}^{vacuum}$ is zero by definition, there are correlated vacuum fluctuations, that is

$$\langle\phi|\hat{T}_{\mu\nu}^{vacuum}(x)\hat{T}_{\lambda\sigma}^{vacuum}(y)|\phi\rangle \neq 0. \quad (6.12)$$

Now, in order to derivate the Friedmann–Robertson–Walker metric equation

$$ds^2 = -dt^2 + a(t)^2(dr^2 + r^2 d\Omega). \quad (6.13)$$

Santos considered the quantum metric

$$d\hat{s}^2 = \hat{g}_{\mu\nu}dx^\mu dx^\nu \tag{6.14}$$

and imposed it to be close to Minkowski metric. In this way, Santos expressed the quantum coefficients of the metric (in polar coordinates) as

$$\hat{g}_{00} = -1 + \hat{h}_{00}, \quad \hat{g}_{11} = 1 + \hat{h}_{11}, \quad \hat{g}_{22} = r^2(1 + \hat{h}_{22}),$$
$$\hat{g}_{33} = r^2 \sin^2 \vartheta (1 + \hat{h}_{33}), \quad \hat{g}_{\mu\nu} = \hat{h}_{\mu\nu} \quad \text{for } \mu \neq \nu, \tag{6.15}$$

where

$$\hat{h}_{\mu\nu} = 0$$

except

$$\langle \hat{h}_{00} \rangle = \frac{8\pi G}{3}(\rho_{\text{mat}} + \rho_{\text{DE}})r^2$$

and

$$\langle \hat{h}_{11} \rangle = \frac{8\pi G}{3}\left(\rho_{\text{DE}} - \frac{1}{2}\rho_{\text{mat}}\right)r^2, \tag{6.16}$$

where $\langle \hat{h}_{\mu\nu} \rangle$ stands for $\langle \phi | \hat{h}_{\mu\nu} | \phi \rangle$, ρ_{DE} is the density owed to a possible existence of dark matter. On the basis of Eqs. (6.7)–(6.16), one can conclude that the metric of space determined by the energy density of quantum vacuum is quantized as in (6.14) and that the action of the quantum vacuum fluctuations is direct, as it is expressed by a two-point correlation function that, in the case of equal times, depends only on the distance between those two points. Moreover, at the atomic scale quantization is essential, but at a cosmic scale the quantum metric may be approximated by a classical metric which is the expectation value (long distance average) of the former. The relations between the metric coefficients and the matter stress-energy tensor are nonlinear (for example here a quadratic dependence on r may be possible) and, as a consequence, the expectation of the metric (6.14) is not the same as the metric of the expectation of the matter tensor, given by Eq. (6.10). The difference gives rise to a contribution of the vacuum fluctuations mimicking the effect of Einstein's cosmological constant.

From the metric equation (6.15) by means a straightforward calculation, Santos obtained the components of the quantum Einstein equations

$$\hat{G}_{\mu\nu} = \frac{8\pi G}{c^4}\hat{T}_{\mu\nu}, \qquad (6.17)$$

assuming that they are similar to the classical counterparts. In particular, by writing the (operator) metric parameter $\hat{h}_{11}$ in the form

$$\hat{h}_{11} = G\hat{h}^{(1} + G^2\hat{h}^{(2}, \qquad (6.18)$$

one obtains

$$\langle\Phi|h_{11}|\Phi\rangle = \langle\Phi|G\hat{h}^{(1} + G^2\hat{h}^{(2}|\Phi\rangle_{\text{mat}} + G^2\langle\Phi|\hat{h}^{(2}|\Phi\rangle_{\text{vac}}, \qquad (6.19)$$

namely the expectation value is the sum of two expressions, one containing the matter density, ρ_{mat}, and the other one the vacuum density, ρ_{vac}. Modelling the matter density of the universe by a constant, the matter term is

$$\langle\Phi|h_{11}|\Phi\rangle_{\text{mat}} \cong \frac{2GM}{r} + \frac{2G^2M^2}{r^2}, \qquad (6.20)$$

where $M = \frac{4}{3}\pi\rho_{\text{mat}}r^3$ which agrees with the second order expansion of the well known Schwarzschild solution

$$g_{11} = \left(1 - \frac{2GM}{r}\right)^{-1}, \qquad (6.21)$$

while the vacuum contribution is

$$\langle\Phi|h_{11}|\Phi\rangle_{\text{vac}} \cong 600\, G^2 r^2 \int_0^\infty C(s)s\, ds, \qquad (6.22)$$

where r is a typical cosmic distance, which in the second equality is estimated to fulfil $r/s \approx 10^{40}$. In this way, the total expectation value (6.19) becomes

$$\langle\Phi|h_{11}|\Phi\rangle \cong \frac{8\pi G}{3}\rho_{\text{mat}}r^2 + 600G^2 r^2 \int_0^\infty C(s)s\,ds. \quad (6.23)$$

Hence, comparison with the Friedmann equations

$$\left[\frac{\dot{a}}{a}\right]^2 = \frac{8\pi G}{3}(\rho_{\text{mat}} + \rho_{\text{DE}}), \quad \frac{\ddot{a}}{a} = \frac{8\pi G}{3}\left(\frac{1}{2}\rho_{\text{mat}}t + \rho_{\text{DE}}\right) \quad (6.24)$$

leads to the equivalence of the curvature of space-time produced by vacuum vacuum fluctuations and the one determined by a constant dark energy density, expressed by Eq. (6.6). Thus, one can say that in Santos' approach, the quantum vacuum fluctuations characterized by the result (6.22) plus the matter density lead to the Friedmann–Robertson–Walker metric equation

$$ds^2 = -dt^2 + a(t)^2(dr^2 + r^2 d\Omega). \quad (6.25)$$

In synthesis, one can say that Santos' model of quantized general relativity, with plausible hypotheses, predicts that the quantum vacuum fluctuations give rise to a curvature of space-time similar to the curvature produced by a "dark energy" density. Crucial for this result is the fact that Einstein's equation involves a nonlinear relation between the stress-energy tensor and the metric tensor. In fact, if the relation was linear, then the vanishing of the vacuum expectation of the quantum matter density operator would imply Minkowski space (in the absence of baryon or dark matter). The main result of Santos' calculation is equation (6.6), which states that the possible value of the "dark energy" density is the product of Newton constant, G, times the integral of the two-point correlation of the vacuum fluctuations (6.8). Moreover, dimensional analysis leads to Zeldovich's formula [437]

$$\rho_{\text{DE}}c^2 \approx \frac{Gm^2}{r_C} \cdot \frac{1}{r_C^3}, \quad (6.26)$$

($r_C = \hbar/mc$ being Compton's radius) which involves a parameter, m, with dimensions of a mass. If in Zeldovich's original model, Eq. (6.26) reproduces the observed value of the dark energy density for a mass of $m \approx 7,6 \cdot 10^{-29}$ Kg that is about 1/20 times the proton mass or about 80 times the electron mass, Santos' approach does not allow to obtain the value of m, but seems suggest the possibility that vacuum fluctuations of high energy, involving very massive particles, would not be probable. If this is the case, m should be of order the mass of a not too heavy particle. Indeed a mass of order one third the mass of the lightest hadron, the pion, is able to reproduce the observed value of the dark energy.

On the other hand, as regards general relativity, the gravitational effect of a vacuum energy resulting from zero-point energies, virtual particles (higher order vacuum fluctuations), QCD condensates, fields of spontaneously broken theories, and possible other, at present, unknown fields, might curve space-time beyond recognition. It is usually assumed that the vacuum energy density of general relativity ($\langle \rho_{\text{vac}} \rangle$) is equivalent to a contribution to the 'effective' cosmological constant in Einstein equations

$$\Lambda_{\text{eff}} = \Lambda_0 + \frac{8\pi G}{c^4} \langle \rho_{\text{vac}} \rangle, \qquad (6.27)$$

where Λ_0 denotes Einstein's own 'bare' cosmological constant which in itself leads to a curvature of empty space, i.e. when there is no matter or radiation present. Once Eq. (6.27) is established, it follows that anything which contributes to the quantum field theory vacuum energy density is also a contribution to the effective cosmological constant in general relativity.

From the quantum field theories which describe the known particles and forces one can derive various contributions to the vacuum energy density. The vacuum energy density associated with these theories, which has experimentally demonstrated consequences and is thus taken to be physically real, has cosmological implications which follow when certain assumptions are made about the relation between general relativity and quantum field theory. On the basis of the fundamental theories, one can infer that the total vacuum energy density has at least the following three contributions,

$$\begin{pmatrix} Vacuum \\ energy \\ density \end{pmatrix} = \begin{pmatrix} VACUUM \\ ZERO-POINT-ENERGY \\ +FLUCTUATIONS \end{pmatrix}$$
$$+ \begin{pmatrix} QCD \\ gluon-and-quark \\ condensates \end{pmatrix} + \begin{pmatrix} The \\ Higgs \\ field \end{pmatrix} + \cdots$$
(6.28)

namely the fluctuations characterizing the zero-point field, the fluctuations characterizing the quantum chromodynamic level of subnuclear physics and the fluctuations linked with the Higgs field, and the dots represent contributions from possible existing sources outside the Standard Model (for instance, GUT's, string theories, and every other unknown contributor to the vacuum energy density). There is no structure within the Standard Model which suggests any relations between the terms in Eq. (6.28), and it is therefore customary to assume that the total vacuum energy density ρ_{vac} is, at least, as large as any of the individual terms. In order to reconcile the vacuum energy density estimate within the Standard Model with the observational limits on the cosmological constant $|\Lambda| < 10^{-56}\,\text{cm}^{-2}$, the usual programme is to "fine-tune": for example, if the vacuum energy is estimated to be at least as large as the contribution from the QED sector then Λ_0 has to cancel the vacuum energy to a precision of at least 55 orders of magnitude.

As regards the role of the different contributions to the vacuum energy density, the reader can find a detailed analysis in several current research. In particular, in the paper [438] Rugh and Zinkernagel studied the connection between the vacuum concept in quantum field theory and the conceptual origin of the cosmological constant problem. In the paper [439] Timashev examined the possibility of considering the physical vacuum as a unified system governing the processes taking place in microphysics and macrophysics, which manifests itself on all space-time scales, from subnuclear to cosmological.

Even more fascinating is perhaps Solà's paper *Cosmological constant and vacuum energy: old and new ideas* [440] which, in the analysis of the cosmological constant problem and of its significance and possible phenomenological implications in relation to the vacuum energy, focused on the notion of the vacuum energy in quantum field theory both in flat and curved space-time, showing that the term which is usually interpreted as the vacuum energy density (originating from the quantum vacuum fluctuations of the matter fields) is the same both in flat and curved space-time. The zero-point-energy is a pure quantum effect which is linked with vacuum fluctuations inside a field theory. As regards the zero-point energy of the vacuum in flat space-time, the final result is thus that it only depends of a list of parameters P(masses and coupling constants), which can only enter virtually in the loop propagators. If we count the loop order of perturbation theory with the corresponding power of $\hbar$, the loopwise expansion can be presented as a power series in $\hbar$:

$$V_{ZPE}(P) = \hbar V_P^{(1)} + \hbar^2 V_P^{(2)} + \hbar^3 V_P^{(3)} + \cdots \qquad (6.29)$$

where each term has to renormalized in order to obtain meaningful physical conclusions. In particular, upon removing the cutoff one expects that the renormalized result should be of order

$$V_P^{(1)} = -\frac{m^4}{64\pi^2} \ln \frac{\mu^2}{m^2} + \cdots \qquad (6.30)$$

where μ is some arbitrary substraction scale that must remain after renormalization. From the above formula (6.30) it derives that the zero-point energy becomes zero if we set $m = 0$ in it while if we would instead compute the contribution, not from the electromagnetic field, but from a scalar field of mass m, the renormalized contribution should be proportional to m^4. In the minimal substraction scheme, the renormalized vacuum energy of the free field at one loop is

$$\rho_{\text{vac}}^{(1)} = \rho_\Lambda(\mu) + V_{ZPE}^{(1)} = \rho_\Lambda(\mu) + \frac{m^4 \hbar}{64\pi^2} \left(\ln \frac{m^2}{\mu^2} - \frac{3}{2} \right). \qquad (6.31)$$

On the basis of Eq. (6.31), the main message of the renormalization group is the following: the sum of the various μ-dependencies must cancel in the renormalized effective action, and also in the renormalized S-matrix elements (when they can be defined). As a consequence, computing the logarithmic derivative $d/d\ln\mu = \mu d/d\mu$ on both sides of (6.31) and taking into account that this gives $d\rho_{\text{vac}}^{(1)}/d\ln\mu = 0$ on the left-hand side, one immediately finds

$$\mu\frac{d\rho_\Lambda(\mu)}{d\mu} = \frac{\hbar m^4}{32\pi^2}. \tag{6.32}$$

The Renormalization Group has thus the following useful consequence about the explicit μ-dependence affecting incomplete structures of the effective action. While we may not know the full structure of the effective action in a particular complicated situation, our educated guess in associating μ with some relevant dynamical variable of the system can give relevant information on physics. This is the case with an effective charge (or "running coupling constant"), say the QED or QCD renormalized gauge coupling $g = g(\mu)$, which is explicitly μ-dependent even though the full effective action or S-matrix element is not.

Considering now the zero-point energy in curved space-time, the total effective action (classical plus one-loop corrections) can be conveniently expressed as follows:

$$\Gamma = S[\phi_c] + S_{\text{HD}}^{(1)} + S_{\text{EH}}^{(1)}, \tag{6.33}$$

where

$$S[\phi_c] = \int d^4x \sqrt{-g(x)}\left(\frac{1}{2}g^{\mu\nu}\partial_\mu\phi\partial_\nu\phi + \frac{1}{2}\xi\phi_c^2 R - V_c(\phi)\right) \tag{6.33a}$$

is the classical action for the scalar field into a curved space-time,

$$S_{HD}^{(1)} = \int d^4x\sqrt{-g}(\alpha_1^{(1)}C^2 + \alpha_2^{(2)}R^2 + \cdots) \tag{6.34}$$

is the classical vacuum action which contains standard high derivative part representing short distance effects which have no impact at low energies and

$$S_{EH}^{(1)} = -\int d^4x \sqrt{-g} \left(\frac{1}{16\pi G^{(1)}} R + \rho_{\text{vac}}^{(1)} \right) \quad (6.35)$$

is the long-distance (namely low energy) Einstein–Hilbert action, where

$$C^2 = R_{\mu\nu\rho\sigma} R^{\mu\nu\rho\sigma} - 2R_{\mu\nu} R^{\mu\nu} + \frac{1}{3} R^2 \quad (6.36)$$

is the square of the Weyl tensor, $\xi \phi_c^2 R$ is a nonminimal coupling term (ξ being a dimensionless coefficient necessary for renormalizability), $\alpha_1^{(1)}$ and $\alpha_2^{(1)}$ are one-loop coefficients of the HD action and $G^{(1)}$ and $\rho_{\text{vac}}^{(1)}$ are the one-loop Newton's constant and cosmological constant term of the low energy Einstein–Hilbert action.

The final results read

$$\alpha_1^{(1)} = \alpha_1(\mu) - \frac{\hbar}{32\pi^2} \frac{1}{120} \left(\ln \frac{m^2}{\mu^2} + finite\ const. \right) \quad (6.37)$$

$$\alpha_2^{(1)} = \alpha_2(\mu) - \frac{\hbar}{64\pi^2} \left(\frac{1}{6} - \xi \right)^2 \left(\ln \frac{m^2}{\mu^2} + finite\ const. \right) \quad (6.38)$$

$$\frac{1}{16\pi G^{(1)}} = \frac{1}{16\pi G(\mu)} + \frac{\hbar m^2}{32\pi^2} \left(\frac{1}{6} - \xi \right) \left(\ln \frac{m^2}{\mu^2} + finite\ const. \right) \quad (6.39)$$

$$\rho_{\text{vac}}^{(1)} = \rho_\Lambda(\mu) + \frac{m^4 \hbar}{64\pi^2} \left(\ln \frac{m^2}{\mu^2} + finite\ const. \right). \quad (6.40)$$

The parameters $\alpha_1^{(1)}$, $\alpha_2^{(1)}$, $G^{(1)}$ and $\rho_{\text{vac}}^{(1)}$ on the left-hand side of these relations are the final "one-loop parameters"; they are finite at one loop because counterterms are used to cancel the divergences coming from the one-loop contributions to them. Mathematically, the μ-running is determined by the corresponding renormalization group equation for each of the parameters, and follows from setting the total derivatives of the one-loop parameters $\alpha_1^{(1)}$, $\alpha_2^{(1)}$, $G^{(1)}$ and $\rho_{\text{vac}}^{(1)}$ with respect to μ to zero. By computing the logarithmic derivatives of each one of them, from Eqs. (6.37)–(6.40) one can immediately obtain the following renormalization group equations:

$$\frac{d\alpha_1(\mu)}{d\ln\mu} = -\frac{\hbar}{120(4\pi)^2}, \quad (6.41)$$

$$\frac{d\alpha_2(\mu)}{d\ln\mu} = -\frac{\hbar}{2(4\pi)^2}\left(\frac{1}{6}-\xi\right)^2, \quad (6.42)$$

$$\frac{d}{d\ln\mu}\left(\frac{1}{16\pi G(\mu)}\right) = \frac{\hbar m^2}{(4\pi)^2}\left(\frac{1}{6}-\xi\right), \quad (6.43)$$

$$\frac{d\rho_\Lambda(\mu)}{d\ln\mu} = \frac{\hbar m^4}{2(4\pi)^2}. \quad (6.44)$$

Here, the most important point lies in fact that the renormalization group equation for $\rho_\Lambda(\mu)$, namely Eq. (6.44), coincides exactly with Eq. (6.32), the flat space-time result. Sola's treatment based on Eqs. (6.29)–(6.44) introduces thus the relevant perspective that the vacuum energy density (originating from the quantum vacuum fluctuations of the matter fields) is the same both in flat and curved space-time.

On the other hand, as shown clearly by Solà in [440], in the Standard Model of particle physics there are other contributions to the vacuum energy that emerge from the spontaneous symmetry breaking of the electroweak gauge symmetry, specifically from the Higgs mechanism. These contributions are associated to the vacuum expectation value of the Higgs potential and increase as $\langle V \rangle \approx v^2 M_H^2 \approx 10^8\,\text{GeV}^4$, where $v = O(100)\,\text{GeV}$ is the vacuum expectation value that defines the electroweak scale, and $M_H \cong 125\,\text{GeV}$ is the presumed physical mass of the Higgs boson [441, 442]. While every single vacuum energy source alone is already vastly worrisome, the combination of them all amounts to a devastating fine tuning problem, which is further aggravated at the quantum level when one considers the many higher loop effects involved.

Of the two main sources of difficulties with the vacuum energy in quantum field theory, namely the zero-point energy and the spontaneous simmetry breaking, the reality of the former has been disputed sometimes by the inconclusive interpretations about the origin of the Casimir effect as being a pure quantum field theory vacuum effect or something else; whereas the latter O remained in the limbo of the theoretical ideas because of its critical link with the existence of the Higgs mechanism. On the other hand, in the electroweak domain there are also other sorts of vacuum fluctuations, which are linked to the fact that quantum corrections to high precision electroweak observables have reached a level of certainty that is beyond any possible doubt. In this regard, an example is provided by the famous Δr parameter from the electroweak theory that allows the determination of the $W^\pm$ gauge boson mass, M_W, in the on-shell renormalization scheme with quantum precision, namely including the quantum output from radiative corrections. Here, an important challenge lies in finding out to which extent we can attest that the "genuine" electroweak quantum effects have been measured when we compare the theoretical value of M_W and the experimentally measured one. In fact, the electroweak radiative corrections represent a manifestation of the properties of the electroweak vacuum at the quantum level, namely the same vacuum that is supported by the likely existence of the Higgs particle. The answer is known, and is astounding: we know unmistakably that they are there with a confidence level of 25σ! And, overall, the overwhelming evidence of the electroweak vacuum effects becomes even more defiant and challenging for cosmology.

Moreover, in [440] Solà discussed a possible reinterpretation of the results obtained in the calculation of the vacuum energy density ρ_Λ in quantum field theory in curved space-time and suggested that, although the resolution of the cosmological constant problem cannot be addressed from a rigorous computation of ρ_Λ in an expanding Friedmann universe, at least some consistency relations seem to hint at the possible form of the correct dynamical dependence of that important quantity as a function of the Hubble rate. For example, if both the vacuum energy and the gravitational coupling are assumed to depend of the Hubble rate, H, the Bianchi identity leads to the possible form for the running of the vacuum density $\rho_\Lambda = \rho_\Lambda(H)$. In this way, one obtains the following behavior for the low energy regime:

$$\rho_\Lambda = c_0 + \beta M_p^2 H^2, \qquad (6.45)$$

where M_p is the Planck mass, c_0 a constant (close to the current cosmological constant density value $\rho_\Lambda^0 = 2,5 \cdot 10^{-47} \text{GeV}^4$) and β is a dimensionless coefficient parameterizing the cosmological constant running. This form for the evolving vacuum energy density has been profusely tested against the latest cosmological data [443–445]. At the same time, a nonvanishing value of β leads to a dynamical vacuum behavior which may effectively appear as quintessence, or as phantom energy, which could eventually provide — Solà claims — a strong phenomenological evidence in favor of the vacuum energy being a serious candidate for dynamical dark energy.

Solà emphasizes that the cosmological constant problem can only be solved as a physical problem, in the sense that only through phenomenological tests it should be possible to disentangle the most difficult theoretical aspects of the cosmological constant problem [440]. In this regard, another potentially interesting aspect of the dynamical vacuum models is the fact that they could provide an explanation for a possible variation of the so-called fundamental constants of nature [446]. Currently plenty of experimental activity, both in the lab and from observations in the astrophysical domain, could be very helpful for effectively testing the proposed vacuum ideas providing interesting news on this field [447–452].

According to Sola's point of view, the idea of a running vacuum energy in an expanding universe, i.e. $\rho_\Lambda = \rho_\Lambda(H)$, appears as a most natural one [440]. It would be very hard to admit the existence of a tiny and immutable constant from the early times to the present days, maybe even ruling the entire future evolution of the universe. Nonetheless, the ultimate value that $\rho_\Lambda(H)$ takes at present, i.e. $\rho_\Lambda^0 = 2,5 \cdot 10^{-47}$ GeV^4, cannot be predicted within these models and hence can only be extracted from observations. The ability to predict this value corresponds just to solve the old cosmological constant problem (in this regard, the reader can find details, for example, in the references [453–457]). In Solà's approach, however, the running vacuum paradigm suggests new perspective of solution to this problem: it conceives the cosmological term as a time evolving variable that underwent a dramatic reduction from the inflationary time till the present days. A generic model, motivated by the general running vacuum framework, and able to provide a complete description of the cosmic history [458–460], that implements this idea is the following:

$$\rho_\Lambda(H) = c_0 + c_2 H^2 + c_4 H^4, \qquad (6.46)$$

in which the highest power of the Hubble rate, H^4, would only be relevant for the early universe, namely for values of H of order of the inflationary expansion rate. In contrast to the standard cosmological model, in which the two opposite poles of the cosmic history (inflation and dark energy) are completely disconnected, the running vacuum models offer a clue for interconnecting them and let the present dark energy appear as a "quantum fossil" from the inflationary universe, thus providing a justification of the small value of the present vacuum energy density ρ_Λ^0 with respect to the inflationary scale [461]. In this picture, the two de Sitter epochs, viz. the primordial inflationary one and the late dark energy epoch, can be thought of as two vacuum dominated stages of the cosmic evolution smoothly interpolated (within a single unified model) by the radiation and matter dominated epochs.

In Solà's approach, the main task that remains to face is to understand why ρ_Λ^0 has the concrete value, that we have measured. In other words, here one is left with the question of why the mass scale $m_\Lambda = [\rho_\Lambda^0]^{1/4}$ associated to the cosmological term is of order of a millielectronvolt rather than, say, ten times or a hundred times bigger. While such values cannot be admitted, as they would obviously be incompatible with the observations, however Sola's view aims to address the following matter of principle: is there a fundamental physical theory which can explain the value of the vacuum energy that we measure at present? Or, in other words, should the millielectronvolt energy scale m_Λ be ultimately predictable, for example from the value of the Planck scale M_p?

This is a very interesting question, but is not at all an obvious one (as shown, for example, by Padmanabhan in [462, 463]).

Solà underlines that, when we face the possibility to explain the mass scale m_Λ in our universe, we have to note that in particle physics essentially all physical scales remain unexplained. For example, we cannot explain why the value of the electron mass is $m_e \cong 0,511$ MeV in our universe, since we do not understand why its Yukawa coupling, λ_e, takes the value it takes in order to yield the precise value of the electron mass from its product with the vacuum expectation value of the Higgs doublet: $m_e = \lambda_e v$.

Whether we can ultimately predict or not the scale of the vacuum energy in the present universe can be a debatable question, however a fundamental point in the Standard Model of particle physics is to understand the values of the two basic "order parameters" that set the fundamental scales of the model, namely: i) the electroweak scale $v \approx G_F^{-1/2}$ (associated to the Higgs mechanism and linked to Fermi's constant); and ii) the value of the QCD scale Λ_{QCD} (associated to the strong interactions and determined by the nonperturbative confinement dynamics of QCD). In both cases an experimental input is needed in order to make contact with physics. For the electroweak sector (which provides the dominant contribution to the vacuum energy of the Standard Model) one needs to perform a precise measurement of Fermi's constant from muon decay. Similarly, in the domain of strong interactions one has to measure Λ_{QCD} to account for the QCD vacuum contribution.

By virtue of its particular position in the realm of the physical quantities, and because of the general covariance of Einstein's equations, a tight connection of the cosmological constant with the remaining scales of the universe seems natural. For this reason the help of the phenomenological input appears as indispensable. Is this not a most fundamental physical requirement, even for the glory of the cosmological constant problem? In this sense, according to Solà's view, if the vacuum energy density is treated as a running quantity in an expanding universe, it would be natural to input its value at a given cosmic time (say, now) and then focus our efforts on finding its past and future evolution [440]. In this regard — Solà claims — some clues in order to open interesting perspectives are provided by the renormalization group approach: after one inputs m_Λ (or, equivalently, c_0) the cosmological constant problem becomes a problem of explaining why the dynamical term in the low energy expression $\rho_\Lambda(H) = c_0 + \beta M_p^2 H^2$ is just the soft (and completely harmless) $\approx H^2$ contribution in the context of quantum field theory in an expanding space-time. According to Solà, this renormalization group approach provides a very attractive option, achieving an appealing scenario for $\rho_\Lambda(H)$ within quantum field theory in curved spacetime, although much work (and thought) is of course needed to tackle the most difficult angles of the cosmological constant problem face to face.

Here, we want now to focus our attention in detail on the link between quantum vacuum and gravitation as well as on the perspectives opened by this link. As regards the idea of a quantum vacuum as the fundamental medium of gravity, since 1994 Haisch and Rueda and Puthoff suggested that the inertial mass can be interpreted as an effect of the electromagnetic quantum vacuum: it is a change in the momentum of the radiation field associated with the electromagnetic zero-point energy instantaneously transiting through an object of a given volume (and interacting with the quarks and electrons in that object) that creates the inertial mass of the object [464]. In this model, inertia emerges thus as a kind of acceleration-dependent electromagnetic quantum vacuum drag force acting upon electromagnetically interacting elementary particles (electrons and quarks). Successively, Rueda and Haisch showed that the result for inertial mass can be extended to passive gravitational mass. In their article of 2005 *Gravity and the quantum vacuum inertia hypothesis*, they say: "If one assumes, as in stochastic electrodynamics, that the zero-point radiation field carries energy and momentum in the usual way, and if that radiation interacts with the particles comprising matter in the usual way, it can be shown that a law of inertia can be derived for matter comprised of electromagnetically interacting particles that are *a priori* devoid of any property of mass. In other words, the $\vec{F} = m\vec{a}$ law of classical mechanics — as well as its relativistic counterpart — can be traced back not to the existence in matter of mass (either innate or due to a Higgs mechanism), but to a purely electromagnetic effect (and possibly analogous contributions from other vacuum fields). It can be shown that the mass-like properties of matter reflect the energy-momentum inherent in the quantum vacuum radiation field. We call this the quantum vacuum inertia hypothesis. There are additional consequences that make this approach of assuming real interactions between the electromagnetic quantum vacuum and matter appear promising. It can be shown that the weak principle of equivalence — the equality of inertial and gravitational mass — naturally follows. In the quantum vacuum inertia hypothesis, inertial and gravitational mass are not merely equal, they prove to be the identical thing. Inertial mass arises upon acceleration through the electromagnetic quantum vacuum, whereas gravitational mass — as manifest in weight — results from what may in a limited sense be viewed as acceleration of the electromagnetic quantum vacuum past a fixed object. The latter case occurs when an object is held fixed in a gravitational field and the quantum vacuum radiation associated with the freely-falling frame instantaneously comoving with the object follows curved geodesics as prescribed by general relativity." [413].

Now, in order to analyse the formalism which lies at the basis of the contribution of the electromagnetic quantum vacuum to the inertial mass of a material object in the form of an inertial force of opposition to acceleration, and in particular the zero-point field Poynting vector that an extended accelerating object sweeps through, we follow Rueda's and Haisch's treatment in the papers [465–468, 413]. In these papers, the properties of the electromagnetic quantum vacuum as experienced in a Rindler, or constant proper acceleration, frame were investigated, and the authors calculated a resulting force of opposition on material objects fixed in such a Rindler noninertial frame and interacting with the electromagnetic zero-point radiation field. Owing to the symmetry and Lorentz invariance of zero-point radiation in unaccelerated reference frames, the Rindler frame force vanishes at constant velocity, i.e. in inertial frames. Since the resulting Rindler frame force proves to be proportional to acceleration, Rueda and Haisch furthermore hypothesized that some frequency-dependent fraction, $\eta(\omega)$, of the radiative momentum of the radiation field associated with the electromagnetic zero-point energy instantaneously transiting through a material object of a given volume will, via interactions, be conferred upon the particles comprising the object. In this way, Newton's equation of motion can be derived from electrodynamics, and the relativistic version of the equation of motion also naturally follows.

The electromagnetic field inside a cavity of perfectly reflecting can be expressed as an expansion of infinite different modes where each mode corresponds to an independent oscillation which always, even in the case of free space, can be represented as a harmonic oscillator. When quantized, the harmonic oscillator provides an energy of the form $\hbar\omega\left(n + \frac{1}{2}\right)$, where ω is the characteristic frequency of the mode. Therefore, on the basis of the Heisenberg uncertainty relation, the electromagnetic field has a minimum quantum energy state consisting of zero-point fluctuations having an average energy per mode of $\hbar\omega/2$. The density of modes is

$$N(\omega)d\omega = \frac{\omega^2}{\pi^2 c^3}d\omega, \qquad (6.47)$$

which leads to the following expression for the spectral energy density for the zero-point fluctuations

$$\rho(\omega)d\omega = \frac{\hbar\omega^3}{2\pi^2 c^3}d\omega. \qquad (6.48)$$

In the semi-classical approach known as stochastic electrodynamics (SED) — in which the physical effects of the electromagnetic quantum vacuum are included by adding to ordinary Lorentzian classical electrodynamics a random fluctuating electromagnetic background constrained to be homogeneous and isotropic and to look exactly the same in every Lorentz inertial frame of reference — the quantum fluctuations of the electric and magnetic fields are treated as random plane waves summed over all possible modes with each mode having this $\hbar\omega/2$ energy. The electric and magnetic zero-point fluctuations in the SED approximation are thus

$$\vec{E}_\tau^{zp}(\vec{r},t) = \sum_{\lambda=1}^{2} \int d^3k(\hbar\omega/2\pi^2)^{1/2}\hat{\varepsilon}(\vec{k},\lambda)$$
$$\cos[\vec{k}\cdot\vec{r} - \omega t - \theta(\vec{k},\lambda)], \qquad (6.49)$$

$$\vec{B}^{zp}(\vec{r},t) = \sum_{\lambda=1}^{2} \int d^3k(\hbar\omega/2\pi^2)^{1/2}[\hat{k}\times\hat{\varepsilon}(\vec{k},\lambda)]$$
$$\cos[\vec{k}\cdot\vec{r} - \omega t - \theta(\vec{k},\lambda)], \qquad (6.50)$$

where the sum is over two polarization states, $\hat{\varepsilon}$ is a unit vector and $\theta(\vec{k},\lambda)$ is a random variable uniformly distributed in the interval $(0,2\pi)$. Now, by considering the transformation from a stationary frame to a Rindler frame (characterized by a constant proper acceleration) which experiences an asymmetric event horizon leading to a nonzero electromagnetic energy and momentum flux and which is defined by the following velocity and Lorentz factor

$$\frac{v}{c} = \tanh\left(\frac{a\tau}{c}\right), \qquad (6.51)$$

$$\gamma_\tau = \cosh\left(\frac{a\tau}{c}\right), \qquad (6.52)$$

where a is the object's proper acceleration and τ its proper time, one obtains

$$\vec{E}_\tau^{zp}(0,\tau) = \sum_{\lambda=1}^{2} \int d^3k(\hbar\omega/2\pi^2)^{1/2}$$
$$\times \left\{\hat{x}\hat{\varepsilon}_x + \hat{y}\cosh\left(\frac{a\tau}{c}\right)\left[\hat{\varepsilon}_y - \tanh\left(\frac{a\tau}{c}\right)(\hat{k}\times\hat{\varepsilon})_z\right]\right.$$
$$\left. + \hat{z}\cosh\left(\frac{a\tau}{c}\right)\left[\hat{\varepsilon}_z + \tanh\left(\frac{a\tau}{c}\right)(\hat{k}\times\hat{\varepsilon})_y\right]\right\}$$
$$\times \cos\left[k_x\frac{c^2}{a}\cosh\left(\frac{a\tau}{c}\right) - \frac{\omega c}{a}\sinh\left(\frac{a\tau}{c}\right) - \theta(\vec{k},\lambda)\right], \qquad (6.53)$$

$$\vec{B}_\tau^{zp}(0,\tau) = \sum_{\lambda=1}^{2} \int d^3k (\hbar\omega/2\pi^2)^{1/2}$$

$$\times \left\{ \hat{x}(\hat{k}\times\hat{\varepsilon})_x + \hat{y}\cosh\left(\frac{a\tau}{c}\right)\left[(\hat{k}\times\hat{\varepsilon})_y - \tanh\left(\frac{a\tau}{c}\right)\hat{\varepsilon}_z\right] \right.$$

$$\left. + \hat{z}\cosh\left(\frac{a\tau}{c}\right)\left[\hat{\varepsilon}_z + \tanh\left(\frac{a\tau}{c}\right)(\hat{k}\times\hat{\varepsilon})_y\right]\right\}$$

$$\times \cos\left[k_x\frac{c^2}{a}\cosh\left(\frac{a\tau}{c}\right) - \frac{\omega c}{a}\sinh\left(\frac{a\tau}{c}\right) - \theta(\vec{k},\lambda)\right],$$
(6.54)

where $\hat{\varepsilon}_x$ is the scalar projection of the $\hat{\varepsilon}$ unit vector along the x-direction, and similarly for $\hat{\varepsilon}_y$ and $\hat{\varepsilon}_z$.

In order to calculate the rate of change of the momentum applied by the zero-point field on the electromagnetically interacting, accelerating object, one observes that each individual inertial frame can be associated with its own (or proper) random electromagnetic zero-point field background and thus with its proper zero-point field vacuum spectral energy density distribution. For each inertial reference frame its proper electromagnetic vacuum is homogeneously and isotropically distributed. This means that the components of the net Poynting vector of the zero-point field of a given frame when observed in that same frame all should vanish. Moreover, of the components of the 4 × 4 electromagnetic energy-momentum stress tensor only the diagonal components survive as all other components should be zero.

Both the momentum density and the total momentum of the zero-point field inside the proper volume V_0 of the body display a time rate of change and have a nonvanishing time derivative. Thus, although both the Poynting vector and the momentum of the zero-point field inside the body instantaneously vanish, their time derivatives at that coincidence time do not vanish. This can also be visualized as a manifestation of a reduction in symmetry of the event horizon that passes from the perfect 3D symmetry of a sphere to a lower two-dimensional (2D) symmetry that is merely axial. This results in a special situation that may be visualized as a stress or tension in the vacuum field which is manifested on the accelerated particle as the Rindler frame force. For uniform acceleration of the object under consideration in the x-direction, the Poynting vector is

$$\vec{N}^{zp}(\tau) = -\hat{x}\frac{2c}{3}\sinh\left(\frac{2\alpha\tau}{c}\right)\int \rho(\omega)d\omega, \qquad (6.55)$$

where $\rho(\omega) = \frac{\hbar\omega^3}{2\pi^2 c^3}$ is the spectral energy density of the quantum vacuum fluctuations.

The amount of radiative momentum carried by the zero-point fluctuations that are passing through and instantaneously contained in the accelerated object which is undergoing uniform proper acceleration $\vec{\alpha} = \alpha\hat{x}$ and that has proper volume V_0, is

$$\vec{p}^{zp}(\tau) = -\hat{x}\frac{4}{3c^2}V_0 v_\tau \gamma_\tau \int \rho(\omega)d\omega. \qquad (6.56)$$

The interacting fraction of the zero-point momentum instantaneously passing through the object is expressed by

$$\vec{p}^{zp}(\tau) = -\hat{x}\frac{4}{3c^2}V_0 v_\tau \gamma_\tau \int \eta(\omega)\rho(\omega)d\omega, \qquad (6.57)$$

where $\eta(\omega)$ is the fraction of the energy that interacts with the material particles contained in V_0. By taking the time derivative of (6.57) one gets

$$\frac{d\vec{p}^{zp}}{dt} = \vec{f}^{zp} = -\left[\frac{4}{3c^2}V_0 \int \eta(\omega)\rho(\omega)d\omega\right]\vec{a}. \qquad (6.58)$$

Equation (6.58) indicates that in order to maintain the acceleration of such an object, a motive force $\vec{f}$ must continuously be applied to balance the vacuum electromagnetic counteracting reaction force (namely the Rindler frame force $\vec{f}^{zp}$) which is given by

$$\vec{f} = -\vec{f}^{zp} = \left[\frac{V_0}{c^2}\int \eta(\omega)\rho(\omega)d\omega\right]\vec{a}, \qquad (6.59)$$

where the factor has been suppressed the factor 4/3 in order to obtain the proper correct relativistic four-vector force expression. On the basis of Eq. (6.59), one can define the zero-point field contribution to the inertial mass contained in V_0 as

$$m_i = \left[\frac{V_0}{c^2} \int \eta(\omega)\rho(\omega)d\omega \right]. \quad (6.60)$$

In other words, in Haisch's and Rueda's quantum vacuum inertia hyphotesis, some of the apparent inertial mass of an object originates in the interacting fraction of the zero-point energy instantaneously contained in an object. In this view, the apparent momentum of the object can be traced back to the momentum of the zero-point radiation field. The rest mass in Eq. (6.60) exactly corresponds to the amount of zero-point field mass equivalent enclosed within the volume of the body and that, thanks to the $\eta(\omega)$ spectral factor, interacts with the body. A stationary observer would come to the conclusion that the mass of an accelerating object is steadily increasing as $\gamma_\tau m$, but in the accelerating frame no mass change is evident. One's momentum in one's own reference frame is always zero. Physical consequences emerge only in correspondence to a change in the momentum. In the approach of the quantum vacuum inertia hypothesis, it becomes physically evident how the Lorentz factor γ_τ, which characterizes a space-time geometry relationship, acquires the physical role of the relativistic mass increase parameter. Moreover, one can justify also why the relativistic mass increase must become infinite at the speed of light: the effect of any forces due to the zero-point field at speeds faster than c cannot be propagated.

A relevant result that the quantum vacuum inertia hypothesis can obtain is based on the possibility to explain the origin of the equivalence between inertial mass and gravitational mass. While in the standard theoretical framework of general relativity and related theories, the equality (or proportionality) of inertial mass to gravitational mass has to be assumed and remains unexplained, instead Rueda's and Haisch's approach of electromagnetic quantum vacuum leads naturally and inevitably to this equality. The interaction between the electromagnetic quantum vacuum and the electromagnetically-interacting particles constituting any physical object (quarks and electrons) is identical for the following two situations: acceleration with respect to constant velocity inertial frames or remaining fixed above some gravitating body with respect to freely-falling local inertial frames. The existence of a Rindler frame force in an accelerating reference frame translate and correspond exactly to a reference frame fixed above a gravitating body. In the same manner that light rays are deviated from straight-line propagation by a massive gravitating body, the other forms of electromagnetic radiation, including the electromagnetic zero-point field rays (in the SED approximation), are also deviated from straight-line propagation. In the papers [465, 466], Rueda and Haisch interpreted the drag force exerted by the electromagnetic quantum vacuum radiation as the inertia reaction force of an object that is being forced to accelerate through the electromagnetic quantum vacuum field. This leads to the following associated nonrelativistic form of the inertia reaction force

$$\vec{f}_*^{zp} = -m_i \vec{g}_\omega, \qquad (6.61)$$

where $\vec{g}_\omega$ is the acceleration with which ω appears in a local inertial frame I_* and the coefficient m_i is given by (6.60). Here, the coupling function $\eta(\omega)$ — that indicates the relative strength of the interaction between the zero-point field and the massive object, interaction which acts to oppose the acceleration — of the inertial mass (6.60) has different shapes for the electron, a given quark, a composite particle like the proton, a molecule, a homogeneous dust grain or a homogeneous macroscopic body. However, what appears as inertial mass, m_i, to the observer in a local Lorentz frame $I_{*,L}$ is what corresponds to passive gravitational mass, m_g, and thus one has

$$m_g = \left[\frac{V_0}{c^2} \int \eta(\omega) \rho(\omega) d\omega \right]. \qquad (6.62)$$

In synthesis, in Rueda's and Haisch's approach, the physical basis for the principle of equivalence is the fact that accelerating through the electromagnetic quantum vacuum is identical to remaining fixed in a gravitational field and having the electromagnetic quantum vacuum fall past on curved geodesics. In both situations the observer will experience an asymmetry in the radiation pattern of the electromagnetic quantum vacuum which results in a force — either the inertia reaction force or weight — which becomes then within this more general Einsteinian perspective the same thing.

As regards the quantum vacuum inertia hypothesis, significant results have been also obtained recently by Sunahata, Rueda and Haisch in the paper *Quantum vacuum and inertial reaction in nonrelativistic QED* [469]. In this paper, by following the approach originally outlined in the papers [465–467], Sunahata, Rueda and Haisch analyzed the interaction between the zero-point field and an object under hyperbolic motion (constant proper acceleration) within a formulation that uses the low energy version of quantum electrodynamics, also called nonrelativistic quantum electrodynamics with the creation and annihilation operators for all the averaging calculations, finding that a reaction force arises which is proportional in magnitude, and opposite in direction, to the acceleration (and thus which can be interpreted as inertia). The three authors also pointed out that the equivalence principle between inertial mass and gravitational mass follows naturally in this picture and showed that QED leads to the same results reported in the papers [465–467] mentioned above, and thus that there is a contribution to the inertia reaction force coming from the electromagnetic quantum vacuum.

By following Sunahata's, Rueda's and Haisch's treatment in the paper [469], let us consider an object, at rest at time $t_* = 0$ in the laboratory frame I_*, which is uniformly accelerated by an external force which produces a rectilinear motion along the x-axis with constant proper acceleration $\vec{a} = a\hat{x}$. Consider a noninertial frame — namely a Rindler frame — S such that its x-axis coincides with that of I_* and let the body be located at coordinates $(c^2/a, 0, 0)$ in S at all times. So this point of S performs hyperbolic motion. The acceleration of the body point in I_* is $\vec{a}_* = \gamma_\tau^{-3}\vec{a}$ at the body proper time τ. Consider also an infinite collection of inertial frames $\{I_\tau\}$ such that at body proper time τ, the body is located at point $(c^2/a, 0, 0)$

of I_τ. The I_τ frames have all axes parallel to those of I_* and their x-axes coincide with that of I_*. By setting the proper time τ such that at $\tau = 0$ the corresponding I_τ coincides with I_*, one has $I_{\tau=0} = I_*$ and thus the hyperbolic motion assures that

$$x_* = \frac{c^2}{a}\cosh\left(\frac{a\tau}{c}\right), \tag{6.63}$$

$$t_* = \frac{c}{a}\sinh\left(\frac{a\tau}{c}\right), \tag{6.64}$$

$$\beta_\tau = \frac{u_x(\tau)}{c} = \tanh\left(\frac{a\tau}{c}\right), \tag{6.65}$$

$$\gamma_\tau = \left(1 - \beta_\tau^2\right)^{-1/2} = \cosh\left(\frac{a\tau}{c}\right). \tag{6.66}$$

By using Lorentz transformations from the laboratory frame I_* into an instantaneously comoving frame I_τ, the electromagnetic zero-point field vectors $\vec{E}_{zp}$ and $\vec{B}_{zp}$ of I_* as represented in I_τ are given by relations

$$\vec{E}_\tau^{zp}(0,\tau) = \sum_{\lambda=1}^{2}\int d^3k H_{zp}(\omega)$$

$$\times \left\{\hat{x}\hat{\varepsilon}_x + \hat{y}\cosh\left(\frac{a\tau}{c}\right)\left[\hat{\varepsilon}_y - \tanh\left(\frac{a\tau}{c}\right)\left(\hat{k}\times\hat{\varepsilon}\right)_z\right]\right.$$

$$\left. + \hat{z}\cosh\left(\frac{a\tau}{c}\right)\left[\hat{\varepsilon}_z + \tanh\left(\frac{a\tau}{c}\right)(\hat{k}\times\hat{\varepsilon})_y\right]\right\}$$

$$\times\{\vec{\alpha}(\vec{k},\lambda)e^{i\theta} + \vec{\alpha}^+(\vec{k},\lambda)e^{-i\theta}\}, \tag{6.67}$$

$$\vec{B}_\tau^{zp}(0,\tau) = \sum_{\lambda=1}^{2}\int d^3k H_{zp}(\omega)$$

$$\times \left\{\hat{x}(\hat{k}\times\hat{\varepsilon})_x + \hat{y}\cosh\left(\frac{a\tau}{c}\right)\left[(\hat{k}\times\hat{\varepsilon})_y - \tanh\left(\frac{a\tau}{c}\right)(\hat{\varepsilon})_z\right]\right.$$

$$\left. + \hat{z}\cosh\left(\frac{a\tau}{c}\right)\left[(\hat{k}\times\hat{\varepsilon})_z + \tanh\left(\frac{a\tau}{c}\right)(\hat{\varepsilon})_y\right]\right\}$$

$$\times\{\vec{\alpha}(\vec{k},\lambda)e^{i\theta} + \vec{\alpha}^+(\vec{k},\lambda)e^{-i\theta}\}, \tag{6.68}$$

where

$$\theta = \hat{k}_x \frac{c^2}{a}\cosh\left(\frac{a\tau}{c}\right) - \omega\frac{c}{a}\sinh\left(\frac{a\tau}{c}\right), \qquad (6.69)$$

$\hat{\varepsilon}_i$ are the polarization components, $\vec{k}$ is the polarization wave vector, λ is the polarization index, $\vec{a}(\vec{k},\lambda)$ and $\vec{a}^+(\vec{k},\lambda)$ are annihilation and creation operators and the spectral function in QED is $H^2_{zp}(\omega) = \frac{\hbar\omega}{4\pi^2}$. We assume these fields as seen in I_τ also correspond to the fields as instantaneously seen in S at proper time τ. The two fields are the same at a given space-time point; however, the time evolution and space distribution of the field in S and those of the field in I_τ are not the same. If one considers now the ZPF radiation background of I_* in the act of being swept through by the object, the corresponding ZPF perceived in I_τ is obtained by fixing the attention on the point of the observer at $(c^2/a, 0, 0)$ of I_* (that instantaneously coincides with the object at the object initial proper time $\tau = 0$) and by considering that point as referred to another inertial frame I_τ that instantaneously will coincide with the object at a future generalized object proper time $\tau > 0$. Hence, one can compute the I_τ-frame Poynting vector, as instantaneously evaluated at the $(c^2/a, 0, 0)$ space point of the I_* inertial frame, namely in I_τ at the I_τ space-time point

$$ct_\tau = -\frac{c^2}{a}\sinh\left(\frac{a\tau}{c}\right), \quad x_\tau = -\frac{c^2}{a}\cosh\left(\frac{a\tau}{c}\right), \quad y_\tau = 0, \quad z_\tau = 0, \qquad (6.70)$$

where the time in I_τ, called t_τ, is set to zero at the instant when S and I_τ (locally) coincide, which happens at proper time τ. Then, the ZPF Poynting vector that enters the body of the accelerating object in the instantaneous comoving frame I_τ is

$$\begin{aligned}\vec{S}^{zp}_* &= \frac{c}{4\pi}\langle 0|\vec{E}^{zp}_\tau \times \vec{B}^{zp}_\tau|0\rangle_*\\ &= \frac{c}{4\pi}\{\hat{x}\langle 0|E_y B_z - E_z B_y|0\rangle + \hat{y}\langle 0|E_z B_x - E_x B_z|0\rangle\\ &\quad + \hat{z}\langle 0|E_x B_y - E_y B_x|0\rangle\}. \end{aligned} \qquad (6.71)$$

The star in Eq. (6.71) implies that the quantity needs to be evaluated in the laboratory inertial frame I_*. It turns out that only the two terms of the x-component of the ZPF Poynting vector are nonvanishing and the other seven components are zero; so, exact calculations lead to the following result

$$\vec{S}_*^{zp} = -\hat{x}\frac{2c}{3}\sinh\left(\frac{2a\tau}{c}\right)\int\frac{\hbar\omega^3}{2\pi^2 c^3}d\omega. \qquad (6.72)$$

The quantity (6.72) represents the energy flux, namely the ZPF energy that enters the uniformly accelerating object's body per unit area per unit time from the viewpoint of the observer at rest in the inertial laboratory frame I_*. The radiation entering the body is that of the ZPF centered at the I_* frame of the observer. The energy flux (6.72) determines a momentum of the ZPF background the object has swept through after a time duration t_*, which, as judged again from the I_*-frame viewpoint, is

$$p_* = g_* V_* = \frac{S_*}{c^2}V_* = -\hat{x}\frac{1}{6\pi c}\gamma_\tau^2 \beta_\tau \langle E_*^2 + B_*^2 \rangle V_*, \qquad (6.73)$$

where S_* represents energy flux producing a parallel, x-directed momentum density, namely a field momentum growth per unit time and per unit volume as it is incoming towards the object position, $(c^2/a, 0, 0)$ of S, at object proper time τ and as estimated from the viewpoint of I_*. In particular, if V_0 is the proper volume of the object, from the viewpoint of I_*, because of Lorentz contraction such volume is then $V_* = \frac{V_0}{\gamma_\tau}$ and thus the amount of momentum (6.21) due to the field inside the volume of the object according to I_* becomes

$$p_* = g_* V_* = \frac{S_*}{c^2}\frac{V_0}{\gamma_\tau} = -\hat{x}\frac{4V_0}{3}c\gamma_\tau \beta_\tau \left(\frac{1}{c^2}\int \eta(\omega)\frac{\hbar\omega^3}{2\pi^2 c^3}d\omega\right). \qquad (6.74)$$

At proper time $\tau = 0$, the $(c^2/a, 0, 0)$ point of the laboratory inertial system I_* instantaneously coincides and comoves with the object point of the Rindler frame S in which the object is fixed. Since the observer located at $x_* = \frac{c^2}{a}$, $y_* = 0$, $z_* = 0$ at $t_* = 0$ coincides and comoves with the object because the latter is accelerated with constant proper acceleration $\vec{a}$, the object receives a time rate of change of incoming ZPF momentum of the form:

$$\frac{d\vec{p}_*}{dt_*} = \frac{1}{\gamma_\tau} \frac{d\vec{p}_*}{d\tau}|_{\tau=0}, \qquad (6.75)$$

and thus the radiation from the ZPF exerts on the object a force with respect to I_* at $t_* = 0$ is given by relation

$$\vec{F}_* = \frac{d\vec{p}_*}{dt_*} = -\left(\frac{4V_0}{3c^2} \int \eta(\omega) \frac{\hbar \omega^3}{2\pi^2 c^3} d\omega\right) \vec{a}, \qquad (6.76)$$

where

$$m_i = \frac{V_0}{c^2} \int \eta(\omega) \frac{\hbar \omega^3}{2\pi^2 c^3} d\omega, \qquad (6.77)$$

is an invariant scalar having the dimension of a mass which corresponds to the fraction of the energy of the ZPF radiation enclosed within the object that interacts with the object as parametrized by the $\eta(\omega)$ factor in the integrand. Sunahata's, Rueda's and Haisch's approach based on nonrelativistic QED and synthesized in Eqs. (6.63)–(6.77) shows thus that inertial mass and gravitational mass emerge as the identical thing viewed from two complementary perspectives. On one hand, an object accelerating through the electromagnetic zero-point field experiences resistance from the field. On the other hand, in the case of an object in a gravitational field, the electromagnetic zero-point field propagates on curved geodesics, accelerating with respect to the fixed object, thereby generating weight. This means, in other words, that the equivalence principle does not need to be independently postulated.

Already since the second half of the 1960s several authors have proposed the idea that a vacuum energy is ultimately responsible for gravitation. It was Sakharov [411] that, on the basis of the work of Zeldovich [437], originally proposed a model in which a connection is drawn between Hilbert–Einstein action and a quantum vacuum intended as a fundamental arena of physical reality. This model introduced a view of gravity as "a metric elasticity of space".

Following Sakharov's idea and using the techniques of stochastic electrodynamics, Puthoff proposed that gravity could be construed as a (long range) form of the van der Waals force [470]. Although interesting and stimulating in some respects, Puthoff's attempt to derive a Newtonian inverse square force of gravity proves to be unsuccessful [471–474]. An alternative approach has been successively developed by Puthoff in 2002 [414] on the basis of earlier work of Dicke [475] and of Wilson [476]: a polarizable vacuum model of gravitation. In this representation, gravitation comes from an effect by massive bodies on both the permittivity, ε_0, and the permeability, μ_0, of the vacuum and thus on the velocity of light in the presence of matter. Puthoff's model clearly provides a theory alternative to general relativity, since it does not involve an actual curvature of space-time. In the weak field limit, Puthoff's polarizable vacuum model of gravitation replicates the results of general relativity, including the classic tests (gravitational redshift, bending of light near the Sun, advance of the perihelion of Mercury). Differences appear in the strong-field regime, which should lead to interesting tests between the two theories.

By following the paper [414], here we want now to analyse the main features of Puthoff's approach. While the principles of general relativity are generally formulated in terms of tensor formulations in curved space-time, the basic postulate of Puthoff's polarized vacuum approach in order to reproduce the curvature of space-time is that the polarizability of the vacuum in the vicinity of a mass (or other mass-energy concentrations) differs from its asymptotic far-field value because of vacuum polarization effects caused by the presence of the mass. That is, here one postulates the following relation for the variable polarization of the vacuum

$$\vec{D} = K\varepsilon_0 \vec{E}, \quad (6.78)$$

where $\vec{E}$ is the electric field, K is the (altered) dielectric constant of the vacuum (typically a function of position) due to (general relativity-induced) vacuum polarizability changes under consideration. Equation (6.78) establishes that the presence of electromagnetic energy or massive objects modulates the vacuum polarization in a linear fashion. The vacuum dielectric constant K constitutes the ultimate visiting card of Puthoff's model. It acts as a variable refractive index under conditions in which the vacuum polarizability is assumed to change in response to phenomena characteristic of general relativity. In particular, its effects on the various measurement processes that characterize general relativity are the following: the velocity of light changes from c to c/K, the time intervals change from Δt_0 to $\Delta t_0 \sqrt{K}$ (which indicates that for $K > 1$, namely in a gravitational potential, the time intervals between clock ticks is increased, that is the clock runs slower), the lengths of rods change from Δr_0 to $\Delta r_0/\sqrt{K}$. In Puthoff's model, the curvature of space — for example in the vicinity of a planet or a star — is associated with the effects on measurement processes of lengths and time intervals that take place for $K > 1$. Such an influence on the measuring processes due to induced polarizability changes in the vacuum near the body under consideration leads to the general relativistic concept that the presence of the body "influences the metric".

Moreover, Puthoff's model developed in the paper [420] suggests some fundamental equations of motion of a massive body in the polarized quantum vacuum. Let us consider a test particle of mass m_0 and charge q within a relatively weak gravity potential. Variation of the total lagrangian density for matter-field interactions in the polarized vacuum given by relation

$$L_d = -\left(\frac{m_0 c^2}{\sqrt{K}}\sqrt{1-\left(\frac{v}{c/K}\right)^2} + q\Phi - q\vec{A}\cdot\vec{v}\right)\delta^3(\vec{r}-r_0)$$

$$-\frac{1}{2}\left(\frac{B^2}{K\mu_0} + K\varepsilon_0 E^2\right)$$

$$-\frac{\lambda}{K^2}\left((\nabla K)^2 + \frac{1}{(c/K)^2}\left(\frac{\partial K}{\partial t}\right)^2\right) \quad (6.79)$$

(where $(\Phi, \vec{A})$ are the electromagnetic potentials, $\vec{B}$ is the magnetic field and $\lambda = \frac{c^4}{32\pi G}$) with respect to the test particle variables leads to the following equation of motion in a variable dielectric (and thus polarized) vacuum (with $K \neq 1$):

$$\frac{d}{dt}\left[\frac{\left(m_0 K^{3/2}\right)\vec{v}}{\sqrt{1-\left(\frac{v}{c/K}\right)^2}}\right] = m_0(c^2\vec{E} + \vec{v}\times\vec{B}) + \frac{m_0 c^2}{K}$$

$$\cdot \frac{1+\left(\frac{v}{c/K}\right)^2}{2\sqrt{1-\left(\frac{v}{c/K}\right)^2}} \cdot \frac{\nabla K}{K} \qquad (6.80)$$

Equation (6.80) shows that there are two forces acting onto the test particle of mass m_0: the Lorentz force and the dielectric force proportional to the gradient of the vacuum polarization. The importance of this second term lies in the fact that thanks to it one can account for the gravitational potential, either in Newtonian or general relativistic form. It might be interesting to note that with $m_0 \to 0$ and $v \to \frac{c}{K}$, as would be the case for a photon, the deflection of the trajectory is twice as the deflection of a slow moving massive particle. This is an important indication of conformity with general relativity.

In analogous way, variation of the total lagrangian density with regard to the K variable leads to the expression of the generation of the shadowing of space within general relativity, owed to the presence of both matter and fields. The equation has three right-hand side terms:

$$\nabla^2 \sqrt{K} - \frac{1}{(c/K)^2} \cdot \frac{\partial^2 \sqrt{K}}{\partial t^2} = \frac{-K}{4\lambda}[P(K) + Q(K) + R(K)]. \qquad (6.81)$$

Here, $P(K)$ represents the change of the vacuum dielectric constant caused by the mass density:

$$P(K) = \frac{m_0 c^2}{\sqrt{K}} \cdot \frac{1+\left(\frac{v}{c/K}\right)^2}{\sqrt{1-\left(\frac{v}{c/K}\right)^2}} \cdot \delta^3(\vec{r}-\vec{r}_0). \qquad (6.82)$$

$Q(K)$ is the change of the vacuum dielectric constant caused by the electromagnetic field energy density:

$$Q(K) = \frac{1}{2}\left(\frac{B^2}{K\mu_0} + K\varepsilon_0 E^2\right). \tag{6.83}$$

$R(K)$ is the change of the vacuum dielectric constant caused by the vacuum polarization energy density itself:

$$R(K) = -\frac{\lambda}{K^2}\left((\nabla K)^2 + \frac{1}{(c/K)^2}\left(\frac{\partial K}{\partial t}\right)^2\right). \tag{6.84}$$

Equations (6.80) and (6.81), together with Maxwell's equations for propagation in a medium with variable dielectric constant, constitute the master equations of Puthoff's polarized model of the vacuum that must be used in discussing matter-field interactions.

In the case of a static gravity field of a spherical mass distribution (a planet or a star), the appropriate expression of the vacuum dielectric constant, solution of Eq. (6.81), has a simple exponential form:

$$\sqrt{K} = e^{GM/rc^2}. \tag{6.85}$$

The solution (6.86) can be approximated by expanding it into a series:

$$K = e^{2GM/rc^2} = 1 + \frac{2GM}{rc^2} + \frac{1}{2}\left(\frac{2GM}{rc^2}\right)^2 + \cdots \tag{6.86}$$

where G is the gravitational constant, M is the mass, and r is the distance from the origin located at the centre of the mass M. For comparison with expressions derived by conventional general relativity techniques, it is sufficient to focus the attention to a weak-field approximation expressed by expansion of the exponential to second order. This solution reproduces (to the appropriate order) the usual general-relativistic Schwarzschild metric predictions in the weak field limit conditions (i.e. Solar system).

The obtained solution (6.65) or (6.86) lead to explore the classical experimental tests of general relativity in Puthoff's polarized model of gravitation. In particular, as regards the gravitational red shift, if in a gravitation-free part of space a photon emission from an excited atom takes place at some frequency ω_0, in a gravitational field the emission frequency of the photon turns out to be altered (redshifted) to $\omega = \omega_0/\sqrt{K}$. A photon emitted by an atom located on the surface of a body of mass M and radius R will experience a redshift by the following amount:

$$\frac{\Delta\omega}{\omega_0} = \frac{\omega - \omega_0}{\omega_0} \approx -\frac{GM}{Rc^2}, \qquad (6.87)$$

where $\frac{GM}{Rc^2} \ll 1$. The photon, after having climbed up the gravity potential of the star, will retain its acquired frequency unchanged, and the change in frequency can be tested locally by comparing it with photons emitted by the same type of atoms at the same temperature, but within the weak gravity field of the laboratory. On the basis of Puthoff's research, measurement of the redshift of the sodium D_1 line emitted on the surface of the sun and received on earth verified Eq. (6.87) to a precision of 5%, and experiments carried out on the surface of the earth involving the comparison of photon frequencies at different heights improved the accuracy of verification still further to a precision of 1%.

It is also interesting to remark that, if both the conventional approach and Puthoff's polarized vacuum approach to general-relativistic problems lead to the same results for weak gravity fields, namely for small departures from flatness, instead for strong fields, namely for increasingly larger departures from flatness, the two approaches, although initially following similar trends, begin to diverge with regard to specific magnitudes of effects. In Puthoff's polarized vacuum approach the solution for the static gravitational case leads to a metric tensor that is exponential in form, while in the conventional general relativity approach one obtains a more complex Schwarzschild solution. For example, a major difference between the predictions of Puthoff's polarized vacuum model and the standard general relativity for strong fields lies in the fact that whereas the standard general relativistic Schwarzschild solution limits neutron stars (or neutron star mergers) to $\approx 2, 8$ solar masses because of black hole formation, no such constraint exists for the exponential metric characterizing Puthoff's model. This raises the possibility that such anomalous observations as the enormous radiative output of the gamma ray burster GRB990123 might be interpreted as being associated with collapse of a very massive star (hypernova), or the collision of two high density neutron stars. Thus the perspective is opened to find astrophysical evidence able to provide discriminants between the standard general relativity and polarized vacuum approaches.

As regards the epistemology underlying the polarizable-vacuum approach as compared with the standard general relativity, one can draw some interesting considerations in the following sentences by Atkinson [477]: "It is possible, on the one hand, to postulate that the velocity of light is a universal constant, to define "natural" clocks and measuring rods as the standards by which space and time are to be judged, and then to discover from measurement that space-time, and space itself, are "really" nonEuclidean; alternatively, one can define space as Euclidean and time as the same everywhere, and discover (from exactly the same measurements) how the velocity of light, and natural clocks, rods, and particle inertias "really" behave in the neighbourhood of large masses. There is just as much (or as little) content for the word "really" in the one approach as in the other; provided that each is self-consistent, the ultimate appeal is only to convenience and fruitfulness, and even "convenience" may be largely a matter of personal taste..." The true nature of the polarizability of the vacuum has probably to be found and possibly defined by more fundamental physical processes, nonetheless, according to the author of this book, the polarized vacuum model provides an intuitive, physical appeal that can be useful in bridging the gap between flat-space Newtonian physics and the curved space-time formalisms of general relativity.

As regards the interpretation of gravitational interaction as a result of density fluctuations of a polarized quantum vacuum, very interesting results and perspectives have been suggested recently by YE Xing-Hao and LIN Qiang [478] and by Maurizio Consoli [415–417]. Xing-Hao's and Qiang's research and Consoli's results may be considered as relevant developments of the fundamental ideas which are at the basis of Puthoff's polarizable model of gravitation.

In the article *Inhomogenous vacuum: an alternative interpretation of curved space-time* [478], Xing-Hao and Qiang analysed the strong similarities between the light propagation in a curved space-time and in a medium with graded refractive index, demonstrating that a curved space-time is equivalent to an inhomogeneous vacuum for light propagation. On the light of these similarities, Xing-Hao and Qiang suggested to see the vacuum around the gravitational matter as a special optical medium with graded refractive index, which may be the physical interpretation of the curved space-time in general relativity and provided a general method for calculating the corresponding refractive index of the vacuum. Xing-Hao and Qiang found that, for a weak gravitational field, the refractive index of the vacuum turns out to be a unified exponential function of the gravitational potential for both outside and inside the gravitational matter system and showed that this result provides a simple optical way to analyse the gravitational lensing. The equivalence between the inhomogeneous vacuum and the curved space-time of general relativity emerges from the similarity between the general-relativistic angular displacement $d\phi$ of the coordinate radius R in a curved space-time given by relation

$$d\phi = \frac{dR}{R/\sqrt{A(R)}\sqrt{\left[\frac{R/\sqrt{B(R)}}{R_0/\sqrt{B(R_0)}}\right]^2 - 1}}, \qquad (6.88)$$

where R_0 is the coordinate radius at the point P_0 of the light ray nearest to the gravitational matter M, $A(R)$ and $B(R)$ come from a static and spherically symmetric metric of the standard form

$$dT^2 = B(R)c^2 dt^2 - A(R)dR^2 - R^2(d\theta^2 + \sin^2\theta d\phi^2) \qquad (6.89)$$

and the angular displacement of the radius in a plat space-time for the light propagation in a medium of a spherically symmetric refractive index n, given by Fermat's principle

$$d\phi = \frac{dr}{r\sqrt{\left[\frac{nr}{n_0 r_0}\right]^2 - 1}}, \qquad (6.90)$$

where r_0 and n_0 represent the radial distance and refractive index at the point closest to the centre, respectively. From Eqs. (6.88) and (6.90) and the boundary conditions at infinity, the refractive index of this inhomogeneous vacuum can be written as

$$n = \frac{R}{r/\sqrt{B(R)}}, \qquad (6.91)$$

where

$$\frac{R}{r} = \frac{1}{\sqrt{A(R)}} \frac{dR}{dr}. \qquad (6.92)$$

Equations (6.91)) and (6.92) introduce a general method for finding the vacuum refractive index profile of a static spherically symmetric gravitational field, where the coefficients $A(R)$ and $B(R)$ can be obtained from the Schwarzschild solutions. In particular, taking account of the Schwarzschild exterior and interior solutions, Xing-Hao and Qiang found, if the gravitational field is not extremely strong, the following unitary expression for the vacuum refractive index

$$n = \exp\left(-\frac{2P_r}{c^2}\right), \qquad (6.93)$$

where P_r is the gravitational potential at position r ($P_r = -\frac{GM}{r}$ outside the gravitational matter system and $P_r = -\{\frac{GM(r_M)}{r_M} + \int_r^{r_M} \left[\frac{GM(r)}{r^2}\right] dr\}$ inside the gravitational matter system, r_M being the radial coordinate in flat space-time corresponding to R_M in curved space-time). For a many-body system, the vacuum refractive index (6.93) can be generalized as

$$n = \exp\left(-\frac{2}{c^2}(P_{r1} + P_{r2} + P_{r3} + \cdots)\right) = n_1 n_2 n_3 \ldots \qquad (6.94)$$

where P_{r1}, P_{r2}, P_{r3} and n_1, n_2, n_3 are the gravitational potential and the corresponding refractive index caused by each gravitational body respectively.

By following the philosophy that underlies Puthoff's polarizable vacuum model, Consoli [18, 415–417] has introduced a physical vacuum intended as a superfluid medium — a Bose condensate of elementary spinless quanta — whose long-range fluctuations, on a coarse-grained scale, resemble the Newtonian potential, allowing thus the first approximation to the metric structure of classical general relativity to be obtained. In Consoli's model, taking into consideration the long-wavelength modes, gravity is induced by an underlying scalar field $s(x)$ describing the density fluctuations of the vacuum, that in weak gravitational fields, on a coarse-grained scale, is identified with the Newtonian potential

$$s \approx U_N = -G_N \sum_i \frac{M_i}{|\vec{r} - \vec{r}_i|}, \qquad (6.95)$$

namely

$$g_{\mu\nu} = g_{\mu\nu}[s(x)]. \qquad (6.96)$$

The effect of this scalar field (and thus of the density fluctuations of the vacuum) is to determine a re-scaling of the masses, namely under its action any mass scale M (or binding energy) is replaced by an effective mass, $\frac{M}{\lambda(s)}$. The effective metric structure that re-absorbs the local, isotropic modifications of space-time into its basic ingredients (value of the light speed and space-time units) has the form

$$g_{\mu\nu}(s) \equiv diag\left(\frac{1}{\lambda^2(s)}, -\lambda^2(s), -\lambda^2(s), -\lambda^2(s)\right), \qquad (6.97)$$

from which, by expanding in powers of $s(x)$ with the natural condition $\lambda(0) = 1$ and choosing the units so that

$$\frac{1}{\lambda(s)} = 1 + s + O(s^2), \qquad (6.98)$$

one arrives to the weak-field line element

$$g_{\mu\nu}(s) \equiv diag((1+2s), -(1-2s), -(1-2s), -(1-2s)). \qquad (6.99)$$

By using the basic two-valued nature of the zero-momentum connected scalar propagator in the broken-symmetry phase and the 'triviality' of contact, point-like interactions in $3+1$ dimensions, one could naturally explain the origin of a hydrodynamic coupling G_∞ that is infinitesimally weak in units of the physical coupling strength set by the Fermi constant G_F. If this coupling is identified with the Newton constant G_N, one has a toy model of gravity, restricted to a definite class of phenomena, that could be used as a first approximation to reduce the conceptual gap with particle physics. Moreover, Consoli's model advocates a phenomenon of symmetry breaking consisting in the spontaneous creation from the empty vacuum of elementary spinless scalar quanta and provides a simple unified picture of the scalar field in terms of the density n of the elementary quanta (treated as hard spheres) and of their scattering length a. About this model, Consoli recently remarked: "At the present the approach has some interesting aspects [...]. The interesting features consist in a potentially simple explanation of the hierarchical pattern of scales that is observed in nature. By describing the scalar condensate as a true physical medium, and by approaching the continuum limit of the underlying scalar quantum field theory, one could handle, at the same time, an extremely small scattering length $a \approx 10^{-33}$ cm, a typical elementary particle scale $\frac{1}{\sqrt{na}} \approx 10^{-17}$ cm and a macroscopic mean free path $\frac{1}{na^2} \approx 10^{-1}$ cm that can be used to mark the onset of a hydrodynamic regime. At the same time, one can easily understand basic properties of the gravitational interaction. For instance, the equality of both passive gravitational mass with the same inertial mass generated by the vacuum structure is straightforward. Finally, the relevant curvature effects will be orders of magnitude smaller than those expected by solving Einstein's equations with the full energy-momentum tensor as a source term." [284].

The gravitational effects of the quantum vacuum are a subject of questions and puzzlement, in particular because they are expected to function as an effective cosmological constant which generate an accelerated expansion of the universe; but simple estimates of its magnitude are very large, exceeding the observed value by between 60 and 120 orders of magnitude [479–481]. The vacuum energy issue indicates a profound discrepancy between General Relativity and Quantum Field Theory — a major problem for theoretical physics. The field theory view of gravity as a spin-2 field, and with the quantum vacuum contributing to the cosmological constant, is a halfway house between general relativity and a full quantum gravity theory (which will need to have both general relativity and spin-2 quantum field theory as appropriate limits), and it is needed to resolve that discrepancy no matter what final quantum gravity theory is adopted. As regards the problem of the gravitational effects of the vacuum energy, in the recent paper *The gravitational effect of the vacuum* [482] Ellis, Murugan and Van Elst make an interesting proposal, based on the use of Trace-Free Einstein Equations instead of the standard Einstein's Field Equations [483, 484], that while does not give a unique value for the effective cosmological constant, opens the perspective to solve the huge discrepancy between theory and observation, namely resolves the problem of the discrepancy between the vacuum energy density and the observed value of the cosmological constant. Ellis', Murugan's and Van Elst's approach assumes both the trace-free Einstein equations

$$R_{ab} - \frac{1}{4} R g_{ab} = \frac{8\pi G}{c^4} \left(T_{ab} - \frac{1}{4} T g_{ab} \right), \qquad (6.100)$$

and the matter conservation equations

$$\nabla_b T^{ab} = 0. \qquad (6.101)$$

Ellis', Murugan's and Van Elst's approach based on the trace-free Einstein equations (6.100) makes unchanged the solutions of all the Einstein's Field Equations of the vacuum, so results from the Schwarzschild space-time and black hole solutions are still valid. However, it no longer has a cosmological constant problem, as Λ does not affect space-time curvature. In particular, for a perfect fluid the matter source term is the explicitly trace-free stress tensor

$$\left(T_{ab} - \frac{1}{4} T g_{ab} \right) = \left(\rho + \frac{p}{c^2} \right) c^2 \left(u_a u_b + \frac{1}{4} g_{ab} \right), \qquad (6.102)$$

which implies that matter enters the field equations only in terms of the inertial mass density $\left(\rho + \frac{p}{c^2} \right)$, which vanishes for a vacuum. By starting from Eqs. (6.100), through integration one obtains the classic Einstein's field equations but with a new effective cosmological constant

$$R_{ab} - \frac{1}{2}Rg_{ab} + \hat{\Lambda}g_{ab} = \frac{8\pi G}{c^4}T_{ab}. \quad (6.103)$$

Thus, in Ellis', Murugan's and Van Elst's model, the trace-free Einstein equations (6.100) together with the differential relations (6.101) are functionally equivalent to the classic Einstein's field equations (6.103), with the cosmological constant an arbitrary integration constant $\hat{\Lambda}$ unrelated to the vacuum energy. Ellis', Murugan's and Van Elst's approach based on the trace-free Einstein equations (6.100) leads to the same experimental predictions of the classic Einstein field equations, so no experiment can tell the difference between them, except for one fundamental feature: the classic Einstein field equations (confirmed in the solar system and by binary pulsar measurements to high accuracy) together with the quantum field theory prediction for the vacuum energy density (confirmed by Casimir force measurements) seem to provide a wrong answer by many orders of magnitude; the trace-free Einstein equations do not suffer this problem, reproducing cosmology without the vacuum energy problem. In cosmology, the trace-free Einstein's equations (6.100) lead to the trace-free equation

$$\frac{\ddot{a}}{a} - \left(\frac{\dot{a}}{a}\right)^2 - \frac{kc^2}{a^2} = -4\pi G\left(\rho + \frac{p}{c^2}\right), \quad (6.104)$$

where $a(t)$ is a universal time-dependent scale factor and the space-time Ricci scalar R is given by relation

$$Rc^2 = -\frac{8\pi G}{c^2}T + 4\Lambda c^2 = 8\pi G\left(\rho - 3\frac{p}{c^2}\right) + 4\Lambda c^2$$

$$= 6\left[\frac{\ddot{a}}{a} + \left(\frac{\dot{a}}{a}\right)^2 + \frac{kc^2}{a^2}\right]. \quad (6.105)$$

The time derivative of this equation has the form

$$0 = 6\frac{d}{dt}\left[\frac{\ddot{a}}{a} + \left(\frac{\dot{a}}{a}\right)^2 + \frac{kc^2}{a^2} - \frac{4\pi}{3}\left(\rho - 3\frac{p}{c^2}\right)\right], \quad (6.106)$$

which, being a vanishing total time derivative, can be easily integrated to yield an integration constant $\frac{2}{3}\hat{\Lambda}$. By using Eqs. (6.104) and (6.106) one obtains the Raychaudhury equation

$$3\frac{\ddot{a}}{a} + 4\pi G\left(\rho + 3\frac{p}{c^2}\right) + \hat{\Lambda}c^2 = 0, \qquad (6.107)$$

and the Friedmann equation

$$3\left(\frac{\dot{a}}{a}\right)^2 = -3\frac{kc^2}{a^2} + 8\pi G\rho + \hat{\Lambda}c^2, \qquad (6.108)$$

with effective cosmological constant $\hat{\Lambda}$, an arbitrary integration constant, unrelated to the vacuum energy. Thus, by applying Ellis', Murugan's and Van Elst's approach to cosmology, one may solve the basic discrepancy between quantum field theory estimates of the energy density of the quantum vacuum and the disastrous result if one assumes this is a source term in the Raychaudhuri equation in an obvious way.

In another recent paper, Consoli, Pluchino and Rapisarda [485] considered the idea that the quantum vacuum is characterized by quantum fluctuations that might reflect the existence of an 'objective randomness', namely a basic property of the vacuum state which is independent of any experimental accuracy of the observations or limited knowledge of initial conditions. According to these authors, the objective randomness of the quantum vacuum, besides being responsible for the observed quantum behaviour, might introduce a weak, residual form of 'noise' which is an intrinsic feature of natural phenomena and could be important for the emergence of complexity at higher physical levels. In this paper Consoli, Pluchino and Rapisarda, by adopting stochastic electrodynamics as a heuristic model, suggest a picture of the vacuum as a form of highly turbulent ether, which is deep-rooted into the basic foundational aspects of both quantum physics and relativity. Although the analysis of the most precise ether-drift experiments (operating both at room temperature and in the cryogenic regime) shows that, at present, there is some ambiguity in the interpretation of the data, according to Consoli, Pluchino and Rapisarda such an experimental evidence for the stochastic nature of the underlying vacuum state would represent an important step forward in order to take seriously the idea (and start to explore the implications) of the basic randomness of nature.

If vacuum is defined as the state of minimum (but seemingly nonzero) energy in quantum mechanics, it is also an active component in quantum field theory. For all its significance, however, vacuum is not a well-defined concept, and, though no one has found a way of eliminating it from quantum theories, the reason why nature requires it at all has never been made clear. In the recent paper *What is vacuum?*, Peter Rowlands [486] introduces a new way to interpret the physical vacuum. In Rowlands' approach the physical vacuum can be defined with exact mathematical precision as the state which remains when a fermion, with all its special characteristics, is created out of absolutely nothing. It is an existence condition, like conservation of energy, which must apply independently on our knowledge of how it is constructed, and it can be defined with exact mathematical precision. Vacuum, in Rowlands' understanding, is the 'hole' in the zero state produced by the creation of the fermion, or, from another point of view, the 'rest of the universe' that the fermion sees and interacts with. So, if one defines a fermion with interacting field terms, then the 'rest of the universe' needs to be 'constructed' to make the existence of a fermion in that state possible. According to Rowlands, this definition leads to a special form of relativistic quantum mechanics, which only requires the construction of a creation operator. On the basis of Rowlands' results, this definition of the physical vacuum leads to a form of quantum mechanics which is especially powerful for analytic calculation, and can at the same time explain, from first principles, many aspects of the Standard Model of particle physics. In particular, the characteristics of the weak, strong and electric interactions can be derived from the structure of the creation operator itself.

Finally, as regards the interpretation of the physical vacuum, another fascinating recent research — also for its atemporal perspectives — is represented by Chiatti's and Licata's model of transactions where the vacuum, besides to be an eigenstate of minimal energy, is the network of all the possible transactions of the field modes in an "unidivided oneness" and it must be considered as a radical nonlocal and event-symmetric state. Chiatti's and Licata's approach of transactions implies that the physical vacuum as a fundamental arena of the universe is an archaic, atemporal manifold that constrains and conveys the dynamical processes we observe in nature. It is a fabric from which patterns emerge by reduction-state processes and such patterns influence the vacuum activity, in a quantum feedback. According to Chiatti's and Licata's approach the only truly existent "things" in the physical world are the events of creation and destruction (or, in other words, physical manifestation and demanifestation) of certain qualities [487]. Such properties' measurement is all that we know of the physical world from an operational view point [488]. In this picture, any other construction in physics — like the continuous space-time notion itself or the evolution operators — has the role to causally connect the measured properties: they are "emergent" with respect to the network of events. The connection between the events of creation or destruction is possible because these events derive from the transformation of the same atemporal vacuum. In this picture, it looks as if every quantum process and time itself emerges from this invariant atemporal substratum and is re-absorbed within it.

In Chiatti's and Licata's approach, the two extreme events of a transaction correspond to two reductions of the two state vectors which describe the evolution of the quantum process in the two directions of time. They are also called "**R** processes" (**R** stands for *reduction*) and here are the only real physical processes. They are constituted of interaction vertices in which real elementary particles are created or destroyed (however, one must rekark that these interactions are not necessarily acts of preparation or detection of a quantum state in a measurement process). Both the probability amplitudes in the two directions of time and the time evolution operators which act on them are mathematical constructions whose only purpose is to describe the causal connection between the extreme events of the transaction, i.e. between **R** processes. This connection is possible because the two extreme events of a transaction derive from the transformation of the same aspatial and atemporal vacuum. As a specific consequence of this assumption, all virtual processes contained in the expansion of the time evolution operator are deprived of physical reality. According to this approach, therefore, the history of the universe, considered at the basic level, is given neither by the application of forward causal laws at initial conditions nor by the application of backward causal laws at final conditions. Instead, it is assigned as a complete network of past, present and future **R** processes deriving from the same invariant atemporal, archaic vacuum. In this atemporal, archaic substratum, at a fundamental level only the transactions between field modes takes place, and the quantum-mechanical wavefunction simply emerges as a statistical coverage of a great amount of elementary transitions. In the light of the atemporal "archaic vacuum" inside an archaic universe which derives from Licata's and Chiatti's model of transactions, the old Big Bang conception as a "thermodynamic balloon" seems to be irreparably compromised as basic cosmological theory and must be definitely abandoned [489, 490].

6.2 The Quantum Vacuum Energy Density and the RS Processes as the Fundamental Arena of Physics

Several phenomena, such as Planck's radiation law, the spontaneous emission of a photon by a particle in an excited state, Casimir's effect, van der Waals' bonds, Lamb's shift, Davies–Unruh's effect, indicate that the vacuum energy density which describes the universal space should be nonzero. In the recent fascinating paper *Gedanken experiments with Casimir forces, vacuum energy and gravity* [491], G. Jordan Maclay suggested some gedanken experiments in order to explore properties of vacuum energy that are currently challenging or impossible to explore experimentally. By considering systems in which vacuum energy and other forms of energy are exchanged, Maclay demonstrated that a change ΔE in vacuum energy, whether positive or negative with respect to the free field, corresponds to an equivalent inertial mass and gravitational mass $\Delta M = \Delta E/c^2$. A first gedanken experiment demonstrated that there is a constant, finite lateral force at $0\ K$ between two parallel, finite plates that overlap. The force tries to maximize the amount of overlap. Other gedanken experiments show that changes in vacuum energy formally couple to gravity like ordinary forms of energy. Otherwise, according to Maclay's results, it is possible to design gedanken experiments in which an arbitrary amount of energy can be extracted from a physical system without changing the state, violating our usual form of the law of conservation of energy. Specifically, changes in vacuum energy correspond to equivalent active and passive gravitational masses. Maclay's results show that extracting energy from the quantum vacuum is possible if there is a change in the state of the system.

On the other hand, some approaches indicate that the vacuum energy density is somehow related to the average number of elementary fluctuations in terms of a rate of massless particle/antiparticle pairs production/annihilation. In this regard, already in 1951 Schwinger studied the question of the production of charged particle/antiparticle pairs out of the vacuum by a classical electric field and computed the rate associated with vacuum breakdown via pair production per unit volume for the case of a constant electric field [492]. By treating the electric field classically — namely the formal limit of $q \to 0$, $E \to \infty$ with qE fixed — in Schwinger's approach the vacuum persistence probability is

$$P_{\text{vac}}(t) = \exp(-wVt), \qquad (6.109)$$

where

$$w = \frac{(qE)^2}{4\pi^3 \hbar^2 c} \sum_{n=1}^{\infty} \exp\left(-\frac{n\pi m^2 c^3}{qE\hbar}\right), \qquad (6.109a)$$

where V is the spatial volume of the system and w is the rate of vacuum decay per unit volume.

Over the years, the Schwinger mechanism has taken place in a vast literature. Invoked to gain insights on topics as diverse as the string breaking rate in QCD [493, 494] and on black hole physics [495], this mechanism has become a textbook topic in quantum field theory [496]. Topics such as back reaction [497] and finite size effects [498] have moreover been addressed. Although Schwinger's interpretation that Eq. (6.109a) as measuring the local rate of production per unit volume of fermion-antifermion pairs by the electric field has been widely accepted, however in 1970 Nikishov found that the rate of the pair fermion/antifermion production per unit volume out of the quantum vacuum by a constant electric field is given by relation

$$\Gamma = \frac{(qE)^2}{4\pi^3 \hbar^2 c} \exp\left(-\frac{\pi m^2 c^3}{qE\hbar}\right), \qquad (6.110)$$

where q is the charge of the fermions and m is their mass and thus that the rate of pair productions is given by the first term in the series for w [499]. Moreover, recently (2008) T. D. Cohen and D. A. McGady rederived Nikishov's result in a physically transparent context which clearly illustrates why (6.110) rather than (6.109) is the rate of pair creation per unit volume out of the quantum vacuum [500]. According to Cohen's and McGady's approach, as regards the pair production rate of Eq. (6.110), the crucial point is that the exponential factor appearing in it is very small for static macroscopic electric fields realizable in the labs (and this is also the reason for which the Schwinger mechanism has never been tested experimentally). It only becomes of order unity when the electric field is large enough so that qE times the electron's Compton wavelength is greater than mc^2: this requires an electric field of order $10^{16} V/cm$. Instead, one might test in a coherent way equation (6.110) in a condensed matter system which simulates light or massless, electrically charged, relativistic fermions. In this context, the condensed matter systems act as an analog computer to test the underlying result from relativistic field theory.

As regards the quantum vacuum energy density, significant results have also been obtained recently by Klinkhamer and Volokik in the papers [501–506]. In particular, in the *Dynamics of the quantum vacuum: cosmology as relaxation to the equilibrium state* [501], these two authors showed that the gravitating vacuum energy density in an expanding universe follows the steadily decreasing matter energy density, in other words that there is a hierarchy of different contributions to the vacuum energy density from different matter fields. According to Klinkhamer's and Volokik's research, the contributions to the vacuum energy density can be separated into sub-Planckian and trans-Planckian contributions with respect to the Planck energy scale

$$E_{\text{Planck}} = [\hbar c^5/8\pi G_N]^{1/2} = 2,44 \cdot 10^{18} \text{GeV}. \quad (6.111)$$

The sub-Planckian contributions are described by relativistic quantum fields propagating over a classical space-time manifold. The trans-Planckian contributions come from the fundamental microscopic degrees of freedom of the "deep vacuum" which, in the perfect equilibrium Minkowski vacuum, compensate the contribution of the sub-Planckian quantum fields to the gravitating vacuum energy density ρ_{vac} (and thus to the effective cosmological constant λ_{eff}). Cosmology can be viewed here as the process of relaxation towards the equilibrium vacuum state, where the vacuum energy density drops from its initial large value (of order $\rho_{\text{vac}} \approx (E_{\text{Planck}})^4$) to its present small value ($\rho_{\text{vac}} \approx (2\text{meV})^4$). During the expansion of the Universe, one deals with an energy hierarchy of the different contributions to the vacuum energy density from the different quantum fields which gives rise to a time sequence, since each epoch is characterized by one dominant contribution to the vacuum energy density and, thus, by one particular value of the effective cosmological constant. A step-wise relaxation of the effective cosmological constant emerges therefore which, on average, follows the matter energy density. The initial relaxation dynamics of the vacuum energy density is dominated by microscopic (trans-Planckian) degrees of freedom, leading to $1/t^2$ decay of the average vacuum energy density or to exponential decay from dissipative effects. For later times, starting at the electroweak epoch and following with quantum chromodynamics, the dynamics of the vacuum energy density turns out to be ruled by the low-energy contributions of the corresponding quantum fields and can be described by standard-model physics. In Klinkhamer's and Volokik's approach, the hierarchical structure of the quantum vacuum implies a complicated spectral function of the vacuum energy density. However, the single contributions to the spectral function of the vacuum energy density cannot be determined in the static Minkowski vacuum, since, in equilibrium, the contributions of the bosonic and fermionic quantum fields to the low-energy part of the spectral function are compensated by the trans-Planckian part of the spectrum.

In the model developed by the author and Sorli in the recent papers [507–509], the Planckian metric (the system of units introduced by Max Planck in 1899 by combining the speed of light c, the Planck constant $\hbar$, and the Newton's universal gravitational constant G) suggests an interesting way to define a quantum vacuum energy density as a fundamental property of space. This can be shown by starting with a semi-classical argument derived from orbital mechanics, and then by imposing on it the necessary constraints of relativity and of quantum mechanics. A body within a circular orbit experiences a centripetal acceleration of v^2/r, derived from a gravitational force per unit mass, Gm/r^2 where G is universal gravitation constant. If the velocity v becomes the largest velocity possible, namely $v \to c$, then the maximum acceleration possible is $c^2/r = Gm/r^2$. From this equality the Schwarzschild's radius for a given mass m can be easily obtained:

$$r_S = \frac{Gm}{c^2}. \qquad (6.112)$$

Now, the Heisenberg's uncertainty principle requires that the position x of an object and its instantaneous momentum p cannot be known at the same time to a precision better than the amount imposed by the quantum of action, $\Delta x \Delta p \geq \hbar$. By expressing Δp as mc, we obtain in the limiting case the Compton's radius:

$$r_C = \frac{\hbar}{mc}, \qquad (6.113)$$

which is understood as the minimum possible size of a quantum object of mass m.

The laws of physics should be the same, regardless of the size of an object, so it makes sense to assume that the limits $r_C = \frac{\hbar}{mc}$ and $r_S = \frac{Gm}{c^2}$ must apply for all limiting cases, whether we are dealing with black holes or elementary particles. By equating this minimum quantum size for an object of mass m with the Schwarzschild's radius for that same object, $\frac{Gm}{c^2} = \frac{\hbar}{mc}$, we obtain the Planck's mass limit:

$$m_P = \sqrt{c\hbar/G}. \qquad (6.114)$$

The Compton's radius of the Planck's mass is then the shortest possible length in space, the Planck's length:

$$l_P = \sqrt{\frac{\hbar G}{c^3}}. \qquad (6.115)$$

This means that, because of the uncertainty relation, the Planck's mass cannot be compressed into a volume smaller than the cube of the Planck's length. A Planck's mass contained within a Planck's volume l_p^3 therefore represents the maximum density of matter that can possibly exist:

$$\rho_P(m) = \frac{c^5}{\hbar G^2}. \qquad (6.116)$$

Following Einstein's most popular equation:

$$W = mc^2, \qquad (6.117)$$

this mass density is equivalent to an energy density of:

$$\rho_P(W) = \frac{c^7}{\hbar G^2}. \qquad (6.118)$$

Likewise, the maximum energy that can be contained within a volume l_p^3 is the Planck energy:

$$W_P = m_P c^2 \approx 1,98 \cdot 10^9 J. \qquad (6.119)$$

Consequently, the speed of light limit, c, together with the minimum quantized space l_p^3 constrain the highest oscillations that can be sustained in space:

$$\omega_P = \frac{W_P}{\hbar} = \sqrt{\frac{c^5}{\hbar G}} \approx 1,9 \cdot 10^{43} \, rad/s. \qquad (6.120)$$

All these relations represent the fundamental limits of nature, and thus the natural units of measure. By using these units it is possible to calculate the total average volumetric energy density, owed to all the frequency modes possible within the visible size of the universe. Because we are averaging over all possible oscillating modes, it is proper to express it as the square root of the squared energy density given by (6.118):

$$\rho_P = \sqrt{\frac{c^{14}}{\hbar^2 G^4}} \approx 4,641266 \cdot 10^{113} \, J/m^3. \quad (6.121)$$

Therefore, Planck's metric implies that the maximum energy density characterizing the minimum quantized space constituted by Planck's volume is given by relation (6.121) and this defines a universal property of space. By starting from the Planckian metric, in the recent papers [507–509] the author and Sorli proposed a model of a timeless 3D isotropic quantum vacuum composed by elementary packets of energy having the size of Planck's volume and whose most universal physical properties are the energy density and **RS** processes (of creation-annihilation of quantum particles) analogous to Chiatti's and Licata's transactions. By following the treatment made in the papers [507–509] here we want to analyze the fundamental features of this model.

In the free space, in the absence of matter, the energy density of the 3D quantum vacuum is at its maximum and is given by the total volumetric energy density (6.121), which coincides with the Planck energy density

$$\rho_{pE} = \frac{m_p \cdot c^2}{l_p^3} = 4,641266 \cdot 10^{113} \frac{Kg}{ms^2}. \quad (6.122)$$

In the ordinary space-time we perceive, the appearance of matter corresponds to a given change of the quantum vacuum energy density. In analogy with Rueda's and Haisch's interpretation of the inertial mass as an effect of the electromagnetic quantum vacuum in which the presence of a particle with a volume V_0 expels from the vacuum energy within this volume exactly the same amount of energy as is the particle's internal energy (equivalent to its rest mass), here the presence of a given material object of mass m in our level of physical reality derives from a diminishing of the energy density of quantum vacuum. The energy density of quantum vacuum at the centre of this object is given by relation:

$$\rho_{qvE} = \rho_{pE} - \frac{m \cdot c^2}{V}, \qquad (6.123)$$

where

$$V = \frac{4}{3} \cdot \pi \cdot R^3 \qquad (6.124)$$

is the volume and R is the radius of the massive particle under consideration.

In analogy with Chiatti's and Licata's approach as proposed in the papers [487, 488, 490], in the 3D quantum vacuum model of the author the events of preparation of the initial state (creation of a particle or object from the 3D quantum vacuum) and of detection of the final state (annihilation or destruction of a particle or object from the 3D quantum vacuum) are assumed to be as the two real primary physical events. The connection between these two events is ensured by their common origin in the timeless background represented by the 3D quantum vacuum. In this picture, the process amplitudes are mathematical devices suited to calculate the statistics of the connection itself. Accordingly, the virtual processes obtained by expanding these amplitudes into partial amplitudes are also mathematical devices. Moreover, the evolution of a massive particle — which is originated from the 3D quantum vacuum — is determined by appropriate waves of the vacuum associated with a spinor which describes the amplitude of creation or destruction events. The waves of the vacuum act in a nonlocal way through an appropriate quantum potential of the vacuum (which, so to speak, guides the occurring of the processes of creation or annihilation in the 3D quantum vacuum). In this approach, just like the spinors describing the amplitudes of creation and destruction events, also the quantum potential of the vacuum must be considered as a mathematical entity. More precisely, the quantum potential of the vacuum can be interpreted as the primary mathematical reality which emerges from the very real extreme primary physical realities, namely from the processes of creation and annihilation. In virtue of the primary physical reality of the processes of creation and annihilation and of the nonlocal action of the quantum potential which is associated with the amplitudes of them, in the 3D quantum vacuum the duration of the processes from the creation of a particle or object till its annihilation has not a primary physical reality but exists only in the sense of numerical order. In other words, in the 3D quantum vacuum time exists merely as a mathematical parameter measuring the dynamics of a particle or object.

The two only primary extreme physical events of the 3D quantum vacuum (namely the creation and annihilation of an elementary particle) are each corresponding to a peculiar reduction of a state vector. They can be also called "**RS** processes" and from our perspective, are the only real primary physical processes. These two quantum events of reduction, these two **RS** processes correspond, in our level of physical reality, to the evolution of the quantum process in the two directions of time (namely forward-time evolution and time-reversed evolution respectively, which can be associated with the opportune operators S and S^+ respectively). Each **RS** process is a self-connection of the 3D timeless quantum vacuum. A **RS** process that begins with the creation of quality q and ends with the destruction of quality r can be represented simply by the form q^+r^- and can be described as the concurrence of two distinct transformations. In the first transformation, the 3D quantum vacuum divides into two pairs of opposites: $q+$, $q-$ and $r+$, $r-$; the first pair forms "event A", the second pair forms "event B". The second transformation is the real generation of the transaction having the events A and B as its extremities.

The appearance of a massive particle in a given region of the ordinary space-time derives from an opportune change of the energy density of quantum vacuum $\Delta\rho_{qvE} = \rho_{qvE} - \rho_{pE}$ on the basis of equation

$$m = \frac{4\pi R^3 \Delta\rho_{qvE}}{3c^2}. \tag{6.125}$$

The probability of the occurrence of a creation/destruction event for a quantum particle Q of mass (6.125) in a point event x is linked with the probability amplitudes $\psi_{Q,i}(x)$ (for creation events) and $\phi_{Q,i}(x)$ (for destruction events) of a spinor $C = \begin{pmatrix} \psi_{Q,i} \\ \phi_{Q,i} \end{pmatrix}$ at two components. The generic spinor $C = \begin{pmatrix} \psi_{Q,i} \\ \phi_{Q,i} \end{pmatrix}$ satisfies a time-symmetric extension of Klein–Gordon quantum relativistic equation of the form

$$\begin{pmatrix} H & 0 \\ 0 & -H \end{pmatrix} C = 0, \tag{6.126}$$

where $H = (-\hbar^2 \partial^\mu \partial_\mu + m^2 c^2)$. Equation (6.126) corresponds to the following equations,

$$\left(-\hbar^2 \partial^\mu \partial_\mu + \frac{16\pi^2 R^6}{9c^2}(\Delta\rho_{qvE})^2\right)\psi_{Q,i}(x) = 0 \tag{6.127}$$

for creation events and

$$\left(\hbar^2 \partial^\mu \partial_\mu - \frac{16\pi^2 R^6}{9c^2}(\Delta\rho_{qvE})^2\right)\phi_{Q,i}(x) = 0 \tag{6.128}$$

for destruction events respectively, where $m = \frac{4\pi R^3 \Delta\rho_{qvE}}{3c^2}$ is the mass of the quantum particle under consideration. At the nonrelativistic limit, Eq. (6.126) becomes a pair of Schrödinger equations:

$$-\frac{3\hbar^2 c^2}{8\pi R^3 \Delta\rho_{qvE}} \nabla^2 \psi_{Q,i}(x) = i\hbar \frac{\partial}{\partial t} \psi_{Q,i}(x), \tag{6.129}$$

and

$$-\frac{3\hbar^2 c^2}{8\pi R^3 \Delta\rho_{qvE}} \nabla^2 \phi_{Q,i}(x) = i\hbar \frac{\partial}{\partial t} \phi^*_{Q,i}(x). \tag{6.130}$$

Equation (6.129) has only retarded solutions, which classically correspond to a material point with impulse $\mathbf{p}$ and kinetic energy $E = \mathbf{p} \cdot \mathbf{p}/2m > 0$. Equation (6.130) has only advanced solutions, which correspond to a material point with kinetic energy $E = -\mathbf{p} \cdot \mathbf{p}/2m < 0$. On the basis of Eqs. (6.111) and (6.112), thus there are no true causal propagations from the future, in other words there is no possibility to have time travels.

Moreover, in the paper [507] the author and Sorli showed that, by studying Eqs. (6.126)–(6.130) in a bohmian framework, one can introduce a quantum potential of the 3D quantum vacuum as the fundamental mathematical element which guides the occurring of the processes of creation and annihilation of quanta in the different regions of the 3D quantum vacuum. In this regard, by writing the two components of the spinor in polar form

$$\psi_{Q,i} = |\psi_{Q,i}| \exp\left(\frac{iS^{\psi}_{Q,i}}{\hbar}\right), \quad (6.131)$$

$$\phi_{Q,i} = |\phi_{Q,i}| \exp\left(\frac{iS^{\phi}_{Q,i}}{\hbar}\right), \quad (6.132)$$

and decomposing the real and imaginary parts of the Klein–Gordon equation (6.126), for the real part one obtains a couple of quantum Hamilton–Jacobi equations that, by imposing the requirement that they are Poincarè invariant and have the correct nonrelativistic limit, assume the following form

$$\partial_\mu \begin{pmatrix} S^{\psi}_{Q,i} \\ S^{\phi}_{Q,i} \end{pmatrix} \partial^\mu \begin{pmatrix} S^{\psi}_{Q,i} \\ S^{\phi}_{Q,i} \end{pmatrix} = \frac{16\pi^2 R^6}{9c^2} (\Delta\rho_{qvE})^2 \exp\begin{pmatrix} Q^{\psi}_{Q,i} \\ -Q^{\phi}_{Q,i} \end{pmatrix}, \quad (6.133)$$

while the imaginary part gives the continuity equation

$$\partial_\mu \left(\rho\partial^\mu \begin{pmatrix} S^{\psi}_{Q,i} \\ S^{\phi}_{Q,i} \end{pmatrix}\right) = 0, \quad (6.134)$$

where ρ is the ensemble of particles associated with the spinor under consideration and

$$Q_{Q,i} = \frac{9\hbar^2 c^2}{16\pi^2 R^6 (\Delta\rho_{qvE})^2} \left(\frac{\left(\nabla^2 - \frac{1}{c^2}\frac{\partial^2}{\partial t^2}\right)|\psi_{Q,i}|}{|\psi_{Q,i}|} - \frac{\left(\nabla^2 - \frac{1}{c^2}\frac{\partial^2}{\partial t^2}\right)|\phi_{Q,i}|}{|\phi_{Q,i}|} \right) \quad (6.135)$$

is the quantum potential of the vacuum. In the nonrelativistic limit, Eq. (6.133) becomes

$$\frac{3c^2}{8\pi R^3(\Delta\rho_{qvE})}\begin{pmatrix}\left|\nabla S_{Q,i}^{\psi}\right|^2\\ \left|\nabla S_{Q,i}^{\phi}\right|^2\end{pmatrix}+Q_{Q,i}+\begin{pmatrix}V\\-V\end{pmatrix}=-\frac{\partial}{\partial t}\begin{pmatrix}S_{Q,i}^{\psi}\\S_{Q,i}^{\phi}\end{pmatrix},$$

(6.136)

and Eq. (6.134) becomes

$$\frac{\partial}{\partial t}\begin{pmatrix}|\psi_{Q,i}|^2\\|\phi_{Q,i}|^2\end{pmatrix}+\nabla\cdot\begin{pmatrix}\dfrac{|\psi_{Q,i}|^2\nabla S_{Q,i}^{\psi}}{m}\\ \dfrac{|\phi_{Q,i}|^2\nabla S_{Q,i}^{\phi}}{m}\end{pmatrix}=0,\qquad(6.137)$$

where

$$Q_{Q,i}=-\frac{3\hbar^2 c}{8\pi R^3(\Delta\rho_{qvE})}\begin{pmatrix}\dfrac{\nabla^2|\psi_{Q,i}|}{|\psi_{Q,i}|}\\ -\dfrac{\nabla^2|\phi_{Q,i}|}{|\phi_{Q,i}|}\end{pmatrix}\qquad(6.138)$$

is the nonrelativistic quantum potential of the vacuum.

On the basis of equations (6.135) and (6.138), both in the relativistic domain and in the nonrelativistic domain, the quantum potential of the vacuum has a like-space, nonlocal, instantaneous action, both for the processes of creation and for the processes of annihilation. This means, in other words, that the quantum potential is the fundamental mathematical entity which guides the occurring of the processes of creation and annihilation of quanta in the different regions of the 3D quantum vacuum in a nonlocal, instantaneous manner. The first component of the quantum potential regards the processes of creation of quanta, the second component regards the processes of annihilation of quanta in the 3D quantum vacuum. Moreover, the opposed sign of the second component with respect to the first component can be seen as a mathematical translation of the fact that, in the 3D quantum vacuum, time exists only as a measuring system of the numerical order of material changes: the sign of the second component means that it is not possible to go backwards in the physical time intended as a numerical order.

Now, let us indicate with t_1 the numerical order corresponding with the event of the creation–destruction of a quantum particle $\boldsymbol{Q}(|Q\rangle\langle Q|)$ and t_2 the numerical order corresponding with the event of the creation–destruction of a quantum particle $\boldsymbol{R}(|R\rangle\langle R|)$. In analogy with Chiatti's and Licata's approach, these two processes will be linked by a time evolution operator S which acts in the following way: $|Q\rangle$ is transported by S into $|Q^I\rangle$ and projected onto $\langle R|$, $|R\rangle$ is transported by S^+ into $|R^I\rangle$ and projected onto $\langle Q|$. The amplitudes product $\langle R|S|Q\rangle\langle Q|S^+|R\rangle = |\langle R|S|Q\rangle|^2$ is the probability of the entire process.

The evolution of probability amplitudes in the two directions of time constitutes, in Penrose terminology, a **U** process (where **U** stands for *unitary*). From our ontological viewpoint, both the amplitudes and the time evolution operators which act on **U** processes are mathematical devices whose only purpose is to describe the causal connection between the extreme events, namely between **RS** processes (and the fundamental mathematical reality emerging from the amplitudes is the quantum potential of the vacuum). The connection between **RS** processes is possible because the two events derive from the change of the energy density of the same timeless 3D quantum vacuum (and the quantum potential can be considered the ultimate mathematical entity which guides the evolution of the occurring of the processes of creation or annihilation of quanta in the different regions of the 3D quantum vacuum). As a specific consequence of this assumption, all virtual processes contained in the expansion of the time evolution operator, in the ordinary space-time we perceive, are deprived of primary physical reality.

According to this approach, therefore, the history of the Universe, considered at the basic level, is given neither by the application of forward causal laws at initial conditions nor by the application of backward causal laws at final conditions. Instead, it is assigned as a whole as a complete network of **RS** processes that take place in the timeless 3D quantum vacuum. Causal laws are only rules of coherence which must be verified by the network and are *per se* indifferent to the arrow of the time of our level of physical reality.

An interesting merit of this approach is represented by the possibility to obtain, on one hand, an interesting reading of the quantum dynamics as a particular aspect of such general theory and, on the other hand, a suggestive interpretation of gravity as a phenomenon emerging from the timeless 3D quantum vacuum. The philosophy of this model as regards the link between our level of physical reality and the 3D quantum vacuum actually follows someway the program that Bohm had already sketched out when dealing with clarifying the relationship between *implicate* and *explicate order* [510].

6.3 About the Treatment of Gravity in the 3D Quantum Vacuum Energy Density Model

As regards the interpretation of the gravitational interaction, the model developed by the author of this book and Sorli in the papers [507–509] allows us to explore in a suggestive way various perspectives opened by Consoli's model of physical vacuum and Puthoff's polarized vacuum model of gravity. All these models have at the basis the same underlying consideration: an "effective" curvature might originate from the same physical flat-space vacuum, from the same physical flat-space background. However, while in Consoli's model developed in the papers [18, 415–417], the physical vacuum is represented by a superfluid medium, a Bose condensate of elementary spinless quanta, so involving the Higgs mechanism, in the model of a 3D timeless quantum vacuum suggested by the author and Sorli in the papers [507–509], the background of processes, the physical vacuum is a 3D granular structure which is introduced without the need to postulate the consideration of other processes (involving other more fundamental particles): it is based only on the Planckian metric which leads directly to the definition of a Planck energy density (intended as a universal property of nature) as the maximum energy density that can be possibly sustained in space and on **RS** processes of creation-annihilation of quanta. On the other hand, making a parallelism with Puthoff's polarized quantum vacuum model, one can say that the polarization of the quantum vacuum is linked with the changes of the 3D timeless quantum vacuum corresponding to the **RS** processes.

In the model of a 3D timeless quantum vacuum developed by the author and Sorli, the origin of gravitational phenomena lies simply in the changes of the energy density of a 3D background associated with the **RS** processes. In this model, the Planck energy density can be considered as the ground state of the same physical flat-space background; the appearance of material objects and subatomic particles derive from opportune **RS** processes of creation and correspond to opportune changes of the quantum vacuum energy density and thus can be interpreted as the excited states of the same physical flat-space background. The excited state of the background corresponding to the appearance of a material particle of mass m (originated from an opportune **RS** process of creation) is defined by a quantum vacuum energy density (in the centre of this particle) given by Eq. (6.123). Gravity is a phenomenon determined by the changes of the quantum vacuum energy density and thus by the excited states of the same flat quantum vacuum. The appearance of a material particle of mass m corresponds to an excited state of the flat background defined by a change of energy density (with respect to the ground state) given by equation

$$\rho_{qvE} - \rho_{pE} \equiv \Delta\rho_{qvE} = \frac{3m \cdot c^2}{4\pi \cdot R^3}. \qquad (6.139)$$

In other words, each material particle endowed with mass is produced by a change of the energy density of quantum vacuum. It is the change of the quantum vacuum energy density (6.139) which gives origin to the mass (6.125) of a material particle. On the basis of Eq. (6.125), the property of mass derives from a change of the quantum vacuum energy density. Equation (6.125) indicates that the mass of a given massive body or particle is the result of the interaction of that body or particle with space, in particular it corresponds to an opportune change of the quantum vacuum energy density, which can be considered as the fundamental property of the universal space. Each body or particle diminishes energy density of space. We measure this diminishing of the quantum vacuum energy density as "inertial mass" and as "gravitational mass".

The foundational ideas of the model of the 3D timeless quantum vacuum proposed by the author in the papers [507–509] can be synthesized in the following postulates:

1. The medium of space is an isotropic, granular, 3D structure which may be termed "quantum vacuum". It is constituted by energetic packets having the size of Planck's volume and its most universal properties are the energy density and **RS** processes of creation-annihilation of quantum particles.
2. In the free space, without the presence of massive particles, the quantum vacuum energy density is at its maximum and is given by Eq. (6.121) (or the equivalent Eq. (6.122)) which defines the so-called "ground state" of the quantum vacuum.
3. The appearance of matter derives from an opportune excited state of the quantum vacuum which corresponds to an opportune change of the energy density of the quantum vacuum and is determined by opportune **RS** processes of creation/annihilation. The excited state of the 3D quantum vacuum corresponding to the appearance of a material particle of mass m is defined (in the centre of that particle) by the energy density (6.123) (and by the change of the energy density (6.139) with respect to the ground state), depending on the amount of mass m and the volume V of the particle (6.124).

6.3.1 Space as the geometro-hydrodynamic limit of a 3D quantum vacuum condensate

In the model of the 3D timeless quantum vacuum, taking account of Sacharov's assumption that the action of space-time

$$S(R) = -\frac{1}{16\pi G} \int dx \sqrt{-gR}, \qquad (6.140)$$

where R is the invariant Ricci tensor, constitutes a change in the action of quantum fluctuations of vacuum in a curved space and considering the consistent histories approach of quantum mechanics [437, 511, 512], according to which the quantum evolution can be seen as the coherent superposition of virtual fine–grained histories, general relativity can be interpreted as the hydrodynamic limit of an underlying "microscopic" structure, of a 3D quantum vacuum condensate whose most universal physical property is its energy density.

A fine-grained history can be defined by the value of a field $\Phi(x)$ at the point x and has quantum amplitude $\Psi[\Phi] = e^{iS[\Phi]}$, where S is the classical action corresponding to the considered history. The quantum interference between two virtual histories A and B can be quantified by a "decoherence" functional:

$$D_F[\Phi_A, \Phi_B] \approx \Psi[\Phi_A]\Psi[\Phi_B]^* \approx e^{i(S[\Phi_A]-S[\Phi_B])} \quad (6.141)$$

that provides the coarse-grained histories corresponding to the observations in the classical world. The quantum amplitude for a coarse-grained history is then defined by:

$$\Psi[\omega] = \int D_F \Phi e^{iS} \omega[\Phi], \quad (6.142)$$

where ω can be considered as a "filter" function that selects which fine-grained histories are associated to the same superposition with their relative phases. The decoherence functional for a couple of coarse-grained histories is then:

$$D_F[\omega_A, \omega_B] = \int D_F \Phi_A D_F \Phi_B e^{i(S[\Phi_A]-S[\Phi_B])} \omega[\Phi_A]\omega[\Phi_B]^*, \quad (6.143)$$

in which the histories Φ_A and Φ_B assume the same value at a given time instant, where decoherence indicates that the different histories contributing to the full quantum evolution can exist individually, are characterized by quantum amplitude and that the system undergoes an information and predictability degradation [512]: in this sense the system becomes stochastic and dissipative. By applying the formalism (6.143) to hydrodynamics variables [513], Einstein's stress-energy tensor can be expressed through the following operator:

$$\hat{T}_{\mu\nu}(x_A, x_B) = \Gamma_{\mu\nu} \Phi(x_A) \Phi(x_B). \quad (6.144)$$

In Eq. (6.144) $\Gamma_{\mu\nu}$ is a generic field operator defined at two points that leads to the "conservation law":

$$\hat{T}^{;\nu}_{\mu\nu} = 0, \qquad (6.145)$$

meaning that the decohered quantities, showing a classical behavior, are the conserved ones. It can be shown that, for an action $S[\Phi^l] = \Phi^l \Delta_{lm} \Phi^m$, the following relation holds

$$D_F[\hat{T}^A_{\mu\nu}, \hat{T}^B_{\mu\nu}] = \int D_F K^{\mu\nu}_n(x_A, x_B)$$

$$\times \int D_F \Phi^l e^{i\Phi^l [\Delta + K^{\mu\nu}_n(x_A,x_B)\Gamma^n_{\mu\nu}(x_A,x_B)]_{lm} \Phi^m}$$

$$\times e^{iK^{\mu\nu}_n(x_A,x_B)T^n_{\mu\nu}(x_A,x_B)} \approx e^{i\Omega[\hat{T}^A_{\mu\nu}(x_A,x_B), \hat{T}^B_{\mu\nu}(x_A,x_B)]}$$

$$(6.146)$$

in which we have used the integral representation of delta and the CTP indices $l.m.n = 1,2$, Ω being the closed-time path two-particle irreducible action.

The conservation of $\hat{T}_{\mu\nu}$ implies that the decoherence functional has maximum values in correspondence of the hydrodynamic variables (ρ, p) that, in turn, are the most readily decohered and have the highest probability to become classical. By applying the above procedure to Einstein's tensor $G_{\mu\nu}$ an analogy emerges between the conservation law for $\hat{T}_{\mu\nu}$ and the Bianchi identity $G^{;\nu}_{\mu\nu} = 0$ which implies the decoherence and the emergence of the hydrodynamics variables of geometry. In this sense general relativity can be considered as geometro–hydrodynamics and the most readily decohered variables are those associated to the largest "inertia" representing the collective variables of geometry. The expectation value of the stress-energy tensor operator of the quantum fields (6.144) at any point leads to changes of the 3D quantum vacuum energy density. In order to have the correct Friedmann–Robertson–Walker metric, this assumption means that

$$\langle \Psi | \hat{T}^4_4 | \Psi \rangle = \frac{\Delta \rho_{qvE}}{c^2}; \quad \langle \Psi | \hat{T}^\nu_\mu | \Psi \rangle \approx 0 \quad \text{for } \mu\nu \neq 00, \qquad (6.147)$$

Ψ being the quantum state of the universe corresponding to the value of the field $\Phi(x)$ defining a given fine-grained history. This suggests to express the stress-energy tensor (6.144) corresponding to the quantum vacuum fluctuations as

$$\hat{T}^{vac}_{\mu\nu} \equiv \hat{T}_{\mu\nu} - \langle\Psi|\hat{T}_{\mu\nu}|\Psi\rangle\hat{I}, \qquad (6.148)$$

where $\hat{I}$ is the identity operator. The existence of quantum vacuum fluctuations implies that, despite the expectation of $\hat{T}^{vac}_{\mu\nu}$ is zero by definition, one has

$$\langle\Psi|\hat{T}^{vac}_{\mu\nu}(x)\hat{T}^{vac}_{\lambda\sigma}(y)|\Psi\rangle \neq 0 \qquad (6.149)$$

in general.

If general relativity must be regarded as a geometro-hydrodynamic limit of an underlying "microscopic" background and the laws governing macro-classical space-time are expressed in terms of collective variables, the precise characterization of this underlying background and thus the quantization of the general-relativistic metric or the connection variables will only result in the discovery of the excitations in the geometry and not of its quantum micro-structures. If we consider the collective hydrodynamic variables ρ and p appearing in the stress-energy tensor $T_{\mu\nu}$, then the quantization has sense when performed on the field function $\Phi(x)$ from which they are constructed and not on ρ and p themselves. The situation is similar to that regarding condensed matter physics in which the quantization of collective excitations leads to phonons and not to the atomic structure of matter. In the view of general relativity as geometro-hydrodinamic limit of an underlying background, there is therefore an important analogy between quantum to classical transition of gravity and the behavior of condensed matter.

Moreover, given the collective variables (the metric and the connections in general relativity), how can we characterize the microscopic structure of the underlying background, namely what can we say about the quantum micro-structure from which the collective variables derive? In this regard, a possible strategy is of starting from a suitable theory of quantum microscopic structure and studying its previsions in the long wavelength-low energy limit. An approach of this kind is Consoli's one [18, 415–417], where a physical vacuum intended as a superfluid medium — a Bose condensate of elementary spinless quanta — is introduced whose long-range fluctuations, on a coarse-grained scale, resemble the Newtonian potential, allowing thus the first approximation to the metric structure of classical general relativity to be obtained. In analogy with Consoli's model, taking into consideration the long-wavelength modes, in the model of the author gravity is induced by the density fluctuations of the vacuum described by the underlying field $\Phi(x)$, that in weak gravitational fields, on a coarse-grained scale, is identified with the Newtonian potential

$$\Phi \approx U_N = -G_N \sum_i \frac{M_i}{|\vec{r} - \vec{r}_i|} \qquad (6.150)$$

namely one can write

$$g_{\mu\nu} = g_{\mu\nu}[\Phi(x)]. \qquad (6.151)$$

6.3.2 The changes of the 3D quantum vacuum energy density as the origin of the curvature of space-time

The Planck energy density (6.122) is usually considered as the origin of the dark energy and thus of a cosmological constant, if the dark energy is supposed to be owed to an interplay between quantum mechanics and gravity. However the observations are compatible with a dark energy

$$\rho_{\text{DE}} \cong 10^{-26} Kg/m^3, \qquad (6.152)$$

and thus Eqs. (6.122) and (6.152) give rise to the so-called "cosmological constant problem" because the dark energy (6.152) is 123 orders of magnitude lower than (6.122). In order to solve this problem, as mentioned in chapter 6.1, Santos proposed recently an explanation for the actual value (6.152) which invokes the fluctuations of the quantum vacuum: the quantum vacuum fluctuations can be associated with a curvature of space-time similar to the curvature produced by a "dark energy" density, on the basis of equation

$$\rho_{\text{DE}} \cong 70G \int_0^\infty C(s)s\,ds, \quad (6.6)$$

which states that the possible value of the "dark energy" density is the product of Newton constant, G, times the integral of the two-point correlation of the vacuum fluctuations defined by

$$C(|\vec{r}_1 - \vec{r}_2|) = \frac{1}{2}\langle \text{vac}|\hat{\rho}(\vec{r}_1,t)\hat{\rho}(\vec{r}_2,t) + \hat{\rho}(\vec{r}_2,t)\hat{\rho}(\vec{r}_1,t)|\text{vac}\rangle, \quad (6.8)$$

$\hat{\rho}$ being an energy density operator such that its vacuum expectation is zero while the vacuum expectation of the square of it is not zero.

Now, in the approach developed by the author of this book, the curvature of space-time associated with a dark energy density can be interpreted as a consequence of more fundamental changes of the 3D quantum vacuum energy density, in other words it can be physically defined as the mathematical value of the 3D quantum vacuum energy density. In order to analyse the fundamental features of this approach, we base here our discussion on the paper [508].

In analogy with Santos' treatment, inside our model we assume that the underlying quantum vacuum condensate can be characterized by considering the quantized metric of the quantum vacuum defined by relation

$$d\hat{s}^2 = \hat{g}_{\mu\nu}dx^\mu dx^\nu, \quad (6.153)$$

where the coefficients (in polar coordinates) are

$\hat{g}_{00} = -1 + \hat{h}_{00}, \ \hat{g}_{11} = 1 + \hat{h}_{11}, \ \hat{g}_{22} = r^2(1 + \hat{h}_{22}), \ \hat{g}_{33} = r^2 \sin^2\vartheta(1 + \hat{h}_{33}),$

$\hat{g}_{\mu\nu} = \hat{h}_{\mu\nu}$ for $\mu \neq \nu$, \hfill (6.154)

where

$$\langle \hat{h}_{\mu\nu}\rangle = 0 \text{ except } \langle \hat{h}_{00}\rangle = \frac{8\pi G}{3}\left(\frac{\Delta\rho_{qvE}}{c^2} + \rho_{\text{DE}}\right)r^2 \text{ and}$$

$$\langle \hat{h}_{11}\rangle = \frac{8\pi G}{3}\left(\rho_{\text{DE}} - \frac{1}{2}\frac{\Delta\rho_{qvE}}{c^2}\right)r^2, \quad (6.155)$$

where $\langle \hat{h}_{\mu\nu}\rangle$ stands for $\langle\Psi|\hat{h}_{\mu\nu}|\Psi\rangle$, ρ_{DE} is the density owed to a possible existence of dark matter, Ψ is the quantum state of the universe corresponding to the value of the field $\Phi(x)$ defining a given fine-grained history. Due to the fact that the relations between the metric coefficients and the matter stress-energy tensor are nonlinear, the expectation of the quantized metric (6.153) of the vacuum condensate is not the same as the metric of the expectation of the matter tensor (6.144). The difference gives rise to a contribution of the vacuum fluctuations which reproduces the effect of a cosmological constant. Moreover, we will assume that, when the distance $r \to \infty$, one has $\hat{g}_{\mu\nu} \to \eta_{\mu\nu}$, where $\eta_{\mu\nu}$ is the Minkowski metric.

By starting from the quantized metric whose coefficients are defined by relations (6.154) and (6.155), one can obtain the components of the quantum Einstein equations

$$\hat{G}_{\mu\nu} = \frac{8\pi G}{c^4}\hat{T}_{\mu\nu}, \qquad (6.156)$$

on the basis of the assumption that they are similar to the classical counterparts. In particular, the expectation value of the (operator) metric parameter $\hat{h}_{11}$ may be written in the form

$$\langle\Psi|\hat{h}_{11}|\Psi\rangle = \langle\Psi|\hat{h}_{11}|\Psi\rangle_{\text{mat}} + \langle\Psi|\hat{h}_{11}|\Psi\rangle_{\text{vac}}, \qquad (6.157)$$

namely it is the sum of two expressions, one containing the matter density produced by the changes of the quantum vacuum energy density, and the other indicating the vacuum density fluctuations, ρ_{vac}. In Eq. (6.157), the matter term is

$$\langle\Psi|\hat{h}_{11}|\Psi\rangle_{\text{mat}} \cong \frac{2GM}{r} + \frac{2G^2M^2}{r^2} \qquad (6.158)$$

which agrees with the second order expansion of the well known Schwarzschild solution

$$g_{11} = \left(1 - \frac{2GM}{r}\right)^{-1} \qquad (6.159)$$

while the vacuum contribution, to order G^2, is

$$\langle\Psi|\hat{h}_{11}|\Psi\rangle_{\text{vac}} \cong 150\frac{1}{\pi}G^2 r^2 \left(\frac{V}{c^2}\Delta\rho_{qvE}^{DE}\right)^2 \frac{1}{l} \cdot \frac{1}{l^3} \qquad (6.160)$$

where

$$l = \frac{\hbar}{\left(\frac{V}{c^2}\Delta\rho_{qvE}^{DE}\right)c}, \quad (6.161)$$

$\Delta\rho_{qvE}^{DE}$ are opportune fluctuations of the quantum vacuum energy density, which determine the appearance of dark energy on the basis of equation

$$\left(\frac{V}{c^2}\Delta\rho_{qvE}^{DE}\right)^2 \frac{1}{l} \cdot \frac{1}{l^3} = 4\pi \int_0^\infty C(r_{12})r_{12}dr_{12} \quad (6.162)$$

and r is a typical cosmic distance, which is estimated to fulfil $r/s \approx 10^{40}$. In this way, the total expectation value (6.157) becomes

$$\langle\Psi|h_{11}|\Psi\rangle \cong \frac{8\pi G \Delta\rho_{qvE}}{3c^2}r^2 + 150\frac{1}{\pi}G^2 r^2 \left(\frac{V}{c^2}\Delta\rho_{qvE}^{DE}\right)^2 \frac{1}{l}\cdot\frac{1}{l^3}. \quad (6.163)$$

Hence, comparison with the Friedmann equations

$$\left[\frac{\dot{a}}{a}\right]^2 = \frac{8\pi G}{3}(\rho_{\text{mat}} + \rho_{\text{DE}}), \quad \frac{\ddot{a}}{a} = \frac{8\pi G}{3}\left(\frac{1}{2}\rho_{\text{mat}}t + \rho_{\text{DE}}\right), \quad (6.164)$$

taking account of relations (6.6) and (6.162), leads to the following equation

$$\rho_{\text{DE}} \cong \frac{35G}{2\pi}\left(\frac{V}{c^2}\Delta\rho_{qvE}^{DE}\right)^2 \frac{1}{l}\cdot\frac{1}{l^3} \quad (6.165)$$

namely

$$\rho_{\text{DE}} \cong \frac{35Gc^2}{2\pi\hbar^4 V}\left(\frac{V}{c^2}\Delta\rho_{qvE}^{DE}\right)^6 \quad (6.166)$$

which states the equivalence of the curvature of space-time produced by the changes of the quantum vacuum energy density and the one determined by a constant dark energy density. This means that in the approach based on Eqs. (6.153)–(6.166), the changes and fluctuations of the quantum vacuum energy density generate a curvature of space-time similar to the curvature produced by a "dark energy" density. Moreover, it is interesting to observe that, while in Santos' model, the dark energy is associated with the two-point correlation function of the vacuum fluctuations, in the approach suggested by the author of this book, the dark energy is directly determined by fluctuations of the quantum vacuum energy density on the basis of equation (6.166). In other words, one can say that here the fluctuations of the quantum vacuum energy density plays the same role of Santos' two-point correlation function.

Now, introducing Eq. (6.166) into Eq. (6.155), the expectation values of the coefficients of the quantized metric (6.153) become

$$\langle \hat{h}_{\mu\nu} \rangle = 0 \text{ except } \langle \hat{h}_{00} \rangle$$

$$= \frac{8\pi G}{3} \left(\frac{\Delta \rho_{qvE}}{c^2} + \frac{35Gc^2}{2\pi \hbar^4 V} \left(\frac{V}{c^2} \Delta \rho_{qvE}^{DE} \right)^6 \right) r^2 \text{ and}$$

$$\langle \hat{h}_{11} \rangle = \frac{8\pi G}{3} \left(-\frac{\Delta \rho_{qvE}}{2c^2} + \frac{35Gc^2}{2\pi \hbar^4 V} \left(\frac{V}{c^2} \Delta \rho_{qvE}^{DE} \right)^6 \right) r^2. \quad (6.167)$$

namely turn out to depend directly on the changes of the quantum vacuum energy density. As a consequence, one can say that the changes and fluctuations of the quantum vacuum energy density, through the quantized metric (6.153) of the quantum vacuum condensate whose coefficients are defined by Eqs. (6.154) and (6.167) can be considered the origin of the curvature of space-time characteristic of general relativity. In other words, one can say that the curvature of space-time may be considered as a mathematical value which emerges from the quantized metric (6.153) and thus from the changes and fluctuations of the quantum vacuum energy density (on the basis of Eqs. (6.154) and (6.167)). In synthesis, according to the view suggested in this chapter, the quantized metric (6.155) associated with the changes and fluctuations of the quantum vacuum energy density on the basis of Eqs. (6.154) and (6.167) can be considered the ultimate visiting card of general relativity.

6.3.3 About the motion of a material object in the curved space-time

Gravitational interaction emerges as an effect of the changes of the 3D quantum vacuum energy density. The excited state of the 3D quantum vacuum characterized by a change of the energy density given by (6.125) can be described by a gravitational energy density given by equation

$$\rho_{grav}c^2 = -G\left(\frac{V}{c^2}\Delta\rho_{qvE} + \frac{35Gc^2}{2\pi\hbar^4}\left(\frac{V}{c^2}\Delta\rho_{qvE}^{DE}\right)^6\right)^2 \frac{1}{l} \cdot \frac{1}{l^3} \quad (6.168)$$

where

$$l = \frac{\hbar}{\left(\frac{V}{c^2}\Delta\rho_{qvE}\right)c}. \quad (6.169)$$

In this approach, the quantum vacuum energy density (given by Eq. (6.123)) and its corresponding gravitational energy density (given by Eq. (6.168)) can be considered the fundamental elements that determine the gravitational interaction between material objects. Considering the 3D quantum vacuum characterized by a ground state defined by the energy density (6.122) as the fundamental arena of the universe, in this picture of the gravitational space a concept of *curvature/density of space* can be introduced which means that presence of mass (determined by an opportune excited state of the 3D quantum vacuum corresponding to a given change (6.139) of its energy density) decreases the quantum vacuum energy density and increases its effective curvature: gravity is carried by the effective curvature caused by a variable quantum vacuum energy density. A first important result inside this model is that gravitational interaction is an immediate, direct phenomenon: the 3D quantum vacuum (through its fundamental quantity represented by the gravitational energy density of the quantum vacuum (6.168)) acts as a direct medium of gravity. No movement of particle-wave is needed for its acting: gravity is transmitted directly by the quantum vacuum energy density, by means of its corresponding gravitational energy density characterizing the region between the material objects under consideration. Gravity is the result of the geometrical shape of the quantum vacuum energy density.

In order to show in what sense the gravitational energy density (6.168) associated with an opportune excited state of the 3D quantum vacuum defined by the energy density (6.123) (and thus by a change of the energy density (6.139) with respect the ground state (6.122)) acts as a direct medium of gravity, by using Santos' equations (6.7) and (6.8) one may rewrite Eq. (6.140) as

$$-G\left(\frac{V}{c^2}\Delta\rho_{qvE} + \frac{35Gc^2}{2\pi\hbar^4}\left(\frac{V}{c^2}\Delta\rho_{qvE}^{DE}\right)^6\right)^2 \frac{1}{l}\cdot\frac{1}{l^3}$$

$$= -4\pi G \int_0^\infty C(r_{12}) r_{12} dr_{12} \qquad (6.170)$$

Equation (6.170) suggests that an excited state of the 3D quantum vacuum corresponding with a given change of the energy density (6.139) (and with an opportune gravitational energy density (6.140)) is linked with the two-point correlation, $C(r_{12})$ which, on the basis of Eq. (6.8), for equal times, is a function that does not depend explicitly on time: this means, from the physical point of view, that the decreasing of the quantum vacuum energy density associated with an excited state of the 3D quantum vacuum (which determines the presence of a material object endowed with a given mass) is a direct phenomenon. Equation (6.170) (combined with Eq. (6.8)) shows clearly in what sense, in the approach here suggested, gravity-space is a direct phenomenon: it states that (for equal times) the two-point correlation function acts as a direct information channel for the transfer of the gravitational energy density of quantum vacuum from one point to another, in other words it states that there is no time for the transfer of the gravitational energy density of quantum vacuum from one point to another and thus that the gravitational energy density of quantum vacuum acts as a direct medium for the transmission of gravitation.

Now, let us see how the curvature of space-time corresponding to the changes and fluctuations of the quantum vacuum energy density — which characterize a given excited state of the 3D quantum vacuum — acts on a test particle of mass m_0, in other words how the motion of a material object in a background characterized by changes of its energy density can be treated mathematically. When a material object (determined by an excited state of the 3D quantum vacuum and thus corresponding to a given diminishing of the quantum vacuum energy density) moves, this diminishing of the quantum vacuum energy density causes a shadowing of the gravitational space which determines the motion of other material objects present in the region under consideration. We assume that the shadowing of the 3D quantum vacuum can be expressed by equation

$$\vec{D} = \kappa \varepsilon_0 \vec{E}_g \qquad (6.171)$$

where κ is a factor which represents the relatively small amount of the altered permittivity of the free space (with respect to the situation in which the energy density of the quantum vacuum is given by Eq. (6.122)) and

$$\vec{E}_g = -H_{eg} \left(\frac{V}{c^2} \Delta \rho_{qvE} + \frac{35Gc^2}{2\pi\hbar^4} \left(\frac{V}{c^2} \Delta \rho_{qvE}^{DE} \right)^6 \right) \frac{1}{r^2} \hat{r} \qquad (6.172)$$

is the gravitostatic field determined by both density of matter and density of dark energy (here $H_{eg} = \frac{G}{c^2}$ is the basic gravitodynamic constant). The shadowing of the gravitational space, expressed by Eq. (6.171), thus depends on the relatively small amount of the altered permittivity of the free space (with respect to the ground state of the 3D quantum vacuum) and on the field (6.172) associated itself with the changes of the quantum vacuum energy density (corresponding with the excited state under consideration of the quantum vacuum). The total lagrangian density for matter-field interactions in the polarized vacuum is given by relation

$$L_d = -\left(\frac{m_0 c^2}{\sqrt{K}}\sqrt{1-\left(\frac{v}{c/K}\right)^2} + q\Phi - q\vec{A}\cdot\vec{v}\right)\delta^3(\vec{r}-r_0)$$

$$-\frac{1}{2}\left(\frac{B_g^2}{K\mu_0} + K\varepsilon_0 E_g^2\right)$$

$$-\frac{\lambda}{K^2}\left((\nabla K)^2 + \frac{1}{(c/K)^2}\left(\frac{\partial K}{\partial t}\right)^2\right), \quad (6.173)$$

where $(\Phi, \vec{A})$ are the gravitational potentials, $\vec{B}$ is the gravitomagnetic field defined by

$$\vec{B}_g = H_{eg}\frac{\vec{J}}{r^3} \quad (6.174)$$

(where $\vec{J} = \vec{L} + \vec{S}$, $\vec{L} = r \times \left(\frac{V}{c^2}\Delta\rho_{qvE} + \frac{35Gc^2}{2\pi\hbar^4}\left(\frac{V}{c^2}\Delta\rho_{qvE}^{DE}\right)^6\right)\vec{v}$, $\vec{S}$ being the spin angular momentum of the material object determined by the diminishing of the quantum vacuum energy density under consideration) and $\lambda = \frac{c^4}{32\pi G}$. In analogy with Puthoff's polarizable vacuum model of gravitation [414], variation of the action functional with respect to the test particle variables leads to the following equation of motion of a test material object of mass m_0 in the polarized 3D quantum vacuum:

$$\frac{d}{dt}\left[\frac{\left(m_0 \kappa^{3/2}\right)\vec{v}}{\sqrt{1-\left(\frac{v}{c/\kappa}\right)^2}}\right] = m_0\left(c^2\vec{E}_g + \vec{v}\times\vec{B}_g\right)$$

$$+\frac{m_0 c^2}{\kappa}\cdot\frac{1+\left(\frac{v}{c/\kappa}\right)^2}{2\sqrt{1-\left(\frac{v}{c/\kappa}\right)^2}}\cdot\frac{\nabla\kappa}{\kappa}. \quad (6.175)$$

Equation (6.175) shows that there are two forces acting onto the test particle of mass m_0: the Lorentz force due to the quantum vacuum energy density surrounding the object and a second term representing the dielectric force proportional to the gradient of the shadowing of the quantum vacuum (6.171). The importance of this second term lies in the fact that thanks to it one can account for the gravitational potential, either in Newtonian or general relativistic form. It might be interesting to note that with $m_0 \to 0$ and $v \to \frac{c}{\kappa}$, as would be the case for a photon, the deflection of the trajectory is twice as the deflection of a slow moving massive particle. This is an important indication of conformity with general relativity.

Variation of the action functional with regard to the κ variable leads to the expression of the generation of the shadowing of space within general relativity, owed to the presence of both matter and fields. The equation one obtains is the following:

$$\nabla^2 \sqrt{\kappa} - \frac{1}{(c/\kappa)^2} \cdot \frac{\partial^2 \sqrt{\kappa}}{\partial t^2} = \frac{-\kappa}{4\lambda}[P(\kappa) + Q(\kappa) + R(\kappa)]. \quad (6.176)$$

Here $P(\kappa)$ represents the change in space shadowing by the density of matter associated with the object of mass m_0, with the vector $\vec{r}$ as the distance from the system mass centre:

$$P(K) = \frac{m_0 c^2}{\sqrt{K}} \cdot \frac{1 + \left(\frac{v}{c/K}\right)^2}{\sqrt{1 - \left(\frac{v}{c/K}\right)^2}} \cdot \delta^3(\vec{r} - \vec{r}_0). \quad (6.177)$$

$Q(\kappa)$ is the change caused by the energy density of the fields (6.172) and (6.174) determined by the diminishing of the quantum vacuum energy density:

$$Q(\kappa) = \frac{1}{2}\left(\frac{B_g^2}{\kappa \mu_0} + \kappa \varepsilon_0 E_g^2\right). \quad (6.178)$$

$R(\kappa)$ is the change caused by the shadowing of the quantum vacuum energy density itself:

$$R(\kappa) = -\frac{\lambda}{\kappa^2}\left((\nabla \kappa)^2 + \frac{1}{(c/\kappa)^2}\left(\frac{\partial \kappa}{\partial t}\right)^2\right). \quad (6.179)$$

In the case of a static gravity field of a spherical mass distribution (a planet or a star), the solution of Eq. (6.176) has a simple exponential form:

$$\sqrt{\kappa} = e^{GM/rc^2}, \qquad (6.180)$$

where $M = \frac{V\Delta\rho_{qvE}}{c^2}$. The solution (6.180) can be approximated by expanding it into a series:

$$\kappa = e^{2GM/rc^2} = 1 + \frac{2GM}{rc^2} + \frac{1}{2}\left(\frac{2GM}{rc^2}\right)^2 + \cdots. \qquad (6.181)$$

This solution reproduces (to the appropriate order) the usual general-relativistic Schwarzschild metric predictions in the weak field limit conditions (i.e. Solar system).

According to this model, it is important to underline that also particles without mass (for example, photons) have an indirect influence on the quantum vacuum energy density. In fact, because of Eq. (6.178) also a photon will add a contribution to the effective curvature of space associated with the fields (6.172) and (6.174). This result turns out to be also in accordance with general theory of relativity, where both mass and energy cause the curvature of space.

Moreover, with the obtained solution (6.180) or (6.181) regarding the factor κ measuring the polarizability of the quantum vacuum in the presence of matter, one can analyze the gravitational red shift characteristic of general relativity, and find inside this approach a more detailed formula in order to evaluate the frequency shift of the photon emitted by an atom on the surface of a star of mass M and radius R. Just like in Puthoff's model, the photon detected far away from the star will appear red shifted by the following amount:

$$\frac{\Delta\omega}{\omega_0} = \frac{\omega - \omega_0}{\omega_0} \approx -\frac{GM}{Rc^2} \qquad (6.182)$$

where we have assumed $\frac{GM}{Rc^2} \ll 1$. The photon, after having climbed up the gravity potential of the star, will retain its acquired frequency unchanged, and the change in frequency can be tested locally by comparing it with photons emitted by the same type of atoms at the same temperature, but within the weak gravity field of the laboratory.

With that same result it is also possible to analyse the amount of the bending of light rays from a distant star passing near a massive body, like in the classic general relativity test performed by the Eddington's expedition during the solar eclipse in May 1919. The light ray from a distant star, while passing close to the Sun, will experience a gradual slowing of wavefront velocity coming towards the Sun, and a gradual increasing velocity in leaving the Sun's gravity field. Because κ increases closer to a massive body ($\kappa > 1$), the velocity of light will vary as c/κ. The part of the wavefront closer to the Sun will thus experience a greater slowdown than the part of the wavefront passing further away. This is seen from Earth as an apparent shift of the position of the star close to the Sun's disk edge in the outward direction. In general relativity's terms, this deflection is a measure of local space-time curvature. In order to calculate the total bending angle, because in case of the Sun the total deflection is small ($\varphi < 2$ arc-seconds), we can apply the usual low angle approximations throughout the calculation. And because of the same reason, we will not make a big mistake if we approximate the variable velocity of light to the first order term of the series expansion (6.181) of κ:

$$v = \frac{c}{\kappa} \approx \frac{c}{1 + \frac{2GM}{rc^2}} \approx c\left(1 - \frac{2GM}{rc^2}\right) \tag{6.183}$$

In this relation the radius–vector r denotes the distance of the wavefront from the centre of the Sun as it travels by from $-\infty$ to $+\infty$, with the minimum distance of $R+\delta$ where R is the Sun's radius, and δ is the minimum distance from the Sun's surface. By assigning z to the distance of the wavefront along the line of sight (perpendicular to $R+\delta$), the radius-vector becomes $r = \sqrt{(R+\delta)^2 + z^2}$, so Eq. (6.183) can be written as:

$$v \approx c\left(1 - \frac{2GM}{c^2} \cdot \frac{1}{\sqrt{(R+\delta)^2 + z^2}}\right). \tag{6.184}$$

The differential velocity of light, assuming $\delta \ll R$, is then

$$\Delta v = \frac{2GM}{c^2} \cdot \frac{R\delta}{(R^2 + z^2)^{3/2}}. \tag{6.185}$$

As the wavefront travels a distance $dz \approx vdt$, the differential velocity along the path of light results in an accumulated wavefront path difference Δz:

$$\Delta z = \Delta v dt \approx \frac{2GM}{c^2} \cdot \frac{R\delta}{(R^2 + z^2)^{3/2}} dz \tag{6.186}$$

This results in an accumulated tilt angle of:

$$\varphi \approx \Delta z / \delta \approx \frac{2GM}{c^2} \cdot \frac{R}{(R^2 + z^2)^{3/2}} dz \tag{6.187}$$

By integrating Eq. (6.187) over the entire path $-\infty < z < +\infty$ yields:

$$\varphi \approx \frac{4\pi GM}{Rc^2}. \tag{6.188}$$

By inserting $G = 6,672 \cdot 10^{-11} Nm^2 Kg^{-2}$, $M = 1,9891 \cdot 10^{30}$Kg, and $R = 6,96 \cdot 10^8$m, we obtain $\varphi = 1,75$ arc-seconds, which is exactly the value predicted by Einstein's general theory of relativity in 1915, and experimentally verified by Eddington in 1919 (between 1.2 and 1.9 arc-seconds, mainly because of the imperfect optics of the portable telescopes used).

In synthesis, in the model suggested by the author and Sorli in the papers [507–509], one can say that the equation of motion (6.175) for a test material object in the 3D quantum vacuum, characterized by a decreasing of the energy density (associated with a given excited state of the quantum vacuum determined by peculiar RS processes of creation/annihilation) on one hand, and the Eq. (6.176) describing the shadowing of the space produced by the presence of matter and fields on the other hand, allow us to obtain results that are coherent and compatible with general theory of relativity. The perspective is thus opened that the curvature of space of general relativity can be associated with the diminishing of the quantum vacuum energy density (corresponding to a given excited state of the quantum vacuum and determining the presence of material objects), and that the material objects follow the geodesic paths within the shadowed gravitational space (determined just by the change of the quantum vacuum energy density and thus associated with an excited state of the quantum vacuum). Moreover, as regards the equations of motion (6.175) and (6.176), it is important to emphasize that, according to this approach, the modification of the quantum vacuum energy density (corresponding to a given excited state of the quantum vacuum and determining both the matter density and dark energy density) as well as the action of the shadowed quantum vacuum on another material object are phenomena directly determined by the fields (6.174), (6.175), (6.178) and (6.179). This implies that no time is needed to transmit the information from a material object to the surrounding region in order to shadow the gravitational space because the change of the quantum vacuum energy density is already there as it is associated with the fields (6.174), (6.175), (6.178) and (6.179) (what propagates from point to point is just the effective consequences of this change); and, on the other hand, no time is needed to transmit the information from the shadowed space to another material object in order to cause its movement.

Finally, at the end of this paragraph, it is important to emphasize that the view proposed by the author and Sorli of a 3D quantum vacuum as a direct medium for the transmission of gravitation established by Eqs. (6.174), (6.175), (6.178) and (6.179) can express in an elegant mathematical way the perspective about the non existence of gravitational waves. In this regard, it seems compatible with some recent results of Loinger according to which gravitational waves are only hypothetical and do not exist in the physical world [514, 515]. On the other hand, despite several attempts of research about gravitational waves performed since the 1960s (see for example the reference [516]), gravitational waves have not yet been detected. As underlined by Schorn in the paper [517], "To search for gravitational waves in a laboratory, classical or quantum mechanical detectors can be used. Despite the experiments of Weber (1960 and 1969) and many others (Braginskij et al., 1972; Drever et al., 1973; Levine and Garwin, 1973; Tyson, 1973; Maischberger et al., 1991;

Abramovici *et al.*, 1992; and Abramovici *et al.*, 1996) and theoretical calculations and estimations (Braginski and Rudenko, 1970; Harry *et al.*, 1996; and Schutz, 1997), gravitational waves have never been observed directly in laboratory".

6.4 About Quantum Mechanics in the 3D Quantum Vacuum Energy Density Model

On the basis of the 3D quantum vacuum model suggested by the author and Sorli in the papers [507, 509], quantum mechanics can be interpreted as a theory which emerges from a more fundamental descriptive level of reality, the 3D quantum vacuum characterized by **RS** processes which evolve according to Eqs. (6.126), (6.127) and (6.128). In fact, nonrelativistic quantum mechanics based on the standard Schrödinger equation derives from the more general time-symmetric extension of Klein–Gordon quantum-relativistic equation (6.126) in the case of creation events.

Moreover, the quantum mechanical formalism in terms of vectors bra and ket emerges naturally by starting from the **RS** processes of the timeless 3D quantum vacuum. In this regard, if one considers a **RS** process of creation of quality q and destruction of quality s which takes place into the two distinct processes $r_k^- q^+$ and $s^- r_k^+$, the probability of the entire process can be seen as the product of the probabilities of the two processes, namely:

$$P(r_k^- q^+) \times P(s^- r_k^+) = |\langle r_k | q \rangle|^2 \times |\langle s | r_k \rangle|^2 = |\langle r_k | q \rangle \langle s | r_k \rangle|^2$$
(6.189)

Therefore, if the amplitude of the entire process is associated with Eq. (6.189) and $(s^- r_k^+)(r_k^- q^+)$ is defined as $\langle s | r_k \rangle \langle r_k | q \rangle$, the standard Born rule according to which the probability of the process is given by the modulus squared of the amplitude remains valid in this case too.

Let us consider now a **RS** process of creation of quality q and destruction of quality s which takes place as the simultaneous actuation of all the processes: $s^- r_k^+$, $r_k^- q^+$ with $k = 1, 2, \ldots, L$. In this more general case, the amplitude of the entire process can be

A Three-Dimensional Timeless Quantum Vacuum   387

expressed as the sum of the partial amplitudes relating the various subsets corresponding to the various r_k:

$$\langle s|q\rangle = \sum_k \langle s|r_k\rangle\langle r_k|q\rangle \qquad (6.190)$$

This result allows the ket to be defined as

$$|q\rangle = \sum_k |r_k\rangle\langle r_k|q\rangle \qquad (6.191)$$

The probability amplitude is thus obtained by "left-multiplying" Eq. (6.191) by $\langle s|$. In analogous way, one can define the bra vector as:

$$\langle s| = \sum_k \langle s|r_k\rangle\langle r_k| \qquad (6.192)$$

The probability amplitude is thus obtained by "right-multiplying" Eq. (6.192) by $|q\rangle$. Since $\langle r_k|r_i\rangle = \delta_{ki}$, bra and ket are vectors defined with respect to a "complete" orthonormalized basis.

Therefore, in the model of a timeless 3D quantum vacuum suggested by the author of this book and Sorli, each generic vector (bra or ket) can be decomposed into a complete basis of orthonormalized vectors. For example,

$$|\Psi\rangle = \sum_i c_i|\varphi_i\rangle \qquad (6.193)$$

where $\langle \varphi_k|\varphi_i\rangle = \delta_{ki}$ and c_i are complex numbers. The probability of the **RS** process $\varphi_k^- \psi^+$ is expressed by Born's rule:

$$P(\varphi_k^- \psi^+) = \frac{|\langle \varphi_k|\psi\rangle|^2}{\langle \psi|\psi\rangle} = \frac{c_k^* c_k}{\sum_i c_i^* c_i} \qquad (6.194)$$

Let us now suppose that a certain physical quantity O, in the realization of the **RS** process $\varphi_k^- \psi^+$, assumes the value $o(k)$. Let us assume that this value is a real number. Let us suppose that the quality ψ is fixed, while the quality φ can freely vary on its support set $\{\varphi_k\}$. Therefore, the *a priori* probability that the outcome is the quality φ_k is (6.194) and the expectation value of O is given by:

$$\langle O\rangle_\psi = \sum_k P(\varphi_k^- \psi^+) o(k) = \frac{\sum_k o(k) c_k^* c_k}{\sum_i c_i^* c_i}. \qquad (6.195)$$

Equation (6.195) is general and derives directly from the formalism of the **RS** processes of the 3D quantum vacuum. A very important special case of Eq. (6.195) is the one in which an appropriate self-adjoint linear operator Ω exists on the rigged space of kets and bras, such that:

$$\Omega|\varphi_i\rangle = o(i)|\varphi_i\rangle, \qquad (6.196)$$

for every value of i. From the self-adjointness of Ω it follows that:

$$\langle\varphi_i|\Omega = \langle\varphi_i|o(i), \qquad (6.197)$$

for every value of i. From the linearity of Ω, it follows immediately that:

$$\Omega|\psi\rangle = \sum_i c_i \Omega|\varphi_i\rangle = \sum_i c_i o(i)|\varphi_i\rangle \qquad (6.198)$$

$$\langle\psi|\Omega|\psi\rangle = \sum_k c_k^* \langle\varphi_k| \left(\sum_i c_i o(i)|\varphi\rangle_i \right) = \sum_k o(k) c_k^* c_k. \qquad (6.199)$$

Hence, taking account of

$$\langle\psi|\psi\rangle = \sum_k c_k^* \langle\varphi_k| \left(\sum_i c_i|\varphi_i\rangle \right) = \sum_i c_i^* c_i \qquad (6.200)$$

one obtains

$$\langle O \rangle_\psi = \frac{\langle\psi|\Omega|\psi\rangle}{\langle\psi|\psi\rangle}. \qquad (6.201)$$

The formula (6.201) is less general but is often used in quantum physics. The operator Ω is thus said to be "associated" with the quantity O. The $o(k)$ values in this case form the "spectrum" of O (or of Ω). The Hermitianity of the operator Ω ensures on the one hand that the $o(k)$ values are real and on the other that the procedure is symmetrical. Indeed, we would have obtained the same result by developing $\langle\psi|\Omega$ and then right-multiplying by $|\psi\rangle$.

Let us consider a generic **RS** process which is constituted by creation of the quality ψ and destruction of the quality φ. These two events, which according to the 3D timeless quantum vacuum model are the primary physical events in the physical universe, are correlated by the amplitude $\langle\psi|S|\psi\rangle$ which can generally be developed as the sum of the partial amplitudes corresponding with different paths joining the two events, S being the evolution operator of the system into consideration. To be explicit, let us consider for example the well-known double-slit experiment. In such a case $|\psi\rangle$ is the ket associated with the "preparation" of the "particle" in the source and $\langle\varphi|$ is the bra associated with the "detection" of the "particle" behind the screen containing the slits. Under the action of S, $|\psi\rangle$ evolves freely up to the interaction with the screen. If the screen is completely absorbent, then, indicating as x the position of a generic point on its rear wall, one has $\langle x|\psi\rangle = 0$, unless x corresponds to one of the two slits. The following time evolution of $|\psi\rangle$ is still free. Therefore, indicating the two slits as f_1, f_2, one can write:

$$\langle\varphi|S|\psi\rangle = \langle\varphi|S|f_1\rangle\langle f_1|S|\psi\rangle + \langle\varphi|S|f_2\rangle\langle f_2|S|\psi\rangle. \quad (6.202)$$

As one can easily remark, here there is the sum of two partial amplitudes, one of which corresponds to a "passage through slit f_1", the other to a "passage through slit f_2". These two amplitudes interfere. Each of the two amplitudes is represented, in turn, as the product of two free propagation amplitudes of the "particle": one corresponding to the propagation from the source to the slit, the other corresponding to the propagation from the slit to the detector.

Hence, in this example of the double-slit experiment, either the "particle" interacts with the atoms which compose the screen, and is absorbed, or it does not interact with the screen at all. Only in this second case it can reach the detector, passing through f_1 or through f_2. The terms f_1, f_2 which appear in the expression of the amplitude are therefore associated with interactions with the atoms of the screen (to be precise, "negative" interactions, namely noninteractions or absence of actual interactions). The correlation between the two events of creation of quality ψ and destruction of quality φ is therefore conditioned by the atoms of the screen. The correlation between these two events is thus affected by other events of the universe, in this specific case by the events characterizing the atoms which constitute the screen. The structure of $\langle \varphi|S|\psi \rangle$ includes these influences.

Here, the fundamental thing to emphasize is that the interaction events which appear in $\langle \varphi|S|\psi \rangle$ have not a physical reality of their own, they *are virtual events* appearing in the partial amplitudes as mere artifices of calculation. For example, the (non) interaction events f_1, f_2 in the example of the double-slit experiment are virtual events. This means that these events are not pairs of emission/re-absorption of qualities by the 3D quantum vacuum, whereas the extreme events corresponding to the creation of quality ψ and the destruction of quality φ are. However, virtual events can become real. For example, a "particle" counter can be placed behind f_1 or f_2. In this case, though, one is no longer dealing with the original **RS** process, rather with other entirely different processes, for example $f_1^- \psi^+$.

The free propagations $\langle f_1|S|\psi \rangle$, $\langle \varphi|S|f_1 \rangle$, $\langle \varphi|S|f_2 \rangle$, $\langle f_2|S|\psi \rangle$ can be represented in the form of the propagator of a particle of energy E, impulse $\vec{p}$ and spin $\vec{s}$. Energy E appears from the timeless 3D quantum vacuum in correspondence with the creation, and vanishes into the vacuum in correspondence with the destruction, as do, in fact, the other quantities which make up the quantum. From a merely numerical standpoint, one can view creation as the release by the vacuum of an E amount of energy to the manifest physical world, while destruction can be seen as the release by the manifest physical world of an amount of energy E to the vacuum. In this regard, one may have the following two different situations:

— First situation: The energy released by the timeless 3D quantum vacuum is positive and that absorbed by the vacuum is negative. In this case, we shall have at creation events the exchange of an energy $E > 0$, and at destruction events the exchange of an energy $-E < 0$.

— Second situation: The energy released by the timeless 3D quantum vacuum is negative and that absorbed by the vacuum is positive. In this case, we shall have the exchange of an energy $E < 0$ at creation events, and the exchange of an energy $-E > 0$ at destruction events.

The first situation corresponds to the forward propagation in time of a quantum of energy $E > 0$, or to the propagation backwards in time of a quantum of energy $E < 0$. The second situation corresponds to the forward propagation in time of a quantum of energy $E < 0$, or to the propagation backwards in time of a quantum of energy $E > 0$. In the timeless 3D quantum vacuum model, both these situations are physically possible and are incompatible. If the first situation occurs, retropropagations in time of quanta with positive energy cannot exist; if the second situation occurs, retropropagations in time of quanta with negative energy cannot exist. This means physically that the direction of time ordering we perceive in our everyday life cannot be considered as a primary physical reality: since both the situations illustrated above may occur, the time arrow has not an existence on its own. Time, in our everyday life, flows in one direction only while the **RS** processes — the most elementary processes of the physical world — are symmetrical in time.

In the model here presented of a timeless 3D quantum vacuum, each elementary particle which emerges from an elementary **RS** process has a corpuscular and a wave nature. For example electron is a particle which moves through a 3D quantum vacuum and it can be associated with an opportune wave of the vacuum which evolves according to the general Eq. (6.127) (which becomes equation (6.129) in the nonrelativistic limit).

More precisely, the behavior of a quantum particle as we know it from ordinary quantum mechanics derives from the equations of the timeless 3D quantum vacuum model regarding the creation events. In particular, in the relativistic domain the fundamental equation of motion is the first of the quantum Hamilton–Jacobi equations (6.122), namely:

$$\partial_\mu S_{Q,i} \partial^\mu S_{Q,i} = \frac{16\pi^2 R^6}{9c^2} (\Delta\rho_{qvE})^2 \exp Q_{Q,i}, \qquad (6.203)$$

where

$$Q_{Q,i} = \frac{9\hbar^2 c^2}{16\pi^2 R^6 (\Delta\rho_{qvE})^2} \frac{\left(\nabla^2 - \frac{1}{c^2}\frac{\partial^2}{\partial t^2}\right)|\psi_{Q,i}|}{|\psi_{Q,i}|} \qquad (6.204)$$

is the quantum potential of the vacuum. In the nonrelativistic limit, Eq. (6.203) becomes

$$\frac{3c|\nabla S_{Q,i}|^2}{8\pi R^3(\Delta\rho_{qvE})} + Q_{Q,i} + V = -\frac{\partial S_{Q,i}}{\partial t} \qquad (6.205)$$

where

$$Q_{Q,i} = -\frac{3\hbar^2 c}{8\pi R^3(\Delta\rho_{qvE})} \frac{\nabla^2|\psi_{Q,i}|}{|\psi_{Q,i}|} \qquad (6.206)$$

is the quantum potential of the vacuum. According to Eqs. (6.203)–(6.206), the following interpretation of subatomic particles becomes permissible: in our world electrons and other elementary particles have precise positions at every time and follow precise trajectories which emerge from the evolution laws regarding creation events (6.203) and (6.205) and the corresponding quantum potentials (6.204) and (6.206) indicate the quantum force exerting on the corpuscle, which guides the corpuscle in the regions where the wavefunction of the vacuum is more intense. Therefore, according to the 3D timeless quantum vacuum model here proposed, the quantum potential of ordinary nonrelativistic quantum mechanics can be considered as a consequence of the more fundamental quantum potential of the 3D quantum vacuum (6.206) (and an analogous result regards the quantum potential of Klein-Gordon's relativistic quantum mechanics which derives from the more fundamental quantum potential of the 3D quantum vacuum (6.204)).

As regards the quantum potential of the 3D quantum vacuum ((6.204) or (6.206)), it is important to emphasize that it must not be considered as an external entity in the vacuum but as an entity which represents the geometrical properties of the 3D quantum vacuum. In other words, the quantum potential can be considered a geometric entity of the vacuum, the information determined by the quantum potential ((6.204) or (6.206)) is a type of geometrodynamic information "woven" into the 3D quantum vacuum. The quantum potential has a geometric nature just because has a contextual nature, contains a global information on the environment in which the experiment (deriving from opportune **RS** processes of creation or destruction of opportune qualities) is performed; and at the same time it is a dynamical entity just because its information about the **RS** processes and their environment is active, determines the behaviour of the quantum particles (created or destroyed in the **RS** processes). On the basis of its mathematical expressions, the action of the quantum potential of the vacuum is like-space, namely creates onto the particles (created or destroyed in the **RS** processes) a nonlocal, instantaneous action. In a double-slit experiment, for example, if one of the two slits is closed the quantum potential of the vacuum changes, and this information arrives instantaneously to the particle created in the corresponding **RS** processes, which behaves as a consequence. This means that, at a fundamental level, the timeless 3D quantum vacuum, through its special state represented by the quantum potential of the vacuum, acts as an immediate information medium in determining the motion of a subatomic particle.

In accordance with the ontology which derives from the 3D quantum vacuum model suggested by the author of this book and Sorli in the papers [507, 509], the primary physical reality is constituted solely by **RS** processes. The causal connection between **RS** processes is not ensured by the propagation of an object in space-time intended as a primary physical reality, be it a wave, a corpuscle or any other entity which acts as an intermediary; rather, it is ensured by the common emergence of **RS** processes from the same timeless underlying 3D quantum vacuum. The nonlocal action of the quantum potential of the vacuum which is responsible of the evolution of the occurring of the creation and destruction events is the ultimate mathematical reality governing the evolution of these processes. By using a Bohmian terminology, one can define the reality constituted by the **RS** processes as the background from whose differentiation the foreground constituted by the events and their evolution (governed by the waves of the vacuum through the nonlocal action of the quantum potential) emerges. By attributing the status of primary physical reality solely to **RS** processes it follows that the space-time coordinates are labels associated with these processes and which express certain properties of relation between them. Space-time as such is only the domain setoff this labels and has no physical reality of its own. It is, so to speak, materialized by **RS** processes.

In analogous way, also the background associated with the quantum potential is materialized by **RS** processes.

Moreover, if in the a-temporal quantum-gravity space theory analysed in chapter 5 one can provide an important unification, in a geometric picture, between the gravitational and quantum aspects of matter and the wavefunction of the quantum-gravity space emerges as the fundamental entity which allows us to obtain this geometric picture, the timeless 3D quantum vacuum model proposed by the author and Sorli introduces the interesting perspective to interpret gravity and quantum behavior as two different aspects of a same source at a more fundamental level, namely the level of the 3D quantum vacuum defined by **RS** processes. In other words, on the basis of this model, the effects of gravity on geometry and the quantum effects on the geometry of space-time are highly coupled and both of them derive from the **RS** processes of the timeless 3D quantum vacuum. By following the treatment provided in the papers [507, 509], if one starts from Bohm's version of Klein-Gordon equation of the generic component of the probability amplitude of the occurrence of creation event for a quantum particle Q based on Eqs. (6.156) and (6.158), one can show that the generic component of the spinor associated with the quantum particle under consideration (and thus the quantum potential associated with it) has an important link with the curvature of the ordinary space-time we perceive. The treatment of a particle moving in a curved background can be done by changing the ordinary differentiating ∂_μ with the covariant derivative ∇_μ and by changing the Lorentz metric with the curved metric $g_{\mu\nu}$ inside Eq. (6.156). In this way one obtains the equations of motion for a change of the energy density (which determines the occurrence of creation event for a quantum particle Q of mass (6.107)) in a curved background:

$$\tilde{g}_{\mu\nu}\tilde{\nabla}_\mu S_{Q,i}\tilde{\nabla}_\nu S_{Q,i} = \frac{16\pi^2 R^6 (\Delta\rho_{qvE})^2}{9c^2\hbar^2}, \qquad (6.207)$$

where $\tilde{\nabla}_\mu$ represents the covariant differentiation with respect to the metric

$$\tilde{g}_{\mu\nu} = g_{\mu\nu}/\exp Q_{Q,i} \qquad (6.208)$$

which is a conformal metric, where

$$Q_{Q,i} = \frac{9\hbar^2 c^2}{16\pi^2 R^6 (\Delta \rho_{qvE})^2} \frac{\left(\nabla^2 - \frac{1}{c^2}\frac{\partial^2}{\partial t^2}\right)_g |\psi_{Q,i}|}{|\psi_{Q,i}|} \qquad (6.209)$$

is the quantum potential of the vacuum.

The important conclusion one can draw from this treatment is that the presence of the quantum potential of the vacuum is equivalent to a curved space-time with its metric being given by (6.208). On the ground of Eqs. (6.207)–(6.209), it becomes so permissible the following reading, the following interpretation of the curvature of space-time (and thus of gravitation) inside the timeless 3D quantum vacuum characterized by **RS** processes. **RS** processes associated with creation events of quantum particles determine a quantum potential of the vacuum which is equivalent to the curvature of the space-time. The quantum potential of the vacuum corresponding to the generic component of the spinor of a quantum particle is tightly linked with the curvature of the space-time we perceive. In other words, one can say that **RS** processes, through the manifestation of the quantum potential of the vacuum (6.162), lead to the generation, in our macroscopic level of reality, of a curvature of space-time and, at the same time, the space-time metric is linked with the quantum potential of the vacuum which influences and determines the behaviour of the particles (corresponding to creation events from the timeless 3D quantum vacuum).

In this way, the model of the 3D timeless quantum vacuum allows us to achieve a significant geometrization of the quantum aspects of matter and the source of this geometrization can be considered the timeless 3D quantum vacuum characterized by **RS** processes. In this picture, one can say that the space-time geometry sometimes looks like what we call gravity and sometimes looks like what we understand as quantum behaviours and both these features of physical geometry emerge from the **RS** processes of the timeless 3D quantum vacuum.

Conclusions

There has been a very long debate in Western philosophy and physics as regards the following three pairs of choices about how to model and understand the phenomena of the universe: 1) the fundamentality of being versus becoming, 2) the fundamentality of monism versus atomism and 3) the fundamentality of algebra versus geometry broadly construed; more generally, which of the myriad formalisms which appear in the literature will be most unifying.

As regards the point 1, from the very beginning the general assumption is that everything can be explained. Perhaps the cosmological argument for the existence of God is the classic example of such thinking. In this regard, Leibniz appeals to a version of the principle of sufficient reason which states that "no fact can be real or existing and no statement true without a sufficient reason for its being so and not otherwise" [518]. Leibniz uses this principle to argue that the sufficient reason for the "series of things comprehended in the universe of creatures must exist outside this series of contingencies and is found in a necessary being that we call God" [518].

While physics dispensed with appeals to God at some point, it did not jettison the principle of sufficient reason, merely replacing God with fundamental dynamical laws, which point in the direction of a Theory of Everything. In agreement with everyday experience a very early assumption of Western physics — reaching its apotheosis with Newtonian mechanics — is that the fundamental phenomena of nature in need of explanation are motion and change in time, and thus that the explanation of processes should involve dynamical laws essentially. This view accepts the reality of change and time, by embracing becoming as a bona fide attribute of the universe.

In the challenge to find a unifying view of the different physical phenomena, the combination of the principle of sufficient reason and the dynamical perspective has in great part motivated the particular kind of unification being sought, i.e., the search for a theory of everything, quantum gravity theory or similar theories. Therefore, today almost all attempts to unify relativity and quantum theory opt for becoming (dynamism) as fundamental in some form or another. Such theories may deviate from the norm by employing radical new fundamental dynamical entities (such as branes, loops, strings, twistors, etc.), but the game is always dynamical, broadly construed (vibrating branes, geometrodynamics, sequential growth process, etc...).

On the other hand, from fairly early on in Western physics there have been adynamical explanations that focused on the role of the future in explaining the past as well as the reverse, such as integral (as opposed to differential) calculus and various least action principles of the sort Richard Feynman generalized to produce the path integral approach to quantum mechanics. And of course there are the various adynamical constraints in physics such as conservation laws and the symmetries underlying them that constrain if not determine the various equations of motion.

As regards the point 2, namely the debate between atomism and monism, despite all the tension that quantum theory has created for atomism as originally conceived, most physicists still assume there is something fundamentally entity like at bottom, however strange it may be by classical lights. But on the process view, potentia, activity, flux or change itself is fundamental, not entities/things changing in time such as particles or strings. In this monistic physics, all talk of such dynamical entities would emerge from, and be derived from, the more fundamental flux together with — and inseparably from — space-time in a background independent fashion. In this context, Bohm's and Hiley's research suggest a nonlocal monistic model wherein "the whole is prior to its parts, and thus views the cosmos as fundamental, with metaphysical explanation dangling downward from the One" [519]. In this view, nonlocality is a characteristic subtended of space-time and quantum particles are seen as vibration modes of the global field which is the dynamical expression of a fundamental level, of the deep geometrical structure associated with the implicate order. Since the first attempts to develop models about the underlying fundamental level of quantum reality represented by the implicate order, there has been a considerable amount of mathematical work exploring a possible deeper structure than space-time of which Bohm was unaware. These attempts have become better known to the physics community under the term 'noncommutative geometry'. As regards the research line about the implicate order, Hiley recently suggested that quantum processes evolve not in space-time but in a more general space called pre-space, which is not subjected to the Cartesian division between res extensa and res cogitans. In this view, the space-time of the classical world would be some statistical approximation and not all quantum processes can be projected into this space without producing the familiar paradoxes, including nonseparability and non-locality. According to Hiley's research, quantum domain is to be regarded as a structure or order evolving in space-time, but space-time is to be regarded as a higher order abstraction arising from this process involving events and abstracted notions of space or space-like points. Moreover, Hiley's monistic approach extends Bohm's interpretation of quantum mechanics to the relativistic regime and unites space-time geometry and material processes, as he does not want things happening in a background space-time but wants to "start from something more primitive from which both geometry and material process unfold together" [520]. That which he considers "more primitive" is elementary process. Hiley calls the fundamental process/potentia the "holomovement" and it has two intertwined aspects, the "implicate order" (characterized algebraically) and all the physics derived from it, such as space-time geometry, the "explicate (or manifest) order". The holomovement is thus the whole ground form of existence, which contains orders that are both implicate and explicate, wherein the latter expresses aspects of the former.

From the perspective of the implicate order, particles and pilot waves are not fundamental but are emergent from the implicate order. In the picture of Hiley's implicate order, there is no unique "now" successively coming into existence. Instead, there are infinite time-like foliations of the Minkowski space-time manifold, each representing a unique global "now" at various values of its foliating time, and a particular spatial hypersurface in a given foliation A (a "now" for a given observer A) contains events on many different spatial hypersurfaces in foliation B (different "nows" for observer B). Events which are simultaneous for observer A are not simultaneous for observer B and this fact implies a "relativity of simultaneity" and negates an objective flow of time. That is to say, there is no objective (frame independent) distinction in space-time between past, present and future events respectively and therefore no objective distinction to be had about the occurrence or nonoccurrence of events.

As regards the problems connected with time, in Hiley's view the idea is that the holomovement can explicate either a small region or a large region of space-time (to include the future) "at once", but never the entire universe. The extent of the explicate domain depends on the properties of the holomovement in each particular case and the process is apparently stochastic. Hiley's approach implies that what happens in the future cannot be made to rewrite the past, but that the future possibilities can influence the unfolding of the present. What is less clear is whether these moments pass in and out of existence or always stay in existence once explicated. All this suggests that each individual shadow manifold is constantly changing in its own time (evolving "now") such that the past is consistent with the present and the future is understood probabilistically.

As regards the point 3, algebra versus geometry broadly construed, there is a dizzying array of formalisms at work in physics. In quantum mechanics alone we have matrix mechanics, Schrödinger dynamics, Clifford algebras and path integrals, to name a few, and in quantum field theory we have canonical quantization, covariant quantization, path integral method, Becchi-Rouet-Stora-Tyupin (BRST) approach, Batalin–Vilkovisky (BV) quantization and stochastic quantization [521]. When we consider quantum gravity and unification the list is even longer and more variegate [522]. Throughout history there have always been differences of opinion, some pragmatic and some regarding principles, about which formalism(s) best models fundamental physical reality. Indeed, one of the striking things about the state of unification is the heterogeneity of formal approaches and the lack of consensus despite the juggernaut of string theory and its progeny. For example, Hiley's approach implies that geometry (space-time) is derived from algebra (process), rather than the other way around [523]. In particular, on the basis of Hiley's approach quantum processes emerge from a more fundamental background, linked with a fundamental non-commutative algebra (represented by the extended Heisenberg algebra, on one hand, or by the Clifford algebra, on the other hand). The extended Heisenberg algebra and the Clifford background suggested by Hiley's models can be considered as two different and equivalent mathematical ways to express the reality which underlies the features of quantum phenomena, namely as attempts to develop a mathematical formalism for the foreground, the implicate order of the quantum processes. The fundamental idea which characterizes Hiley's approaches based on the extended Heisenberg algebra or the Clifford algebra is that, in the quantum world, there is not *a priori* given manifold: the fundamental reality is the algebra and the geometry is then abstracted from the algebra [524].

As regards the debate between the fundamentality of becoming versus being and of algebra versus geometry, however, other approaches proceed towards an opposite direction with respect to Hiley's research. In this regard, for example, Silberstein, Stuckey and McDevitt proposed recently a monistic approach of unification of relativity and quantum theory, called the Relational Blockworld, in which the unity of space, time and matter is postulated at the fundamental level and is used to recover dynamical or process-like classical physics only statistically. Silberstein's, Stuckey's and McDevitt's approach, unlike Hiley's approach, emphasizes being over becoming and has something closer to geometry at bottom (in the form of a discrete graphical structure). According to this model, quantum physics is the continuous approximation of a more fundamental, discrete graph theory whereby the transition amplitude is not viewed as a sum over all paths in configuration space, but is a measure of the symmetry of the difference matrix and source vector of the discrete graphical action for a four-dimensional (4D) process. The graph picture of the Relational Blockworld approach produces divergence-free classical dynamics in the appropriate statistical limit, and provides an acausal global constraint that results in a self-consistent co-construction of space, time and matter that is background independent. Thus, in the Relational Blockworld one has an acausal, adynamical unity of "space-time-matter" at the fundamental level that results statistically in the causal, dynamical "space-time + matter" of classical physics. Silberstein's, Stuckey's and McDevitt's approach provides a wave-function-epistemic account of quantum mechanics with a time-symmetric explanation of interference via acausal global constraints. In this approach, quantum physics simply provides a distribution function for graphical relations responsible for the experimental equipment and process from initiation to termination. So, EPR-correlations and the processes of superluminal information exchange (quantum non-locality), in the Relational Blockworld approach are actually evidence of the deeper graphical unity of space-time-matter responsible for the experimental set up and process, to include outcomes [525–527].

Both Hiley's implicate order approach and Silberstein's, Stuckey's and McDevitt's Relational Blockworld approach want to derive general relativity and quantum theory from something more fundamental in a background independent fashion such that the explanation for quantum entanglement and EPR correlations, rather than creating tensions with space-time and relativity, requires neither nonlocality nor nonseparability in space-time. In both theories such quantum effects are explained by introducing a more fundamental level of physical reality, but the features of this fundamental level is different: it is graphical and implies the fundamentality of being versus becoming (in the Relational Blockworld approach) while it is algebraic and implies the fundamentality of becoming versus being (in the implicate order approach).

According to the Relational Blockworld approach, quantum theory and relativity theory both tell us that every facet of dynamism is false, that dynamism emerges just as a historical contingency based on the fact that all physics must start with experience. This model suggests the following perspective regarding the nature of time: time as part of a fundamental (pre-geometric) regime wherein the notions of space, time and matter are co-defined. In this picture, time, space and matter as stand-alone concepts are not fundamental, emergent or illusions; this model, being fundamentally relational, implies that there are no perspectives "external to the universe".

Another interesting physical model which implies the fundamentality of being versus becoming is Ruth Kastner's possibilist transactional interpretation of quantum mechanics. As known, the transactional interpretation of quantum mechanics was first proposed by John G. Cramer in a series of papers in the 1980s [528–531] by inspiring to previous theories of Wheeler–Feynman and of Hoyle–Narlikar. According to Cramer's approach, a quantum process is a time-symmetric process, in which we have an "absorber" which generates confirmation waves in response to an emitted offer wave: a system emits a field in the form of half-retarded, half-advanced solutions to the wave equation, and the response of the absorber combines with that primary field to create a process that transfers energy from the emitter to the absorber. Cramer's theory leads to a definition of a vacuum which gives form to the anticipated waves; these inform a particle about the global situation in its space-time region and in this way one can reproduce the quantum behaviors, and in particular explain EPR-type experiments, in terms of information — the transaction events — coming from the future owed to the relativistic nature of the vacuum.

In the recent papers [532–535], Ruth Kastner extended Cramer's approach into a more general and fundamental theory, the possibilist transactional interpretation, which can explain the nature of the process leading to an actualized transaction. In Kastner's approach, space-time is not a pre-existing substance, a structured container for events, but rather unfolds as an emergent manifold from a more fundamental collective structure, a "pre-spacetime" characterized by "transactional processes" (constituted by emission and absorption of quanta) involving de Broglie waves. In the picture proposed by Kastner, space-time intervals are measuring systems of realized transactions resulting in transfers of energy from an emitter to an observer, which must be considered as pre-spacetime objects at the micro-level. In Kastner's view, the transactions are processes that somewhat transcend the space-time structure, in other words are the expression of the nonlocal feature of the quantum processes. In this model, space-time is no more — and no less — than the set of actualized transactions. Thus, actualizations of transactions, and thus of emissions and absorptions of quanta, give rise to the set of related events comprising the space-time arena. The actualization of a transaction is an a-spatiotemporal process; it is the coming into being of an event (actually two linked events, the emission and the absorption of quanta).

In an actualized transaction, the emission defines the past and the absorption defines the future. That is, past and future supervene on actualized transactions; there is no "space-time" without actualized transactions. The apparent 4D space-time universe we perceive is not something "already there"; rather, it crystallizes from an indeterminate (but real) pre-spacetime of dynamical possibility. Thus, space-time "grows" but not in the usual "growing universe" sense wherein an advancing "now" proceeds from present to future; rather, events arise from a set of dimensions defining the quantum domain outside space-time. In fact, it is the past that "grows" and is extruded from the present; in the pre-space-time there is no actualized future. Kastner's possibilist transactional interpretation of quantum mechanics implies that only the "now" is the fundamental empirical realm of physical processes and that the changing properties of physical systems are associated with electromagnetic signals that transfer energy from what we are observing to our sense organs (by way of actualized transactions). The "now" is defined by a spatial coordinate (or, in a relational view of space-time, the object(s) with which we are currently in direct contact) and any light signals that have reached our eyes from other objects. Moreover, the empirical realm indeed cannot be "objective" in absolute terms in the sense that it is defined in terms of appearances which can only exist with respect to a given observer. This means that, strictly speaking, everyone has a different "empirical realm". However, we can corroborate our experiences and arrive at a consistent intersubjective consensus about a "larger" world of appearance beyond our individual empirical realms. All of these corroborations are associated with electromagnetic waves. Thus, what is referred to as "the empirical realm" of physical processes can be no more than that corroborated collection of individual empirical "nows".

According to the possibilist transactional interpretation, what is fundamental in the temporal sense is just the now, which changes. The now, and thus being is the ultimate, primary reality of the universe, and the time index is just a way of recording the changes of the now. In other words, one deals here with a sort of "growing universe" theory of time; the past grows and continues to become actualized as it falls away from the present. This approach implies a fundamentality of being versus becoming in the sense that one deals with a blockworld of "nows" originating actualized transactions. The future is a realm of physical possibilities. The present does not "advance" into the future. Rather, the future is a set of possibilities that becomes woven into the created past through the action of now. The origin of time is contained in matter. According to the possibilist transactional interpretation, if there were no matter, there would be no space-time. It is matter that creates the separation between temporal and spatial axes through transactions.

The considerations made in this book suggest new ways in order to face and solve the debate regarding how to model the phenomena of the universe. In the light of the timeless perspectives, explored in this book, in different domains, from special relativity to quantum theory to quantum cosmology to quantum gravity and to the quantum vacuum, it is possible to provide a new suggestive unifying key of reading to the Heraclitean and Parmenedian aspects of contemporary physics and to the debate between monism versus atomism and between algebra versus geometry.

On the basis of the treatment made in this book, the fundamental reality is dynamical inside a timeless background. The themes

Conclusions 405

focused in this book show that it is possible to explore new ways of describing physical processes that do not begin with an *a priori* given space-time arena intended as a primary, fundamental physical reality.

In the light of current relevant research analyzed in Chapter 1, the duration of physical events measured by clocks has not a primary existence but is an emergent mathematical quantity. Palmer's view of a fundamental level of physical reality based on an Invariant Set Postulate, Elze's approach of time, Girelli's, Liberati's and Sindoni's toy model of a nondynamical timeless space as fundamental background, Caticha's approach of entropic time and Prati's model of physical clock time imply that physical time exists only as a parameter measuring the order of the dynamics of processes inside a timeless background. The notion of time as a measuring system of the numerical order of material changes — which can be defined mathematically in three equivalent ways: in terms of Elze's reparametrization incident number (1.132), Caticha's entropic succession of instants (1.158) or Prati's number k_{AB} of states of the Hilbert space of the system of interest whose dynamics is described (and which belong to the interval (1.176) that satisfies an appropriate initial condition $\psi_1(\sigma) = \bar{\psi}_1$ of the subsystem acting as a clock) — introduces unifying interesting perspectives inside the Jacobi–Barbour–Bertotti theory and Rovelli's approach. By replacing time with the concept of numerical order of material changes one can provide a unifying reading both of the Jacobi–Barbour–Bertotti theory of time based on Eqs. (1.9)–(1.10), (1.22)–(1.38), and of Shyam's and Ramachandra's results of canonical quantum gravity based on Eq. (1.52), and of Gryb's approach of Jacobi-Barbour-Bertotti path integrals based on equation (1.66), and finally of Rovelli's results regarding the thermal time hyphotesis. One can conclude that the concept of numerical order of material changes of the system under consideration can be considered as the fundamental element which yields Barbour's ephemeris time and Rovelli's thermal time hypothesys.

In the light of the themes faced in Chapter 2, the fundamental arena of special relativity is a three-dimensional (3D) space: the standard Lorentz transformations of Einstein's special relativity can be derived from a more fundamental 3D Euclid space described by Eqs. (2.21) and (2.22) under the hypothesis that the duration of material change is a quantity that emerges from a more fundamental numerical order on the basis of Eq. (2.24) and that the link between the duration of material change and the numerical order corresponding to the velocity of the moving observer with respect to the rest frame determines a re-scaling of the position of the object into consideration in the moving frame. The treatment made in Chapter 2 introduces the suggestive perspective that the standard einsteinian formalism of special relativity may be obtained by starting from the idea that the fundamental arena of physical processes is a 3D arena where time, at a fundamental level, is a different entity with respect to the spatial coordinates, namely exists only as a numerical order (in the sense that the transformation of the speed of clocks between the two inertial systems does not depend on the spatial coordinates) and the physical duration of material change represents a scaling function of the numerical order.

In the light of the treatment of Chapter 3, the symmetrized quantum potential introduces a significant approach in order to describe the nonlocal geometry characteristic of quantum processes. The symmetrized quantum potential (3.80) (deriving from the symmetrized quantum entropy (3.81) which indicates, in the quantum domain, the degree of order and chaos of the vacuum which supports the space-temporal distribution of the ensemble of particles associated with the wave function under consideration) implies that in the quantum domain a timeless 3D space has a crucial role in determining the motion of a subatomic particle in the sense that the symmetrized quantum potential produces a like-space, instantaneous action on the particles under consideration (which is characterized by a symmetric feature) and contains an active information about the environment and, on the other hand, implies the concept of time as a numerical order of material changes. In EPR-type experiments a 3D timeless space acts as an immediate information medium in the sense that, as a consequence of the symmetrized quantum entropy, the first component of the symmetrized quantum potential makes physical space an "immediate information medium" which keeps two elementary particles in an immediate contact (while the second component of the symmetrized quantum potential reproduces, from the mathematical point of view, the symmetry in time of this communication and the fact that time exists only as a numerical order of material change). Moreover, in the symmetrized quantum potential approach, the nonlocal geometry of the three-dimensional space background can be characterized by introducing opportune symmetrized quantum-entropic lengths (given by Eq. (3.94) for one-body systems and by Eq. (3.95) for many-body systems) which provide a direct measure of the correlation degree (and thus of the degree of the departure from the Euclidean geometry characteristic of classical physics) in a quantum system in a time-symmetric picture.

In Chapter 4, it has been shown that the idea of a timeless background which acts as an immediate information medium — in the form of a symmetrized quantum potential, given by Eq. (4.47), which corresponds to the symmetrized quantum entropy (4.48) — can be embedded also in quantum cosmology in the context of Wheeler-deWitt (WDW) equation.

Finally, the models analyzed in Chapters 5 and 6 illustrate two possible descriptive levels of reality with appropriate mathematical formalisms which introduce interesting a-temporal perspectives towards the unification of gravity with quantum theory. The a-temporal quantum-gravity space — whose mathematical formalism is described by Eqs. (5.28)–(5.137) and which emerges from a universal cosmic space defined by a density of cosmic space and a rotation–orientation of each point of the gravitational space — can be considered as a first deep level of physical reality able to explain and reproduce the gravitational and quantum behavior of matter. The 3D timeless quantum vacuum defined by a energy density and state-reduction (RS) processes of creation–annihilation of quantum particles (and whose mathematical formalism is represented by Eqs. (6.121)–(6.209)) can be considered as a second, more profound level of physical reality which underlies both the gravitational space we perceive and the a-temporal quantum gravity space. In other words, in the light of the model of the timeless 3D quantum vacuum illustrated in Chapter 6, the mathematical formalism regarding the a-temporal quantum-gravity space theory developed in Chapter 5 can be considered as a first intermediate level of reality between the macroscopic level of physical processes and the most profound level (namely, by using a bohmian terminology, between the explicate order and the implicate order of physical processes), while the mathematical formalism of the 3D timeless quantum vacuum developed in Chapter 6 can be considered as a deeper level of physical reality, which is nearer to the real "implicate order". The Planck–Fiscaletti granules of the a-temporal quantum-gravity space defined by the density of cosmic space (5.28) and by the rotation–orientation (5.28a) can be considered as the fundamental quanta of the gravitational space which emerge from the processes of the 3D timeless quantum vacuum, namely the changes of the quantum vacuum energy density and the RS processes of creation–annihilation. In other words, in the light of the models analyzed in Chapter 5 and 6, one can say that the gravitational space we perceive is a projection of the a-temporal quantum-gravity space characterized by Planck–Fiscaletti granules and this a-temporal quantum-gravity space, in turn, emerges from the 3D timeless quantum vacuum.

Since the 3D timeless quantum vacuum defined by a energy density and RS processes of creation–annihilation of quantum particles can be considered as the most profound level which describes the so-called implicate order of processes, in this view, at a fundamental level, being reigns over becoming, in the sense that RS processes are the only primary physical reality in a timeless picture. At the explicate order of the macroscopic domain, time is an emerging quantity and thus the becoming characterizing the events can be considered as an emergent reality. But the emphasizing of being over becoming here gets a different significance with respect to Silberstein's, Stuckey's and McDevitt's relational blockworld approach, in the sense that in the picture of the universe proposed by the author there is not an incompatibility between being and becoming, there is not an incompatibility between the Heraclitean and Parmenedian aspects of the physical processes: here we deal rather with a peculiar unitary complementarity in which the becoming of the events, of motion and evolution derives from the primary physical reality represented by the RS processes of creation–annihilation of quanta in the timeless 3D quantum vacuum. In this picture, being and becoming can be indeed considered as two different aspects of the same coin. Moreover, in the view proposed by the author of this book, it is so possible to clarify in what sense — as established by Kastner's transactional approach — the origin of time is contained in matter. Matter creates the separation between temporal and spatial axes through transactions in the sense that, at a fundamental level, time constitutes a numerical order of material changes and the primary physical reality is represented by RS processes of creation–annihilation of quanta from a timeless 3D quantum vacuum. The apparent 4D space-time universe we perceive in our everyday life is not a fundamental physical reality: rather, it emerges from the timeless 3D quantum vacuum as a consequence of evolutions of RS processes of creation/annihilation of quanta, which define the "nows". The phenomenon of "duration" which is used to analyze the events emerges at a secondary level, as a mathematical ("scaling") parameter of the fundamental numerical order of material changes.

The introduction of a 3D timeless quantum vacuum defined by a energy density and RS processes of creation–annihilation suggests interesting and relevant monistic perspectives in the reading of quantum mechanics and in the interpretation of gravity. On one hand, within this model the shadowing of the gravitational space determined by the change of the quantum vacuum energy density influences the motion of the objects in a way that is coherent with the results of general relativity. On the other hand, on the basis of the mathematical formalism regarding RS processes, the ordinary quantum mechanics emerges directly from the timeless 3D quantum vacuum. In this approach, the fundamental result is obtained that the quantum behavior of matter and gravity appear as two different aspects of the same timeless 3D quantum vacuum. The excited states of the 3D quantum vacuum (corresponding to opportune changes of the quantum vacuum energy density associated with opportune RS processes of creation/annihilation of quanta) can be considered as the real bridges between gravitational effects and quantum effects of matter. Moreover, inside the picture of the 3D timeless quantum vacuum, the algebra regarding quantum mechanics and gravity can be considered as the fundamental mathematical reality which emerges from the RS processes and, by starting from this fundamental mathematical reality, the geometry of physics may be derived. In other words, one can say that in the approach of the 3D timeless quantum vacuum proposed by the author, neither geometry nor algebra may be considered as fundamental; there is only a monistic principle corresponding to the RS processes of creation–annihilation of quanta and the changes of quantum vacuum energy density associated with them.

Finally, it is interesting to make some general considerations about the development of scientific knowledge in the light of the themes and the perspectives suggested in this book. From a general point of view, a physical theory can be defined as a conceptual structure that we develop and use to organize, read and understand the world, and make predictions about it. A successful physical theory is a theory that manages to reach these aims in a consistent and effective way. However, each new theoretical scheme has to take into account what previous theories have already reached. As Rovelli rightly underlined, scientific thinking is not a static entity but is in constant evolution and reorganization: science is itself the evolution process of thinking [23]. There is no reason to believe that our understanding of the universe has not to be in constant evolution. Searching for a fixed point on which to rest our restlessness is naïve, useless and counterproductive for the development of knowledge. It is only by believing our insights and, at the same time, questioning our mental habits, that we can go ahead. This process of cautious faith and self confident doubt is the core of the scientific thinking. Science is the human adventure that consists in exploring possible ways of thinking of the world. Being ready to subvert, if required, anything we have been thinking so far can be considered one of the best of human adventures. The timeless background suggested in this book, by allowing us to open interesting perspectives towards a unitary and holistic view of the universe, can be then considered as a fundamental step of this adventure.

Part of the reflection about science of the last decades has put in evidence the "noncumulative" aspect in the development of scientific knowledge. According to this view, the evolution of scientific theories is marked by large or small breaking points in which empirical results are reorganized within new theories. These new theories would be consequently, to some extent, "incommensurable" with respect to the previous ones.

The considerations that we have made in this book about a timeless background as a fundamental arena of physics provide a different key of reading the evolution of scientific knowledge. In the physical models analyzed in this book, the fundamental ideas are based on the expectation that quantum mechanics, special relativity and general relativity represent in their respective domains our best guide to describe physical world. The timeless approach proposed in this book lets us realize that there is a subtle, but very definite cumulative aspect in the progress of physics, which goes far beyond the growth in the validity and precision of the empirical content of the theories. In moving from a theory to another, more general, theory that incorporates it and extends and supersedes it in such a way to include other domains of nature (or in such a way to provide a new key of reading of reality), we do not save only the verified empirical content of the old theory, but there is more. This "more" regards the spectacular and undeniable predictive power of theoretical physics. In fundamental physics the most significant progresses are at the same time linked with intuitions that lead to great conceptual revolutions and to the confidence in the predictive power of the old theories.

The central core of scientific thinking lies in a cautious faith in the known theories and, at the same time, by means of the intuition, in being open to put questions about nature. Exploring the possible ways of thinking about the world, being ready to subvert, if required, our old prejudices, opening the possibility to investigate new scenarios is among the best human adventures. The view of a timeless background where time is an emergent quantity measuring the numerical order of material changes can be then considered a relevant and significant step of this adventure: it shows how we can have faith in the fundamental theories of physics and, at the same time, how we can modify our view of reality. In virtue of the scenarios and perspectives that the timeless approach opens as regards the interpretation of physical processes in the different domains, the theories analyzed in this book can be considered an important step in the process of improvement in our description and understanding of nature.

References

1. I. Licata, *Osservando la Sfinge*, Di Renzo, Roma (2003).
2. I. Licata, "Visions of oneness. Space-time geometry and quantum physics", in *Visions of Oneness*, I. Licata and A. Sakaji eds., Di Renzo, Roma (2011).
3. I. Newton, *Philosophiae Naturalis Principia Mathematica* (1687); American edition: *Sir Isaac Newton's Mathematical Principles of Natural Philosophy*, University of California Press (1962); Italian edition: *Principi Matematici Della Filosofia Naturale*, UTET, Torino (1954).
4. G. Leibniz, *Discourse on Metaphysics* (1686).
5. E. Mach, *Die Mechanik in Ihrer Entwicklung Historisch-kritsch Dargestellt*, Barth, Leipzig (1883). English translation: *The Science of Mechanics*, Open Court, Chicago (1960).
6. G. J. Whitrow, *The Natural Philosophy of Time*, Clarendon Press, Oxford (1980).
7. D. Fiscaletti and A. Sorli, "Time Measured with Clocks is Exclusively a Mathematical Quantity", *The IUP Journal of Physics* **4**(2), 13–23 (2011).
8. A. Shimony, *Search for a Naturalistic World View*, Cambridge University Press, Cambridge (1993).
9. A. Macias and H. Quevedo, "Time paradox in quantum gravity", in *Quantum Gravity — Mathematical Models and Experimental Bounds*, B. Fauser, J. Tolksdorf and E. Zeidler Eds., Birkhauser, Basel, (2006), pp. 41–60.
10. W. G. Unruh and R. M. Wald, "Time and the Interpretation of Canonical Quantum Gravity", *Physical Review D* **40**, 2598–2614 (1989).
11. J. J. Halliwell, "Derivation of the Wheeler–deWitt Equation from a Path Integral for Minisuperspace Models", *Physical Review D* **38**(8), 2468–2481 (1988).
12. R. Sorkin, "Role of Time in the Sum-Over-Histories Framework for Gravity", *International Journal of Theoretical Physics* **33**(3), 523–534 (1994).
13. C. J. Isham, "Canonical quantum gravity and the problem of time", in *Integrable Systems, Quantum Groups, and Quantum Field Theory*, L. A. Ibort and M. A. Rodriguez eds., Kluwer, Dordrecht, pp. 157–287 (1993).
14. C. Kiefer, *Quantum Gravity*, second edition, Oxford University Press, Oxford (2007).

15. K. V. Kuchař, "Time and interpretations of quantum gravity", in *Proceedings of the 4th Canadian Conference on General Relativity and Relativistic Astrophysics*, G. Kunstatter, D. Vincent and J. Williams eds., World Scientific, Singapore, pp. 211–314 (1992).
16. A. Macìas and A. Camacho, "On the Incompatibility Between Quantum Theory and General Relativity", *Physics Letters B* **663**, 99–102 (2008).
17. A. Sacharov, "Vacuum Quantum Fluctuations In Curved Space and the Theory of Gravitation", *Soviet Physics Doklady* **12**(11), 1040–1041 (1968).
18. M. Consoli, "Ultraweak Excitations of the Quantum Vacuum as Physical Models of Gravity", *Classical and Quantum Gravity* **26**(22), 225008 (2009); e-print arXiv0904.1272 [gr-qc] (2009).
19. I. Licata, *Emergence and Computation at the Edge of Classical and Quantum Systems*, Aracne Editrice, Rome (2008).
20. A. Riotto, "*D*-branes, String Cosmology, and Large Extra Dimensions", *Physical Review D* **61**(12), 123506 (2000).
21. L. Randall and R. Sundrum, "Large Mass Hierarchy from a Small Extra Dimension", *Physical Review Letters* **83**(17), 3370–3373 (1999).
22. S. A. Huggett and K. P. Todd, *An Introduction to Twistor Theory*, Cambridge University Press, Cambridge (1994).
23. C. Rovelli, *Quantum Gravity*, Cambridge University Press, Cambridge (2007).
24. H. Kleinert, P. Jizba and F. Scardigli, "Uncertainty Relation on a World Crystal and Its Applications to Micro Black Holes", *Physical Review D* **81**(8), 084030 (2010).
25. G. Preparata and She Sheng Xue, "Quantum Gravity, the Planck Lattice and the Standard Model", http://arxiv.org/abs/hep-th/9503102 (1994).
26. A. Einstein, *Relativity: The Special and General Theory* (1920).
27. J. A. Wheeler, "Information, physics, quantum: The search for links", in *Complexity, Entropy, and the Physics of Information*, W. Zurek ed., Addison-Wesley, Redwood City, California (1990).
28. P. Mittelstaedt, *Der Zeitbegriff in der Physik*, B.I.-Wissenschaftsverlag, Mannheim, Germany (1976).
29. J. McTaggart, "The Unreality of Time", *Mind: A Quarterly Review of Psychology and Philosophy* **17**, 456–473 (1908).
30. P. Yourgrau, *A World Without Time: The Forgotten Legacy of Godel and Einstein*, Basic Books, New York (2006); http://findarticles.com/p/articles/mi_m1200/is_8_167/ai_n13595656.
31. J. B. Barbour and B. Bertotti, "Mach's Principle and the Structure of Dynamical Theories", *Proceedings of Royal Society A* **382**(1783), 295–306 (1982).
32. E. Anderson, J. Barbour, B. Z. Foster, B. Kelleher and N. O. Murchadha, "The Physical Gravitational Degrees of Freedom," *Classical and Quantum Gravity* **22**, 1795–1802 (2005); e-print arXiv:gr-qc/0407104.
33. J. Barbour, B. Z. Foster and N. O' Murchadha, "Relativity Without Relativity", *Classical and Quantum Gravity* **19**, 3217–3248 (2002); e-print arXiv:gr-qc/0012089.

34. E. Anderson, J. Barbour, B. Foster and N. O'Murchadha, "Scale-invariant Gravity: Geometrodynamics", *Classical and Quantum Gravity* **20**, 1543–1570 (2003); e-print arXiv:gr-qc/0211022.
35. K. Kuchař, *The Problem of Time in Quantum Geometrodynamics*, Oxford University Press, Oxford (1999).
36. E. Anderson, "The Problem of Time in Quantum Gravity", *Annalen der Physik* **524**(12), 757–786 (2012); e-print arXiv:1206.2403v2 [gr-qc] (2012).
37. E. Anderson, "Problem of Time: Facets and Machian Strategy", e-print arXiv:1306.5816v4 [gr-qc] (2014).
38. C. W. Misner, "Minisuperspace", in *Magic without Magic: John Archibald Wheeler*, J.R. Klauder ed., Freeman, San Francisco (1972), p. 441.
39. A. Macìas, O. Obregon and M. P. Ryan, "Quantum Cosmology: The Supersymmetric Square Root", *Classical and Quantum Gravity* **4**(6), 1477–1486 (1987).
40. M. Bojowald, "Loop Quantum Cosmology II. Volume Operators", *Classical and Quantum Gravity* **17**, 1509–1526 (2000).
41. M. Bojowald, "Loop Quantum Cosmology II. Discrete Time Evolution", *Classical and Quantum Gravity* **18**, 1071–1087 (2001).
42. A. Sen, "Time and Tachyon", *International Journal of Modern Physics A* **18**(26), 4869–4888 (2003).
43. K. V. Kuchař and M. P. Ryan, "Is Minisuperspace Quantization Valid?: Taub in Mixmaster", *Physical Review D* **40**, 3982–3996 (1989).
44. A. Ashtekar and M. Pierri, "Probing Quantum Gravity Through Exactly Soluble Midi-Superspaces I", *Journal of Mathematical Physics* **37**(2), 6250–6270 (1996).
45. B. S. DeWitt, "Quantum Theory of Gravity. 1. The Canonical Theory", *Physical Review* **160**(5), 1113–1148 (1967).
46. J. Barbour, "Relative–distance Machian Theories", *Nature* **249**(5455), 328–329 (1974).
47. J. Barbour, "Dynamics of Pure Shape, Relativity and the Problem of Time", *Lecture Notes in Physics* **633**, 15–35 (2003); e-print arXiv/gr-qc/0309089.
48. J. Barbour, *The End of Time*, Oxford University Press, New York (1999).
49. J. Barbour, "Scale-Invariant Gravity: Particle Dynamics", *Classical and Quantum Gravity* **20**, 1543–1570 (2003).
50. J. Barbour and N. Ó. Murchadha, "Classical and quantum gravity on conformal superspace", e-print arXiv:gr-qc/9911071 (1999).
51. J. B. Barbour, "The Timelessness of Quantum Gravity. 1: The Evidence from the Classical Theory", *Classical and Quantum Gravity* **11**, 2853–2873 (1994).
52. C. Kiefer, *Quantum Gravity*, Clarendon, Oxford (2004).
53. E. Anderson, "The problem of time and quantum cosmology in the relational particle mechanics arena", e-print arXiv:gr-qc/1111.1472 (2011).
54. G. M. Clements, *Reviews of Modern Physics* **29**(2) (1957).
55. J. Barbour, "The Nature of Time", http://arxiv.org/abs/ 0903.3489 (2009).

56. R. F. Baierlein, D. Sharp and J. A. Wheeler, "Three Dimensional Geometry as Carrier of Information about Time", *Physical Review* **126**(5), 1864–1865 (1962).
57. E. Anderson and J. Barbour, "Interacting Vector Fields in Relativity without Relativity", *Classical and Quantum Gravity* **19**, 3249–3262 (2002); e-print arXiv:gr-qc/0201092 (2002).
58. J. Barbour and B. Z. Foster, "Constraints and gauge transformations: Dirac's theorem is not always valid", e-print arXiv:0808.1223 (2008).
59. V. Shyam and B. S. Ramachandra, "Presympletic geometry and the problem of time. Part 2", e-print arXiv:1210.5619v5 [gr-qc] (2012).
60. J. B. Hartle and K. V. Kuchař, "The role of time in path integral formulations of parametrized theories", in *Quantum Theory of Gravity: Essays in honor of the 60th birthday of Bryce S. DeWitt*. Bayce S. DeWitt, ed. Adan Hilger Ltd., Bristol, England, (1984).
61. J. D. Brown and J. W. J. York, "Jacobi's Action and the Recovery of Time in General Relativity", *Physical Review D* **40**, 3312–3318 (1989).
62. C. Teitelboim, "The Proper Time Gauge in Quantum Theory of Gravitation", *Physical Review D* **28**(2), 297–309 (1983).
63. M. Henneaux and C. Teitelboim, "Relativistic Quantum Mechanics of Supersymmetric Particles," *Annals of Physics* **143**(1), 127–159 (1982).
64. S. Gryb, "Jacobi's Principle and the Disappearance of Time", *Physical Review D* **81**, 044035 (2010), e-print arXiv: 0804.2900v3 [gr-qc] (2010).
65. L. D. Faddeev, "The Feynman Integral for Singular Lagrangians", *Theoretical and Mathematical Physics* **1**(1), 1–13 (1969).
66. C. Rovelli, "Time in Quantum Gravity: An Hypothesis", *Physical Review D* **43**(2), 442–456 (1991).
67. C. Rovelli, "Quantum Mechanics Without Time: A Model", *Physical Review D* **42**(8), 2638–2646 (1991).
68. C. Rovelli, "Quantum Evolving Constants", *Physical Review D* **44**(4), 1339–1341 (1991).
69. C. Rovelli, "What is Observable in Classical and Quantum Gravity?", *Classical and Quantum Gravity* **8**, 297–316 (1991).
70. C. Rovelli, "Quantum Reference Systems", *Classical and Quantum Gravity* **8**, 317–331 (1991).
71. C. Rovelli, "Is there incompatibility between the ways time is treated in general relativity and in standard quantum mechanics?", in *Conceptual Problems of Quantum Gravity*, A. Ashtekar and J. Stachel eds., Birkhauser, New York (1991).
72. C. Rovelli, "Analysis of the Different Meaning of the Concept of Time in Different Physical Theories", *Il Nuovo Cimento B* **110**(1), 81–93 (1995).
73. C. Rovelli, "Partial Observables", *Physical Review D* **65**, 124013-8 (2002); e-print arXiv:gr-qc/0110035.
74. C. Rovelli, "A note on the foundation of relativistic mechanics. I: Relativistic observables and relativistic states", in *Proceedings of the 15th SIGRAV Conference on General Relativity and Gravitational Physics*, Rome, September 2002; e-print arXiv:gr-qc/0111037.

75. C. Rovelli, "A note on the foundation of relativistic mechanics. II: Covariant hamiltonian general relativity", gr-qc/0202079 (2002).
76. C. Rovelli, "Statistical Mechanics of Gravity and Thermodynamical Origin of Time", *Classical and Quantum Gravity* **10**(8), 1549–1566 (1993).
77. C. Rovelli, "The Statistical State of the Universe", *Classical and Quantum Gravity* **10**(8), 1567–1578 (1993).
78. C. Rovelli, "Forget time", e-print arXiv:0903.3832v3 [gr-qc] (2009).
79. D.-W. Chiou, "Timeless Path Integral for Relativistic Quantum Mechanics", *Classical and Quantum Gravity* **30**(12), 125004 (2013); e-print arXiv:1009.5436v3 [gr-qc] (2009).
80. A. Connes and C. Rovelli, "Von Neumann Algebra Automorphisms and Time-Thermodynamics Relation in Generally Covariant Quantum Theories", *Classical and Quantum Gravity* **11**(12), 2899–2917 (1994).
81. P. Martinetti and C. Rovelli, "Diamond's Temperature: Unruh Effect for Bounded Trajectories and Thermal Time Hypothesis", *Classical and Quantum Gravity* **20**(22), 4919–4931 (2003).
82. R. F. Streater and A. S. Wightman, *PCT, Spin and Statistics and all that*, Benjamin, New York (1964).
83. C. Rovelli and M. Smerlak, "Thermal Time and Tolman-Ehrenfest Effect: Temperature as the Speed of Time", *Classical and Quantum Gravity* **28**(7), 075007 (2011); e-print ArXiv:1005.2985v5 [gr-qc].
84. C. Rovelli, "Why do we remember the past and not the future? The 'time oriented coarse graining hypothesis'," e-print arXiv:14.07.3384v2 [hep-th] (2014).
85. J. F. Woodward, "Killing Time", *Foundations of Physics Letters* **9**(1), 1–23 (1996).
86. T. N. Palmer, "The Invariant Set Hypothesis: A New Geometric Framework for the Foundations of Quantum Theory and the Role Played by Gravity", http://arxiv.org/abs/0812.1148 (2009).
87. H. T. Elze, "Quantum mechanics and discrete time from "timeless" classical dynamics", *Lecture Notes in Physics* **633**(196), (2003); e-print arXiv:grqc/0307014v1.
88. H.-T. Elze and O. Schipper, "Time Without Time: A Stochastick Clock Model", *Physical Review D* **66**, 044020 (2002).
89. H.-T. Elze, "Emergent Discrete Time and Quantization: Relativistic Particle with Extra Dimensions", *Physics Letters A* **310**(2–3), 110–118 (2003).
90. H.-T. Elze, "Quantum mechanics emerging from "timeless" classical dynamics", e-print arXiv:quant-ph/0306096 (2003).
91. G. 'tHooft, "*Quantum Mechanics and Determinism*, in: Proceedings of the Eigth" Int. Conf. on "Particles, Strings and Cosmology", P. Frampton and J. Ng eds. Rinton Press, Princeton (2001), p. 275; e-print arXiv:hep-th/0105105 (2001).
92. G. 'tHooft, *Determinism Beneath Quantum Mechanics*, e-print arXiv:quant-ph/0212095 (2002).
93. C. Wetterich, "Quantum correlations in classical statistics", in *Decoherence and Entropy in Complex Systems*, H.-T. Elze eds., *Lecture Notes in Physics*

Vol. 633, Springer-Verlag, Berlin Heidelberg, New York (2003); e-print arXiv:quant-ph/0212031.
94. H. T. Elze, "Quantum Features in Statistical Observations of Timeless Classical Systems", *Physica A: Statistical Mechanics and its Applications* **344**(3–4), 478–483 (2004); e-print arXiv:gr-qc/0312062v1 (2003).
95. F. Girelli, S. Liberati and L. Sindoni, "Is the Notion of Time Really Fundamental?", *Symmetry* **3**(3), 389–401 (2011); e-print arXiv:0903.4876v1 [gr-qc].
96. S. Weinberg and E. Witten, "Limits on Massless Particles", *Physics Letters B* **96**(1–2), 59–62 (1980).
97. A. Jenkins, "Topics in particle physics and cosmology beyond the standard model", e-print arXiv:hep-th/0607239 (2006).
98. A. Caticha, "Entropic Dynamics, Time and Quantum Theory", *Journal of Physics A: Mathematical and Theoretical* **44**(22), 225303 (2011); e-print arXiv:1005.2357v3 [quant-ph].
99. E. Prati, "Generalized Clocks in Timeless Canonical Formalism", *Journal of Physics: Conference Series* **306**(1), 012013 (2011); original e-print "The nature of time: from a timeless hamiltonian framework to clock time metrology", arXiv:0907.1707v1 [physics.class-ph] (2009).
100. W. H. Oskay, et al., "Single-Atom Optical Clock with High Accuracy," *Physical Review Letters* **97**(2), 020801-4 (2006).
101. J. Kofler and C. Brukner, "Classical World Arising Out of Quantum Physics Under the Restriction of Coarse-Grained Measurements", *Physical Review Letters* **99**(18), 180403 (2007).
102. D. Fiscaletti and A. Sorli, "Perspectives of the Numerical Order of Material Changes in Timeless Approaches in Physics", *Foundations of Physics* **45**(2), 105–133 (2015).
103. M. Pavsic, "Towards the Unification of Gravity and Other Interactions: What Has Been Missed", *Journal of Physics: Conference Series* **222**, 012017-1–9 (2010); e-print arXiv:0912.4836v1 [gr-qc] (2009).
104. R. J. Kennedy and E. M. Thorndike, "Experimental Establishment of the Relativity of Time", *Physical Review* **42**, 400–418 (1932).
105. H. E. Ives and G. R. Stilwell, "An Experimental Study of the Rate of a Moving Atomic Clock", *Journal of the Optical Society of Amercia* **28**(7), 215–219 (1938).
106. B. Rossi and D. B. Hall, "Variation of the Rate of Decay of Mesotrons with Momentum", *Physical Review* **59**(3), 223–228 (1941).
107. M. Kaivola, O. Poulsen, E. Riis and S. A. Lee, "Measurement of the Relativistic Doppler Shift in Neon", *Physical Review Letters* **54**, 255–258 (1985).
108. S. Reinhardt, G. Saathoff, H. Buhr, L. A. Carlson, A. Wolf, D. Schwalm, S. Karpuk, C. Novotny, G. Huber, M. Zimmermann, R. Holzwarth, T. Udem, T. W. Hänsch and Gerald Gwinner, "Test of Relativistic Time Dilation with Fast Optical Atomic Clocks at Different Velocities", *Nature Physics* **3**, 861–864 (2007).

109. J. C. Hafele and R. E. Keating, "Around-the-World Atomic Clocks: Predicted Relativistic Time Gains", *Science* **177**(4044), 166–168 (1972).
110. J. C. Hafele, "Relativistic Time for Terrestrial Circumnavigations", *American Journal of Physics* **40**(1), 81–85 (1972).
111. J. Bailey, K. Borer, F. Combley, H. Drumm, F. Krienen, F. Lange, E. Picasso, W. von Ruden, F. J. M. Farley, J. H. Field, W. Flegel and P. M. Hattersley, "Measurements of Relativistic Time Dilation for Positive and Negative Muons in a Circular Orbit", *Nature* **268**, 301–305 (1977).
112. A. A. Michelson and E. W. Morley, "On the Relative Motion of the Earth and the Luminiferous Ether", *American Journal of Science* **34**, 333–345 (1887).
113. A. Brillet and J. L. Hall, "Improved Laser Test of the Isotropy of Space", *Physical Review Letters* **42**(9), 549–552 (1979).
114. D. Hils and J. L. Hall, "Improved Kennedy-Thorndike Experiment to Test Special Relativity", *Physical Review Letters* **64**(15), 1697–1700 (1990).
115. V. W. Hughes, H. G. Robinson and V. Beltran-Lopez, "Upper Limit for the Anisotropy of Inertial Mass from Nuclear Magnetic Resonance", *Physical Review Letters* **4**(7), 342–344 (1960).
116. R. W. P. Drever, "A Search for Anisotropy of Inertial Mass Using a Free Precession Technique", *Philosophical Magazine* **6**(65), 683–687 (1961).
117. V. A. Kostelecky and C. D. Lane, "Constraints on Lorentz Violation from Clock-Comparison Experiments", *Physical Review D* **60**, 116010 (1999).
118. M. E. Tobar, P. Wolf, Sébastien Bize, G. Santarelli and V. Flambaum, "Testing Local Lorentz and Position Invariance and Variation of Fundamental Constants by Searching the Derivative of the Comparison Frequency Between a Cryogenic Sapphire Oscillator and Hydrogen Maser", *Physical Review D* **81**, 022003 (2010).
119. D. Mattingly, "Modern Tests of Lorentz Invariance," *Living Reviews in Relativity* **8**, 5–84 (2005), http://www.livingreviews.org/lrr-2005-5, accessed on 31 May 2012.
120. T. Alväger, F. J. Farley, J. Kjellman and L. Wallin, "Test of the Second Postulate of Special Relativity in the GeV Region", *Physics Letters* **12**(3), 260–262 (1964).
121. K. Brecher, "Is the Speed of Light Independent of the Velocity of the Source?", *Physical Review Letters* **39**(17), 1051–1054 (1977).
122. T. Roberts and S. Schleif, "What is the experimental basis of Special Relativity?", *Usenet Physics FAQ*, University of California, Riverside (2007), accessed on 01 June 2012.
123. C. Will, "Special relativity: A centenary perspective," in *Einstein. Poincaré Seminar 2005*, T. Damour, O. Darrigol, B. Duplantier and V. Rivasseau, eds. Birkhäuser Verlag, Basel, pp. 33–58 (2005).
124. E. F. Taylor and J. A. Wheeler, *Spacetime Physics: Introduction to Special Relativity*, second edition, W.H. Freeman and Co., New York (1992).
125. S. Bergia, "Strutture e dimensionalità dello spaziotempo: realtà, modello o occasione di formalismo?", in *Dove va la scienza. La questione del realismo*, F. Selleri and V. Tonini eds., Dedalo, Bari, pp. 51-69 (1990).

126. S. Bergia, "Formulari, interpretazioni, ontologie: il caso delle teorie relativistiche", in *Ancora sul Realismo*, G. Giuliani, La Goliardica Pavese eds., Pavia, pp. 47–68 (1995).
127. F. Selleri, "Velocity-symmetrizing synchronization and conventional aspects of relativity", in *Waves and particles in light and matter*, A. van der Merwe *et al.* eds., Plenum Press, London (1994) pp. 439–446.
128. F. Selleri, "The Inertial Tranformations and the Relativity Principle", *Foundations of Physics Letters* **18**(4), 325–339 (2005).
129. F. Selleri, "Non-invariant One-Way Velocity of Light and Particle Collisions", *Foundations of Physics Letters* **9**(1), 43–60 (1996).
130. F. Selleri, "Non-Invariant One-Way Speed of Light and Locally Equivalent Reference Frames", *Foundations of Physics Letters* **10**(1), 73–83 (1997).
131. F. Selleri, "Space and Time Should be Preferred to Spacetime — 1 and 2", *International Workshop Physics for the 21st Century*, 5–9 June 2000.
132. R. Manaresi and F. Selleri, "The International Atomic Time and the Velocity of Light", *Foundations of Physics Letters* **17**(1), 65–79 (2004).
133. G. Rizzi, M. L. Ruggiero and A. Serafini, "Synchronization Gauges and the Principles of Special Relativity", *Foundations of Physics* **34**, 1885 (2005).
134. B. Buonaura, "Electromagnetic Waves, Inertial Transformations and Compton Effect", *Apeiron* **14**(3), 184–213 (2007).
135. R. T. Cahill, "Process Physics: Inertia, Gravity and the Quantum", *General Relativity and Gravitation* **34**(10), 1637–1656 (2002).
136. R. T. Cahill, *Process Physics: From Information Theory to Quantum Space and Matter*, Nova Science Publishers, New York (2005).
137. R. T. Cahill, *The Michelson and Morley 1887 Experiment and the Discovery of Absolute Motion*, *Progress in Physics* **3**, 25–29 (2005).
138. R. T. Cahill, "Dynamical Fractal 3-Space and the Generalised Schrödinger Equation: Equivalence Principle and Vorticity Effects", *Progress in Physics* **1**, 27–34 (2006).
139. R. T. Cahill, "Dynamical 3-Space: A Review", e-print arXiv: 0705.4146 (2007).
140. R. T. Cahill, "Unravelling Lorentz Covariance and the Spacetime Formalism", e-print arXiv:0807.1767v1 [physics.gen-ph] (2008).
141. D. K. Wise, "Holographic special relativity", e-print arXiv:1305.3258v1 [hep-th] (2013).
142. A. Sorli and D. Fiscaletti, "Special Theory of Relativity in a Three-Dimensional Euclid Space", *Physics Essays* **25**(1), 141–143 (2012).
143. D. Fiscaletti and A. Sorli, "About a New Suggested Interpretation of Special Theory of Relativity within a Three-Dimensional Euclid Space", *Annales UMCS Sectio AAA: Physica* **LXVIII**, 36–62 (2013).
144. A. Sorli, D. Fiscaletti, D. Klinar, "Time is a Measuring System Derived from Light Speed", *Physics Essays* **23**(2), 330–332 (2010).
145. A. Sorli, D. Klinar and D. Fiscaletti, "New Insights into the Special Theory of Relativity", *Physics Essays* **24**(2), 313–318 (2011).

146. A. Sorli, D. Fiscaletti, D. Klinar, "Replacing Time with Numerical Order of Material Change Resolves Zeno Problems of Motion", *Physics Essays* **24**(1), 11–15 (2011).
147. J. M. Levi, "The simplest derivation of Lorentz transformation", http://arxiv.org/PS_cache/physics/pdf/0606/0606103v4.pdf (2006).
148. G. Zanella, "Lorentz contraction or Lorentz dilation?", e-print arXiv:1010.5988v2 [physics.gen-ph] (2010).
149. M. Duffy and J. M. Levy eds., *Ether Space-time & Cosmology*, Vol. 3, Apeiron (2009).
150. S. J. Ostro, "Planetary Radar Astronomy", *Reviews of Modern Physics* **65**(4), 1235–1279 (1993).
151. J. von Neumann, *Mathematische Grundlagen der Quantenmeckanik*, Springer, Berlin (1932).
152. W. Heisenberg, *Physics and Philosophy*, Harper and Row, New York (1958).
153. C. F. Von Wezsacker, in *The Physicist's Conception of Nature*, J. Mehra ed., Reidal Publishing Company, Boston (1973).
154. D. Fiscaletti, *I Gatti di Schrödinger. Meccanica Quantistica e Visione Del Mondo*, Muzzio Editore, Roma, 2007.
155. S. Bergia, "La versione di Bohm della meccanica quantistica: variazioni sul tema", in *Quanti Copenaghen? Bohr, Heisenberg e le Interpretazioni Della Meccanica Quantistica*, a cura di I. Tassani. Il Ponte Vecchio, Cesena (2004) pp. 179–199.
156. D. Lewis, *On the Plurality of Worlds*, Blackwell, Oxford (1986).
157. J. Earman, "Reassessing the Prospects for a Growing Block Model of the Universe", *International Studies in the Philosophy of Science* **22**(2), 135–164 (2008).
158. S. McCall, "QM and STR: The Combining of Quantum Mechanics and Relativity Theory", *Philosophy of Science* **67**(4), S535–S548 (2000).
159. J. Earman, "Pruning some branches from 'branching spacetimes'", in *The Ontology of Spacetime II*, D. Dieks ed., Elsevier, Amsterdam, pp. 187–205 (2008).
160. T. Placek, "Stochastic Outcomes in Branching Spacetime: Analysis of Bell's Theorems", *British Journal for the Philosophy of Science* **51**, 445–475 (2000).
161. J. S. Bell, *Speakable and Unspeakable in Quantum Mechanics*, Cambridge University Press, Cambridge (1987); second edition 2004, with an introduction by Alain Aspect.
162. J. Butterfield, "On time in quantum physics", in *The Blackwell Companion to the Philosophy of Time*, A. Bardon and H. Dyke eds., Wiley-Blackwell, (2013), pp. 220–241.
163. D. Deutsch, "Apart from universes", in *Many Worlds? Everett, Quantum Theory and Reality*, S. Saunders et al. eds., Oxford University Press, Oxford, (2010), pp. 542–552.
164. S. Saunders, J. Barrett, A. Kent and D. Wallace eds., *Many Worlds? Everett, Quantum Theory and Reality*, Oxford University Press, Oxford (2010).

165. D. Wallace, *The Emergent Multiverse*, Oxford University Press, Oxford (2012).
166. D. Wallace, "Decoherence and its Role in the Modern Measurement Problem", *Philosophical Transactions of the Royal Society*, http://uk.arxiv.org/abs/1111.2187 (2012).
167. P. Busch, "On the Energy-time Uncertainty Relation: Part I: Dynamical Time and Time Indeterminacy", *Foundations of Physics* **20**(1), 1–32 (1990).
168. P. Busch, "The time-energy uncertainty relation", in *Time in Quantum Mechanics*, G. Muga et al. eds., Springer Verlag, New York (2008).
169. J. Hilgevoord, "The Uncertainty Principle for Energy and Time I", *American Journal of Physics* **64**(12), 1451–1456 (1996).
170. J. Hilgevoord, "Time in Quantum Mechanics", *American Journal of Physics* **70**(3), 301–306 (2002).
171. J. Hilgevoord, "Time in Quantum Mechanics: A Story of Confusion", *Studies in the History and Philosophy of Modern Physics B* **36**(1), 29–60 (2005).
172. J. Hilgevoord and D. Atkinson, "Time in quantum mechanics", in *The Oxford Handbook of Philosophy of Time*, C. Callender et al. eds., Oxford University Press, Oxford (2011), pp. 647–662.
173. Y. Aharonov and D. Bohm, "Time in the Quantum Theory and the Uncertainty Relation for Time and Energy", *Physical Review* **122**(5), 1649–1658 (1961).
174. P. Busch, "On the Energy-Time Uncertainty Relation: Part II: Pragmatic Time Versus Energy Indeterminacy", *Foundations of Physics* **20**, 33–43 (1990).
175. E. Schrödinger, "The present situation in quantum mechanics: a translation of Schrödinger's 'cat paradox'", (transl. J. D. Trimmer), *Proceedings of the Americal Philosophical Society* **124**, 323–338 (1980).
176. E. Schrödinger, "Are There Quantum Jumps?", *The British Journal for the Philosophy of Science* **3**(11), 233–242 (1952).
177. D. Bohm, "A New Suggested Interpretation of Quantum Theory in Terms of Hidden Variables. Part I", *Physical Review* **85**, 166–179 (1952).
178. J. Kijowski, "On the Time Operator in Quantum Mechanics and the Heisenberg Uncertainty Relation for Energy and Time", *Reports in Mathematical Physics* **6**, 362 (1974).
179. P. R. Holland, *The Quantum Theory of Motion*, Cambridge University Press, Cambrige (1993).
180. N. Grot, C. Rovelli and R. S. Tate, "Time of Arrival in Quantum Mechanics", *Physical Review A* **54**, 4676–4690 (1996).
181. V. Delgado and J. G. Muga, "Arrival Time in Quantum Mechanics", *Physical Review A* **56**, 3425–3435 (1997).
182. C. R. Leavens, "Time of Arrival in Quantum and Bohmian Mechanics", *Physical Review A* **58**, 840–847 (1998).
183. P. Kochański and K. Wódkiewicz, "Operational Time of Arrival in Quantum Phase Space", *Physical Review A* **60**, 2689–2699 (1999).

184. J. Leoń, J. Julve, P. Pitanga and F. J. de Urríes, "Time of Arrival in the Presence of Interactions", *Physical Review A* **61**(6), 062101 (2000).
185. J. G. Muga and C. R. Leavens, "Arrival Time in Quantum Mechanics", *Physics Reports* **338**(4), 353–438 (2000).
186. J. G. Muga, R. Sala Mayato and I. L. Egusquiza, (eds.) *Time in Quantum Mechanics*, Springer, Berlin (2002).
187. E. A. Galapon, "Pauli's theorem and quantum canonical pairs: the consistency of a bounded, self-adjoint time operator canonically conjugate to a Hamiltonian with non-empty point spectrum", *Proceedings of Royal Society A* **458**, 451–472 (2002).
188. E. A. Galapon, "Self-adjoint Time Operator is the Rule for Discrete Semi-bounded Hamiltonians", *Proceedings of Royal Society A* **458**(2027), 2671–2689 (2002).
189. E. A. Galapon, "Shouldn't There be an Antithesis to Quantization?", *Journal of Mathematical Physics* **45**, 3180 (2004).
190. E. A. Galapon, R. F. Caballar and R. T. Bahague Jr, "Confined Quantum Time of Arrivals", *Physical Review Letters* **93**, 180406 (2004).
191. E. A. Galapon, R. Caballar and R. T. Bahague, "Confined Quantum Time of Arrival for Vanishing Potential", *Physical Review A* **72**, 062107 (2005).
192. E. A. Galapon, F. Delgado, J. G. Muga and I. Egusquiza, "Transition from Discrete to Continuous Time-of-Arrival Distribution for a Quantum Particle", *Physical Review A* **72**, 042107 (2005).
193. G. C. Hegerfeldt, D. Seidel, J. G. Muga and B. Navarro, "Operator-Normalized Quantum Arrival Times in the Presence of Interactions", *Physical Review A* **70**, 012 110 (2004).
194. G. R. Allcock, "The Time of Arrival in Quantum Mechanics. I. Formal Considerations", *Annals of Physics* **53**, 253–285 (1969).
195. G. R. Allcock, "The Time of Arrival in Quantum Mechanics. II. The Individual Measurement", *Annals of Physics* **53**, 286–310 (1969).
196. G. R. Allcock, "The Time of Arrival in Quantum Mechanics. III. The Measurement Ensemble", *Annals of Physics* **53**, 311–348 (1969).
197. N. Yamada and S. Takagi, "Quantum Mechanical Probabilities on a General Spacetime-Surface", *Progress in Theoretical Physics* **85**(5), 985–1012 (1991).
198. N. Yamada and S. Takagi, "Quantum Mechanical Probabilities on a General Spacetime-Surface. II — Nontrivial Example of Non-Interfering Alternatives in Quantum Mechanics", *Progress in Theoretical Physics* **86**(3), 599–615 (1991).
199. N. Yamada and S. Takagi, "Spacetime Probabilities in Nonrelativistic Quantum Mechanics", *Progress in Theoretical Physics* **87**(1), 77–91 (1992).
200. J. J. Halliwell and E. Zafiris, "Decoherent Histories Approach to the Arrival Time Problem", *Physical Review D* **57**(7), 3351–3364 (1998).
201. A. D. Baute, R. S. Mayato, J. P. Palao, J. G. Muga and I. L. Egusquiza, "Time-of-Arrival Distribution for Arbitrary Potentials and Wigner's Time-Energy Uncertainty Relation", *Physical Review A* **61**, 022118 (2000).

202. E. A. Galapon, "Theory of Quantum Arrival and Spatial Wave Function Collapse on the Appearance of Particle", *Proceedings of Royal Society A* **465**, 71–86 (2009).
203. E. A. Galapon, "Theory of Quantum First Time of Arrival via Spatial Confinement I: Confined Time of Arrival Operators for Continuous Potentials", *International Journal of Modern Physics A* **21**(31), 6351–6381 (2006).
204. D. Bohm, "A New Suggested Interpretation of Quantum Theory in Terms of Hidden Variables. Part II", *Physical Review* **85**, 180–193 (1952).
205. D. Bohm, "Proof that Probability Density Approaches $|\psi|^2$ in Causal Interpretation of the Quantum Theory", *Physical Review* **89**, 458–466 (1953).
206. L. de Broglie, "La Mecanique Ondulatoire et la Structure Atomique de La Matiere et du Rayonnement, *Le Journal de Physique et le Radium* **6**(8), 225–241 (1927).
207. L. de Broglie, *Conseil de Physique Solvay*, octobre 1927, Bruxelles; *Rapports et discussions*, Gauthier-Villars, Paris (1928), pp. 253–256.
208. L. de Broglie, "Interpretation of Quantum Mechanics by the Double Solution Theory", *Annales de la Fondation Louis de Broglie* **12**(4), 399–421 (1987).
209. D. Fiscaletti, "The Geometrodynamic Nature of the Quantum Potential", *Ukrainian Journal of Physics* **57**(5), 560–572 (2012).
210. D. Fiscaletti, A.S. Sorli and D. Klinar, "The Symmetryzed Quantum Potential and Space as a Direct Information Medium", *Annales de la Fondation Louis de Broglie* **37**, 41–71 (2012).
211. D. Fiscaletti and A. Sorli, "Toward an A-temporal Interpretation of Quantum Potential", *Frontier Perspectives* **14**(2), 43–54 (2005/2006).
212. D. Fiscaletti and A. Sorli, "Perspectives Towards the Interpretation of Physical Space as a Medium of Immediate Quantum Information Transfer", *Prespacetime Journal* **1**(6), 883–898 (2010).
213. D. Fiscaletti and A. Sorli, "Timeless Space is a Fundamental Arena of Quantum Processes", *The IUP Journal of Physics* **3**(4), 34–49 (2010).
214. D. Fiscaletti and A. Sorli, "Three-dimensional Space as a Medium of Quantum Entanglement", *Annales UMCS Sectio AAA: Physica* **67**, 47–72 (2012).
215. D. Bohm, *Quantum Theory*, Prentice-Hall, New York (1951).
216. A. Sorli and I. K. Sorli, "From Space-Time to a-Temporal Physical Space", *Frontier Perspectives* **14**(1), 38–40 (2005).
217. M. Novello, J. M. Salim and F. T. Falciano, "On a Geometrical Description of Quantum Mechanics", *International Journal of Geometrical Methods of Modern Physics* **8**(1), 87–98 (2011).
218. D. Fiscaletti and I. Licata, "Weyl Geometries, Fisher Information and Quantum Entropy in Quantum Mechanics", *International Journal of Theoretical Physics* **51**(11), 3587–3595 (2012).
219. G. Resconi, I. Licata and D. Fiscaletti, "Unification of Quantum and Gravity by Non Classical Information Entropy Space", *Entropy* **15**, 3602–3619 (2013); e-print http://arxiv.org/abs/1110.5491.

220. V. I. Sbitnev, "Bohmian Split of the Schrödinger Equation onto Two Equations Describing Evolution of Real Functions", *Kvantovaya Magiya* **5**(1), 1101–1111 (2008); http://quantmagic.narod.ru/volumes/VOL5120 08/p1101.html.
221. D. Fiscaletti, "The Quantum Entropy as an Ultimate Visiting Card of the De Broglie-Bohm Theory", *Ukrainian Journal of Physics* **57**(9), 946–963 (2012).
222. D. Fiscaletti, "A Geometrodynamic Entropic Approach to Bohm's Quantum Potential and the Link with Feynman's Path Integrals Formalism", *Quantum Matter* **2**(2), 122–131 (2013).
223. V. I. Sbitnev, "Bohmian Trajectories and the Path Integral Paradigm. Complexified Lagrangian Mechanics", *International Journal of Bifurcation and Chaos* **19**(7), 2335–2346 (2009); e-print arXiv:0808.1245v1 [quant-ph] (2008).
224. D. Fiscaletti and I. Licata, "Bell Length in the Entanglement Geometry", *International Journal of Theoretical Physics* **54**(7), 2362–2381 (2015).
225. I. Licata and D. Fiscaletti, "Bell Length as Mutual Information in Quantum Interference", *Axioms* **3**, 153–165 (2014).
226. G. Resconi and I. Licata, "Quantum computing in non-Euclidean geometry", e-print arXiv:0911.0842 [quant-ph] (2009).
227. D. Fiscaletti and A. Sorli, "Non-Locality and the Symmetrized Quantum Potential", *Physics Essays* **21**(4), 245–251 (2008).
228. D. Fiscaletti, "The Bohmian Quantum Potential: From the Original to the Timeless, Symmetrized Version", *Galilean Electrodynamics* **24**(5), 83–95 (2013).
229. D. Fiscaletti and A. Sorli, "Non-Local Quantum Geometry and Three-Dimensional Space as a Direct Information Medium", *Quantum Matter* **3**(3), 200–214 (2014).
230. J. Anandan, "Symmetries, Quantum Geometry and the Fundamental Interactions", *International Journal of Theoretical Physics* **41**(2), 199–220 (2002); e-print arXiv:quant-ph/0012011v4 (2001).
231. Y. Aharonov and D. Bohm, "Significance of Electromagnetic Potentials in Quantyum Theory", *Physical Review* **115**(3), 485–491 (1959).
232. K. Bradonjic and J. D. Swain, "Quantum Measurement and the Aharonov-Bohm Effect with Superposed Magnetic Fluxes", *Quantum Information Processing* **13**(2), 323–331; e-print arXiv:1103.1607v2 [quant-ph], 2011.
233. P. O. Mazur, "Spinning Comsic Strings and Quantization of Energy", *Physical Review Letters* **57**(8), 929–932 (1986).
234. J. Samuel and B. R. Iyer, "Comment on 'Spinning Cosmic Strings and Quantization of Energy'", *Physical Review Letters* **59**(20), 2379–2379 (1987).
235. C. Philippidis, D. Bohm and R. D. Kaye, "The Aharonov-Bohm Effect and the Quantum Potential", *Nuovo Cimento* **71B**, 75–87 (1982).
236. E. Sjökvist, "A New Phase in Quantum Computation", *Physics* **1**, 35 (2008).

237. M. Pavsic, *The Landscape of Theoretical Physics: A Global View*, Kluwer Academic Publishers, Boston/Dordrecht, London (2001).
238. C. Rovelli, "Incerto Tempore, Incertisque loci: can We Compute the Exact Time at Which a Quantum Measurement Happens?", *Foundations of Physics* **28**(7), 1031–1043 (1998); e-print arXiv:quant-ph/9802020v3.
239. J. A. Wheeler, "The 'past' and the 'delayed choice' double slit experiment", in *Mathematical Foundations of Quantum Theory*, A. R. Marlow ed., Academic Press, New York (1978), pp. 9–48.
240. M. O. Scully and K. Drühl, "Quantum Eraser: A Proposed Photon Correlation Experiment Concerning Observation and 'Delayed Choice' in Quantum Mechanics", *Physical Review A* **25**, 2208–2213 (1982).
241. P. G. Kwiat, A. Steinberg and R. Y. Chiao, "Observation of a 'Quantum Eraser': A Revival of Coherence in a Two-Photon Interference Experiment", *Physical Review A* **45**, 7729–7739 (1992).
242. Y. Aharonov and M. S. Zubairy, "Time and the Quantum: Erasing the Past and Impacting the Future", *Science* **307**(5711), 875–879 (2005).
243. D. H. Coule, "Quantum Cosmological Models", *Classical and Quantum Gravity* **22**(12), R125–R166 (2005).
244. J. J. Halliwell, "Introductory lectures on quantum cosmology", in *Quantum Cosmology and Baby Universes*, S. Coleman, J. B. Hartle, T. Piran and S. Weinberg eds., World Scientific, Singapore (1991), pp. 159–243.
245. C. Kiefer, *Quantum Gravity*, third edition, Oxford University Press, Oxford, UK (2012).
246. C. Kiefer and B. Sandhoefer, "Quantum cosmology", http://arxiv.org/abs/0804.0672 (2008).
247. D. L. Wiltshire, "An introduction to quantum cosmology", in *Cosmology: The Physics of the Universe*, B. Robson, N. Visvanathan and W. S. Woolcock eds., World Scientific, Singapore (1996), pp. 473–531.
248. G. Montani, M. V. Battisti, R. Benini and G. Imponente, *Primordial Cosmology*, World Scientific, Singapore (2011).
249. P. V. Moniz, "Quantum cosmology — the supersymmetric perspective", in *Lecture Notes in Physics*, Springer, Berlin (2010).
250. M. Bojowald, "Quantum Cosmology", in *Lecture Notes in Physics*, Springer, Berlin (2011).
251. M. Bojowald, C. Kiefer and P. V. Moniz, "Quantum cosmology for the 21st century: A debate", http://arxiv.org/abs/1005.2471 (2010).
252. C. Bastos, O. Bertolami, N. C. Dias and J. N. Prata, "Phase-Space Noncommutative Quantum Cosmology", *Physical Review D* **78**(2), 023516 (2008).
253. O. Bertolami and C. A. D. Zarro, "Hořava-Lifshitz Quantum Cosmology", *Physical Review D* **85**(4), 044042 (2011).
254. S. P. Kim, "Massive Scalar Field Quantum Cosmology", In press.
255. M. Bojowald, "Quantum Cosmology: Effective Theory", *Classical and Quantum Gravity* **29**(21), 213001 (2012).
256. S. W. Hawking, "The Boundary Conditions of the Universe", *Pontificia Academiae Scientarium Scripta Varia* **48**, 563–574 (1982).

257. J. B. Hartle and S. W. Hawking, "Wave Function of the Universe", *Physical Review D* **28**(12), 2960–2975 (1983).
258. S. W. Hawking, "The Quantum State of the Universe", *Nuclear Physics B* **239**(1), 257–276 (1984).
259. J. J. Halliwell and J. Louko, "Steepest-Descent Contours in the Path-Integral Approach to Quantum Cosmology. III. A General Method with Applications to Anisotropic Minisuperspace Models", *Physical Review D* **42**(12), 3997–4031 (1990).
260. C. Kiefer, "On the Meaning of Path Integrals in Quantum Cosmology", *Annals of Physics* **207**(1), 53–70 (1991).
261. A. Vilenkin, "Quantum cosmology and eternal inflation", in *The Future of the Theoretical Physics and Cosmology*, Cambridge University Press, Cambridge (2003), pp. 649–666.
262. A. Vilenkin, "Interpretation of the Wave Function of the Universe", *Physical Review D* **39**, 1116–1122 (1989).
263. C. Kiefer, "Wave Packets in Minisuperspace", *Physical Review D* **38**(6), 1761–1772 (1988).
264. C. Kiefer, "Conceptual problems in quantum gravity and cosmology", *ISRN Mathematical Physics*, 509316 (2013).
265. E. Calzetta and J. J. Gonzalez, "Chaos and Semiclassical Limit in Quantum Cosmology", *Physical Review D* **51**, 6821–6828 (1995).
266. N. J. Cornish and E. P. S. Shellard, "Chaos in Quantum Cosmology", *Physical Review Letters* **81**, 3571–3574 (1998).
267. W. H. Zurek and J. P. Paz, "Quantum Chaos: A Decoherent Definition", *Physica D* **83**, 300–308 (1995).
268. E. Joos, H. D. Zeh, C. Kiefer, D. Giulini, J. Kupsch and I.-O. Stamatescu, *Decoherence and the Appearance of a Classical World in Quantum Theory*, second edition, Springer, Berlin, Germany (2003).
269. E. Calzetta, "Chaos, Decoherence and Quantum Cosmology", *Classical and Quantum Gravity* **29**(14), 143001 (2012); e-print http://arxiv.org/abs/1304.7439.
270. R. W. Carroll, *Fluctuations, Information, Gravity and the Quantum Potential*, Springer, Dordrecht (2006).
271. F. Shojai and A. Shojai, "Pure Quantum Solutions of Bohmian Quantum Gravity", *Journal of High Energy Physics* **2001**, JHEP05 (2001); e-print arXiv:gr-qc/0105102.
272. F. Shojai and A. Shojai, "Constraint Algebra and the Equations of Motion in the Bohmian Interpretation of Quantum Gravity", *Classical and Quantum Gravity* **21**, 1–9 (2004); e-print arXiv:gr-qc/0311076.
273. F. Shojai and A. Shojai, "Causal Loop Quantum Gravity and Cosmological Solutions", *Europhysics Letters* **71**(6), 886–892 (2005); e-print arXiv:gr-qc/0409020.
274. F. Shojai and A. Shojai, "Constraint algebra in causal loop quantum gravity", e-print arXiv:gr-qc 0409035 (2004).
275. N. Pinto-Neto, "Quantum Cosmology: How to Interpret and Obtain Results", *Brazilian Journal of Physics* **30**(2), 330–345 (2000).

276. S. P. Kim, "Problem of Unitarity and Quantum Corrections in Semi-Classical Quantum Gravity", *Physical Review D* **55**, 7511–7517 (1997).
277. A. Ashtekar and J. Lewandowski, "Quantum field theory of geometry", e-print arXiv:hep-th 9603083 (1996).
278. A. Ashtekar, A. Corichi and J. Zapata, "Quantum Theory of Geometry III: Non-Commutativity of Riemannian Structures", *Classical and Quantum Gravity* **15**, 2955–2972 (1998); e-print arXiv:gr-qc/9806041.
279. A. Ashtekar and J. Lewandowski, "Background Independent Quantum Gravity: A Status Report", *Classical and Quantum Gravity* **21**, R53–R152 (2004); e-print arXiv:gr-qc/0404018.
280. A. Ashtekar and C. Isham, "Representations of Holonomy Algebras of Gravity and Non-Abelian Gauge Theories", *Classical and Quantum Gravity* **9**, 1433–1467 (1992).
281. A. Ashtekar, "Gravity and the Quantum", *New Journal of Physics* **7**, 200–232 (2005); e-print arXiv:gr-qc/0410054.
282. J. Baez (ed.), *Knots and Quantum Gravity*, Oxford University Press, Oxford (1994).
283. S. Biswas, A. Shaw and D. Biswas, "Schrödinger Wheeler-DeWitt Equation in Multidimensional Cosmology", *International Journal of Modern Physics D* **10**, 585–594 (2001); e-print arXiv:gr-qc/9906009.
284. B. Dittrich and T. Thiemann, "Testing the Master Constraint Programme for Loop Quantum Gravity I. General Framework", *Classical and Quantum Gravity* **23**, 1025–1046 (2006); e-print arXiv:gr-qc/0411138 (2004).
285. B. Dittrich and T. Thiemann, "Testing the Master Constraint Programme for Loop Quantum Gravity II. Finite Dimensional Systems", *Classical and Quantum Gravity* **23**, 1067–1088 (2006); e-print arXiv: gr-qc 0411139 (2004).
286. B. Dittrich and T. Thiemann, "Testing the Master Constraint Programme for Loop Quantum Gravity III. SL(2R) Models", *Classical and Quantum Gravity* **23**, 1089–1120 (2006); e-printarXiv: gr-qc 0411140 (2004).
287. B. Dittrich and T. Thiemann, "Testing the Master Constraint Programme for Loop Quantum Gravity IV. Free Field Theories", *Classical and Quantum Gravity* **23**, 1121–1142 (2006); e-print arXiv:gr-qc 0411141 (2004).
288. B. Dittrich and T. Thiemann, "Testing the Master Constraint Programme for Loop Quantum Gravity V. Interaction Field Theories", *Classical and Quantum Gravity* **23**, 1143–1162 (2006); e-print arXiv:gr-qc 0411142 (2004).
289. R. Gambini and J. Pullin, *Loops, Knots, Gauge Theories and Quantum Gravity*, A. Ashtekar and J. Lewandowski, Cambridge University Press, Cambridge (1996); e-print arXiv:hep-th 9603083 (1996).
290. A. Ashtekar, A. Corichi and J. Zapata, "Quantum Theory of Geometry III: Non Commutativity of Riemannian Structure", *Classical and Quantum Gravity* **15**(10), 2955 (1998); e-print arXiv:gr-qc/9806041.
291. T. Horiguchi, K. Maeda and M. Sakamoto, "Analysis of the Wheeler–DeWitt Equation Beyond Planck Scale and Dimensional Reduction", *Physics Letters B* **344**(1), 105–109 (1995); e-print arXiv:hep-th/9409152.

292. M. Kenmoku, H. Kubortani, E. Takasugi and Y. Yamazaki, "De Broglie–Bohm interpretation for analytic solutions of the Wheeler–DeWitt equation in spherically symmetric space-time", e-print arXiv:gr-qc/9906056 (1999).
293. S. Kim, "Does Lorentz Boost Destroy Coherence?", e-print arXiv:gr-qc/9703065 (1997).
294. S. Kim, "Quantum Potential and Cosmological Singularities", *Physics Letters A* **236**, 11–15 (1997).
295. F. Markopoulou and L. Smolin, "Quantum Theory from Quantum Gravity", *Physical Review D* **70**(12), 124029-10 (2004); e-print arXiv:gr-qc/0311059 (2003).
296. I. Moss, *Quantum Theory, Black Holes, and Inflation*, Wiley, New York (1996).
297. C. Rovelli, "Covariant hamiltonian formalism for field theory: Hamilton–Jacobi equation on the space G", e-print arXiv:gr-qc 0207043 (2002).
298. L. Smolin, *Three Roads to Quantum Gravity*, Oxford University Press, Oxford (2000).
299. E. A. Giannetto, "On the Quantum Geometry and Quantum Potential of the Universe", *Quantum Matter* **3**(3), 273–275 (2014).
300. G. Preparata, *An Introduction to a Realistic Quantum Physics*, World Scientific, Singapore (2002).
301. A. Shojai and M. Golshani, "Direct particle interaction as the origin of the quantal behaviours", e-print arXiv:quant-ph/9812019 (1998).
302. F. Shojai and A. Shojai, "On the Relation of Weyl Geometry and Bohmian Quantum Gravity", *Gravitation and Cosmology* **9**(3), 163–168 (2003); e-print arXiv:gr-qc/0306099.
303. A. Shojai and F. Shojai, "About Some Problems Raised by the Relativistic form of de Broglie–Bohm Theory of Pilot Wave", *Physica Scripta* **64**(5), 413–416; e-print arXiv:quant-ph/0109025 (2001).
304. L. Smolin, "Could quantum mechanics be an approximation to another theory?", e-print arXiv:quant-ph/0609109 (2006).
305. J. Vink, "Quantum Mechanics in Terms of Discrete Beables", *Physical Review A* **48**(3), 1808–1818 (1993).
306. J. Vink, "Quantum Potential Interpretation of the Wave Function of the Universe", *Nuclear Physics B* **369**, 707–728 (1992).
307. J. A. de Barros, N. Pinto-Neto and M. A. Sagioro-Leal, *Physics Letters A* **241**, 229–239 (1998).
308. Jr. R. Colistete, J. C. Fabris and N. Pinto-Neto, *Physical Review D* **57**, 4707–4717 (1998).
309. Y. V. Shtanov, "Pilot Wave Quantum Cosmology", *Physical Review D* **54**(4), 2564–2570 (1996).
310. J. A. de Barros and N. Pinto-Neto, "The Causal Interpretation of Quantum Mechanics and the Singularity Problem and Time Issue in Quantum Cosmology", *International Journal of Modern Physics D* **7**(2), 201–213 (1998).
311. J. Kowalski-Glikman and J. C. Vink, "Gravity-Matter Mini-Superspace: Quantum Regime, Classical Regime and in Between", *Classical and Quantum Gravity* **7**, 901–918 (1990).

312. E. J. Squires, "A Quantum Solution to a Cosmological Mystery", *Physics Letters A* **162**(1), 35–36 (1992).
313. N. Pinto-Neto, G. Santos and W. Struyve, "Quantum-to-Classical Transition of Primordial Cosmological Perturbations in de Broglie–Bohm Quantum Theory", *Physical Review D* **85**(8), 083506 (2012).
314. N. Pinto-Neto, "The Bohm Interpretation of Quantum Cosmology", *Foundations of Physics* **35**(4), 577–603 (2005); e-print arXiv:gr-qc/0410117 (2004).
315. N. Pinto-Neto, "Perturbations in Bouncing Cosmological Models", *International Journal of Modern Physics D* **13**(7), 1419–1424 (2004); e-print arXiv:gr-qc/0410225 (2004).
316. N. Pinto-Neto and E. Santini, "The Consistency of Causal Quantum Geometrodynamics and Quantum Field Theory", *General Relativity and Gravitation* **34**(4), 505–532 (2002); e-print arXiv:gr-qc/0009080 (2000).
317. N. Pinto-Neto and E. Santini, "The Accelerated Expansion of the Universe as a Quantum Cosmological Effect", *Physics Letters A* **315**, 36–50 (2003); e-print arXiv:gr-qc/0302112.
318. N. Pinto-Neto and J. C. Fabris, "Quantum cosmology from the de Broglie–Bohm perspective", e-print arXiv:1306.0820v1 [gr-qc] (2013).
319. N. Pinto-Neto and E. Santini, "Quantum Cosmology: How to Interpret and Obtain Results", *Brazilian Journal of Physics* **30**(2), 330–345 (2000).
320. N. Pinto-Neto and E. Santini, "Must Quantum Spacetimes be Euclidean?", *Physical Review D* **59**(12), 123517-14 (1999); e-print arXiv:gr-qc/9811067.
321. G. Dautcourt, "Of the ultrarelativistic limit of general relativity", e-print arXiv:gr-gc/9801093 (1998).
322. K. B. Wharton, "Time-Symmetric Quantum Mechanics", *Foundations of Physics* **37**(1), 159–168 (2007).
323. D. Fiscaletti, "Features and Perspectives of the Atemporal Quantum-Gravity Space Theory", *The IUP Journal of Physics* **3**(2), 7–28 (2010).
324. D. Fiscaletti, "About Dirac-Type Equations in an Atemporal Quantum-Gravity Space Theory", *The IUP Journal of Physics* **5**(1), 6–25 (2012).
325. D. Fiscaletti, "The Interpretation of the Gravitational Space Through an Atemporal Quantum-Gravity Space Theory", *Journal of Advanced Physics* **1**(2), 150–160 (2012).
326. D. Fiscaletti, "Wave Dynamics in an Atemporal Quantum-Gravity Space", *Galilean Electrodynamics* **24**(3), 43–52 (2013).
327. D. Fiscaletti, "On the Curvature of Space in an Atemporal Quantum-Gravity Space Theory", *Galilean Electrodynamics* **25**(1), 3–11 (2014).
328. C. Rovelli and L. Smolin, "Discreteness of Area and Volume in Quantum Gravity", *Nuclear Physics B* **442**, 593–619 (1995).
329. J. Lewandowski, "Volume and Quantizations", *Classical and Quantum Gravity* **14**, 71–76 (1997); e-print arXiv:gr-qc/9602035.
330. R. Loll, "Volume Operator in Discretized Quantum Gravity", *Physical Review Letters* **75**, 3048–3051 (1995); e-print arXiv:gr-qc/9506014.
331. E. Bianchi, "The Length Operator in Loop Quantum Gravity", *Nuclear Physics B* **807**, 591–624 (2009); e-print arXiv:gr-qc/0806.4710.

332. Y. Ma, C. Soo and J. Yang, "New Length Operator for Loop Quantum Gravity", *Physical Review D* **81**(12), 124026–124029 (2010); e-print arXiv:1004.1063 [gr-qc].
333. T. Thiemann, "A Length Operator for Canonical Quantum Gravity", *Journal of Mathematical Physics* **39**, 3372–3392 (1998); e-print arXiv:gr-qc/9606092 [gr-qc].
334. S. A. Major, "Operators for Quantized Directions", *Classical and Quantum Gravity* **16**, 3859–3877 (1999); e-print arXiv:gr-qc/9905019 [gr-qc].
335. G. Amelino-Camelia, J. Ellis, N. E. Mavromatos, D. V. Nanopoulos and S. Sarkar, "Potential Sensitivity of Gammaray Burster Observations to Wave Dispersion in Vacuo", *Nature* **293**, 763–765 (1998); e-print astro-ph/9712103 [gr-qc].
336. G. Amelino-Camelia, "Quantum gravity phenomenology", e-print arXiv:0806.0339 [gr-qc] (2008).
337. C. Rovelli, "A New Look at Loop Quantum Gravity", *Classical and Quantum Gravity* **28**(11), 114005 (2011); e-print arXiv:1004.1780v1 [gr-qc].
338. A. Ashtekar and J. Lewandowski, "Quantum Theory of Geometry. I. Area Operators", *Classical and Quantum Gravity* **14**, A55–A81 (1997); e-print arXiv:gr-qc/9602046.
339. C. Rovelli, "A Generally Covariant Quantum Field Theory and a Prediction on Quantum Measurements of Geometry", *Nuclear Physics B* **405**, 797–815 (1993).
340. A. Ashtekar, A. Corichi and J. A. Zapata, "Quantum Theory of Geometry. III. Non-Commutativity of Riemannian Structures", *Classical and Quantum Gravity* **15**, 2955–2972 (1998); e-print arXiv:gr-qc/9806041.
341. L. Freidel, M. Geiller and J. Ziprick, "Continuous Formulation of the Loop Quantum Gravity Phase Space", *Classical and Quantum Gravity* **30**(8), 085013 (2013); e-print arXiv:gr-qc/1110.4833.
342. A. Ashtekar and J. Lewandowski, "Quantum Theory of Geometry. II. Volume Operators", *Advances in Theoretical and Mathematical Physics* **1**, 388–429 (1997); e-print arXiv:gr-qc/9711031.
343. J. Brunnemann and D. Rideout, "Properties of the Volume Operator in Loop Quantum Gravity. I. Results", *Classical and Quantum Gravity* **25**(6), 065001 (2008); e-print arXiv:0706.0469 [gr-qc].
344. J. Brunnemann and D. Rideout, "Properties of the Volume Operator in Loop Quantum Gravity. II. Detailed Presentation", *Classical and Quantum Gravity* **25**(6), 065002 (2008); e-print arXiv:0706.0382 [gr-qc].
345. J. Brunnemann and T. Thiemann, "Simplification of the Spectral Analysis of the Volume Operator in Loop Quantum Gravity", *Classical and Quantum Gravity* **23**, 1289–1346 (2006); e-print arXiv:gr-qc/0405060.
346. R. De Pietri and C. Rovelli, "Geometry Eigenvalues and the Scalar Product from Recoupling Theory in Loop Quantum Gravity", *Physical Review D* **54**, 2664–2690 (1996); e-print arXiv:gr-qc/9602023.
347. K. A. Meissner, "Eigenvalues of the Volume Operator in Loop Quantum Gravity", *Classical and Quantum Gravity* **23**, 617–625 (2006); e-print arXiv:gr-qc/0509049.

348. T. Thiemann, "Closed Formula for the Matrix Elements of the Volume Operator in Canonical Quantum Gravity", *Journal of Mathematical Physics* **39**, 3347–3371 (1998); e-print arXiv:gr-qc/9606091.
349. K. Giesel and T. Thiemann, "Consistency Check on Volume and Triad Operator Quantization in Loop Quantum Gravity. I", *Classical and Quantum Gravity* **23**, 5667–5691 (2006); e-print arXiv:gr-qc/0507036.
350. K. Giesel and T. Thiemann, "Consistency Check on Volume and Triad Operator Quantization in Loop Quantum Gravity. II", *Classical and Quantum Gravity* **23**, 5693–5771 (2006); e-print arXiv:gr-qc/0507037.
351. L. Freidel and S. Speziale, "Twisted Geometries: A Geometric Parametrisation of SU(2) Phase Space", *Physical Review D* **82**(8), 084040 (2010); e-print arXiv:gr-qc/1001.2748.
352. E. Bianchi, P. Donà and S. Speziale, "Polyhedra in Loop Quantum Gravity", *Physical Review D* **83**(4), 044035 (2011); e-print arXiv:gr-qc/1009.3402.
353. E. Bianchi and H. Haggard, "Discreteness of the Volume of Space from Bohr–Sommerfeld Quantization", *Physical Review Letters* **107**(1), 011301, (2011); e-print arXiv:gr-qc/1102.5439.
354. B. Dittrich and T. Thiemann, "Are the Spectra of Geometrical Operators in Loop Quantum Gravity Really Discrete?", *Journal of Mathematical Physics* **50**(1), 012503–012511 (2009); e-print arXiv:gr-qc/0708.1721.
355. C. Rovelli, "Comment on Are the spectra of geometrical operators in loop quantum gravity really discrete? by B. Dittrich and T. Thiemann", e-print arXiv:gr-qc/0708.2481 (2007).
356. F. Girelli, F. Hinterleitner and S. A. Major, "Loop quantum gravity phenomenology: linking loops to observational physics", *Symmetry, Integrability and Geometry: Methods and Applications* **8**(098) (2012); e-print arXiv:1210.1485v2 [gr-qc].
357. R. Gambini and J. Pullin, "Holography in Spherically Symmetric Loop Quantum Gravity", e-print arXiv:0708.0250 [gr-qc].
358. Y. Jack Ng, "Holographic Foam, Dark Energy and Infinite Statistics", *Physics Letters B* **657**, 10–14 (2007).
359. Y. Jack Ng, "Spacetime Foam: From Entropy and Holography to Infinite Statistics and Non-Locality", *Entropy* **10**, 441–461 (2008).
360. Y. Jack Ng, "Holographic quantum foam", e-print arXiv:1001.0411v1 [gr-qc] (2010).
361. Y. Jack Ng, "Various facets of spacetime foam", e-print arXiv:1102.4109.v1 [gr-qc] (2011).
362. N. Margolus and L. B. Levitin, "The Maximum Speed of Dynamical Evolution", *Physica D* **120**, 188–195 (1998).
363. X. Calmet, M. Graesser and S. D. Hsu, "Minimum Length from Quantum Mechanics and Classical General Relativity", *Physical Review Letters* **93**, 211101-1–211101-4 (2004).
364. G. 't Hooft, in *Salamfestschrift*; A. Ali, et al., eds., World Scientific, Singapore (1993).

365. L. Susskind, "The World as a Hologram", *Journal of Mathmatical Physics* **36**, 6377–6396 (1995).
366. J. D. Bekenstein, "Black Holes and Entropy", *Physical Review D* **7**, 2333–2346 (1973).
367. S. Hawking, "Particle Creation by Black Holes", *Communications in Mathematical Physics* **43**, 199–220 (1975).
368. S. B. Giddings, "Black Holes and Massive Remnants", *Physical Review D* **46**, 1347–1352 (1992).
369. R. Bousso, "The Holographic Principle", *Reviews in Modern Physics* **74**, 825–874 (2002).
370. M. Arzano, T. W. Kephart and Y. Jack Ng, "From Spacetime Foam to Holographic Foam Cosmology", *Physics Letters B* **649**, 243–246 (2007).
371. O. W. Greenberg, "Example of Infinite Statistics", *Physical Review Letters* **64**, 705–708 (1990).
372. K. Fredenhagen, "On the Existence of Antiparticles", *Communications in Mathematical Physics* **79**, 141–151 (1981).
373. M. Arzano, "Quantum Fields, Non-Locality and Quantum Group Symmetries", *Physical Review D* **77**, 025013 (2008); e-print arXiv:0710.1083 [hep-th] (2007).
374. A. P. Balachandran, A. Pinzul, B. A. Qureshi and S. Vaidya, "S-Matrix on the Moyal Plane: Locality versus Lorentz Invariance", *Physical Review D* **77**, 025020 (2008); e-print arXiv:0708.1379 [hep-th] (2007).
375. V. Jejjala, M. Kavic and D. Minic, "Fine Structure of Dark Energy and New Physics", *Advances in High Energy Physics*, 21586 (2007); e-print arXiv:0705.4581 [hep-th] (2007).
376. F. Benatti and R. Floreanini, "Non-Standard Neutral Kaons Dynamics from D-Brane Statistics", *Annals Physics* **273**, 58–71 (1999).
377. A. J. M. Medved, "A Comment or Two on Holographic Dark Energy", *General Relativity and Gravitation* **41**(2), 287–303 (2009); e-print arXiv:0802.1753 [hep-th] (2008).
378. S. B. Giddings, "Black Holes, Information, and Locality", *Modern Physics Letters A* **22**(39), 2949–2954 (2007); e-print arXiv:0705.2197 [hep-th] (2007).
379. G. T. Horowitz, "Black Holes, Entropies, and Information", e-print arXiv:0708.3680 [astro-ph] (2007).
380. A. Hamma, F. Markopoulou, S. Lloyd, F. Caravelli, S. Severini and K. Markstrom, "A Quantum Bose–Hubbard Model with Evolving Graph as Toy Model for Emergent Spacetime", *Physical Review D* **81**(10), 104032 (2010); e-print arXiv:0911.5075.
381. F. Caravelli, A. Hamma, F. Markopoulou and A. Riera, "Trapped Surfaces and Emergent Curved Space in the Bose-Hubbard Model", *Physical Review D* **85**, 044046 (2012); e-print arXiv:1108.2013 (2011).
382. F. Caravelli and F. Markopoulou, "Disordered Locality and Lorentz Dispersion Relations: An Explicit Model of Quantum Foam", *Physical Review D* **86**(2), 024019 (2012); e-print arXiv:1201.3206v3 [gr-qc].

383. F. Cardone and R. Mignani, *Deformed Spacetime. Geometrizing Interactions in Four and Five Dimensions*, Springer, Berlin-Heidelberg (2007).
384. R. Mignani, F. Cardone and A. Petrucci, "Metric Gauge Fields in Deformed Special Relativity", *Electronic Journal of Theoretical Physics* **10**(29), 1–21 (2013).
385. University of Wisconsin, "Cosserat Elasticity; micropolar elasticity", http://silver.neep.wisc.edu/~lakes/Coss.html
386. H. Nikolic, "Bohmian Particle Trajectories in Relativistic Fermionic Quantum Field Theory", *Foundations of Physics Letters* **18**(2), 123–138 (2003); e-print arXiv:quant-ph/0302152 v3.
387. H. Nikolic, "Bohmian Particle Trajectories in Relativistic Bosonic Quantum Field Theory", *Foundations of Physics Letters* **17**(4), 363–380 (2004).
388. D. Fiscaletti, "Perspectives Towards a Density Theory of Everything", *Scientific Inquiry* **9**(2), 173–200 (2008).
389. A. Shojai, "Quantum, Gravity and Geometry", *International Journal of Modern Physics A* **15**(12), 1757–1771 (2000); e-print arXiv:gr-qc/0010013.
390. F. Shojai and A. Shojai, "Quantum Einstein's Equations and Constraints Algebra", *Pramana* **58**(1), 13–19 (2002); e-print arXiv:gr-qc/0109052.
391. L. de Broglie, *Non-linear Wave Dynamics — A Causal Interpretation*, transl. A.J. Knodel, Elsevier Publishing Company, New York (1960).
392. F. Shojai and A. Shojai, "Understanding quantum theory in terms of geometry", e-print arXiv:gr-qc/0404102 v1 (2004).
393. Y. M. Cho and D. H. Park, "Higher Dimensional Unification and Fifth Force", *Nuovo Cimento B* **105**(8–9), 817–829 (1990).
394. F. Shojai, A. Shojai and M. Golshani, "Conformal Transformations and Quantum Gravity", *Modern Physics Letters A* **13**(34), 2725–2729 (1998).
395. A. Shojai, F. Shojai and M. Golshani, "Nonlocal Effects in Quantum Gravity" *Modern Physics Letters A* **13**(37), 2965–2969 (1998).
396. F. Shojai and A. Shojai, "Nonminimal Scalar-Tensor Theories and Quantum Gravity", *International Journal of Modern Physics A* **15**(13), 1859–1868 (2000).
397. V. Fock, "Geometrization of Dirac's Theory of Electron", *Zeitschrift Physik* **57**, 261–277 (1929).
398. W. K. Clifford, "On the Space Theory of Matter", in *The World of Mathematics*, Simon and Schuster (1956).
399. W. K. Clifford, "On the Space-Theory of Matter", 2 February 1870, Transactions of the Cambridge Philosophical Society, **2**, 157–158 (1866–1876); reprinted in *William K. Clifford, Mathematical Papers*, R. Tucker ed., Chelsea, New York (1968).
400. H. Weyl, "Elektron und Gravitation I", *Zeitschrift Physik* **56**, 330–352 (1929).
401. M. D. Pollock, "On the Dirac Equation in Curved Space-Time", *Acta Physica Polonika B* **41**(8), 1827–1846 (2010).
402. Th. Kaluza, "Zum Unitatsproblem der Physik", *Sitzungsberichte der: J. Koniglich PreuBisch Akademie der Wissenschaften* **33**, 966–972 (1921).

403. O. Klein, "Meson Fields and Nuclear Interaction", *Arkiv for Matmatik Astronomi Och Fysik* **34A**, 1 (1946).
404. E. Schrödinger, *Sitzungsberichte der: J. Koniglich Preußisch Akademie der Wissenschaften* Berlin, 105–128 (1932).
405. K. T. Compton, E. A. Trousdale, "The Nature of the Ultimate Magnetic Particle", *Physical Review* **5**(4), 315–318 (1915).
406. H. Compton, O. Rognley, "Is the Atom the Ultimate Magnetic Particle", *Physical Review* **16**(5), 464–476 (1920).
407. H. S. Allen, *Philosophical Magazine* **40**(235), 426 (1920).
408. W. Pauli, in *Handbuch der Physik*, Quantentheorie, Springer Verlag, Berlin, pp. 83–272 (1933).
409. W. Pauli, "Relativistic Theories of Elementary Particles", *Reviews of Modern Physics* **13**, 203–232 (1941).
410. A. D. Alhaidari and A. Jellal, "Dirac and Klein-Gordon Equations in Curved Space", e-print arXiv:1106.2236v1 (2011).
411. A. D. Sakharov, "Vacuum Quantum Fluctuations in Curved Space and the Theory of Gravitation", *Doklady Akademii Nauk SSSR* **177**(1), 70–71 (1967).
412. C. W. Misner, K. S. Thorne and J. A. Wheeler, *Gravitation*, Freeman, New York (1971).
413. A. Rueda and B. Haisch, "Gravity and the Quantum Vacuum Inertia Hypothesis", *Annalen der Physik* **14**(8), 479–498 (2005); e-print arXiv:gr-qc0504061v3 (2005).
414. H. E. Puthoff, "Polarizable-Vacuum (PV) Approach to General Relativity", *Foundations of Physics* **32**(6), 927–943 (2002).
415. M. Consoli, "Do Potentials Require Massless Particles?", *Physical Review Letters B* **672**(3), 270–274 (2009).
416. M. Consoli, "On the Low-Energy Spectrum of Spontaneously Broken Phi4 Theories", *Modern Physics Letters A* **26**, 531–542 (2011).
417. M. Consoli, "The vacuum condensates: a bridge between particle physics to gravity?", in *Vision of oneness*, I. Licata and A. Sakaji eds., Aracne Editrice, Roma (2011).
418. P. A. M. Dirac, *Principles of Quantum Mechanics*, fourth edition Oxford University Press, Oxford (1982).
419. G. W. Erickson, "Improved Lamb-shift Calculation for All Values of Z", *Physical Review Letters* **27**, 780–783 (1971).
420. G. L. Klimchitskaya, U. Mohideen and V. M. Mostepanenko, "The Casimir Force between Real Materials: Experiment and Theory", *Reviews of Modern Physics* **81**, 1827–1885 (2009).
421. R. L. Jaffe, "The Casimir Effect and the Quantum Vacuum", *Physical Review D* **72**, 021301, 1–5 (2005); e-print http://arXiv:hep-th/0503158v1.
422. C. Beck and M. C. Mackey, "Zeropoint Fluctuations and Dark Energy in Josephson Junctions", *Fluctuation and Noise Letters* **7**(2), C27–C35 (2007).
423. R. Loudon, *The Quantum Theory of Light*, Clarendon Press, Oxford (1983).
424. L. de la Peña and A. M. Cetto, *The Quantum Dice: An Introduction to Stochastic Electrodynamics*, Kluwer Academic Publishing (1996).
425. P. W. Milonni, *The Space*, Academic Press (1994).

426. P. Davies, T. Dray and C. Manogue, *Physical Review D* **53**, 4382–4387 (1996).
427. A. Bettini, *Introduction to Elementary Particle Physics*, Cambridge University Press, Cambridge (2008).
428. V. A. Rubakov, "Hierarchies of Fundamental Constants", *Physics — Uspekhi. Advances in Physical Sciences* **50**(4), 390–396 (2007).
429. S. M. Carroll, W. H. Press and E. L. Turner, "The Cosmological Constant", *Annual Review of Astronomy and Astrophysics* **30**(1), 499–542 (1992).
430. A. D. Chernin, "Dark Energy and Universal Antigravitation", *Physics — Uspekhi. Advances in Physical Sciences* **51**(3), 253–282 (2008).
431. T. Padmanabhan, "Darker Side of the Universe", *29 International Cosmic Ray Conference Pune* **10**, 47–62 (2005).
432. T. Reichhardt, "Cosmologists Look Forward to Clear Picture", *Nature* **421**, 777 (2003).
433. V. Sahni, "Dark Matter and Dark Energy", *Lecture Notes in Physics* **653**, 141–180 (2004); e-print arXiv.org/abs/astro-ph/0403324v3.
434. Yu. B. Zeldovich, "Vacuum Theory: A Possible Solution to the Singularity Problem of Cosmology", *Physics — Uspekhi. Advances in Physical Sciences* **24**(3), 216–230 (1981).
435. E. Santos, "Quantum vacuum fluctuations and dark energy", e-print arXiv:0812.4121v2 [gr-qc] (2009).
436. E. Santos, "Space–Time Curvature Induced by Quantum Vacuum Fluctuations as an Alternative to Dark Energy", *International Journal of Theoretical Physics* **50**(7), 2125–2133 (2010).
437. Yu. B. Zeldovich, "Cosmological Constant and Elementary Particles", *Zhurnal Eksperimental'noi i Teoreticheskoi Fiziki Pis'ma* **6**, 883–884 (1967).
438. S. E. Rugh and H. Zinkernagel, "The Quantum Vacuum and the Cosmological Constant Problem", *Studies in History and Philosophy of Modern Physics* **33**(4), 673–705 (2001).
439. S. F. Timashev, "Physical Vacuum as a System Manifesting Itself on Various Scales — from Nuclear Physics to Cosmology", e-print arXiv:1107.pdf [gr-qc] (2011).
440. J. Solà, "Cosmological Constant and Vacuum Energy: Old and New Ideas", *Journal of Physics: Conference Series* **453**, 012015 (2013).
441. G. Aad *et al.* (ATLAS Collaboration), "Observation of a New Particle in the Search for the Standard Model Higgs Boson with the ATLAS Detector at the LHC", *Physics Letters B* **716**(1), 1–29 (2012).
442. S. Chatrchyan *et al.* (CMS Collaboration), "Observation of a New Boson of a Mass of 125 GeV with the CMS experiment at the LHC", *Physics Letters B* **716**(1), 30–61 (2012).
443. S. Basilakos, M. Plionis and J. Solà, "Hubble Expansion and Structure Formation in Time Varying Vacuum models", *Physical Review D* **80**(8), 093511 (2009); e-print arXiv:0907.4555 [astro-ph.CO].

444. J. Grande, J. Solà, S. Basilakos and M. Plionis, "Hubble Expansion and Structure Formation in the 'running FLRW model' of the Cosmic Evolution", *Journal of Cosmology and Astroparticle Physics JCAP* **08**(007), 37 pages (2011); e-print arXiv:1103.4632.
445. S. Basilakos, D. Polarski and J. Solà, "Generalizing the running vacuum energy model and comparing with the entropic-force models", *Physical Review D* **86**, 043010 (2012); e-print arXiv:1204.4806.
446. H. Fritzsch and J. Solà, "Matter Non-Conservation in the Universe and Dynamical Dark Energy", *Classical and Quantum Gravity*, **29**(21), 25 (2012); e-print arXiv:1202.5097.
447. J-P. Uzan, "Varying Constants, Gravitation and Cosmology", *Living Reviews in Relativity* **14**(2), 1–155 (2011); e-print arXiv:1009.5514.
448. J-P. Uzan, "The Fundamental Constants and Their Variation: Observational Status and Theoretical Motivations", *Reviews in Modern Physics* **75**(5), 403–455 (2003); e-print arXiv:hep-th/0205340.
449. T. Chiba, "The Constancy of the Constants of Nature: Updates", *Progress in Theoretical Physics* **126**(6), 993–1019 (2011); e-print arXiv:1111.0092; and references therein.
450. E. Reinhold, R. Buning, U. Hollenstein, A. Ivanchik, P. Petitjean and W. Ubachs, "Indication of a Cosmological Variation of the Proton–Electron Mass Ratio Based on Laboratory Measurement and Reanalysis of H2 Spectra", *Physical Review Letters* **96**(15), 151101 (2006).
451. W. Ubachs and E. Reinhold, Highly Accurate H_2 Lyman and Werner Band Laboratory Measurements and an Improved Constraint on a Cosmological Variation of the Proton-to-Electron Mass Ratio", *Physical Review Letters* **92**, 101302 (2004).
452. A. Ivanchik, P. Petitjean, D. Varshalovich, B. Aracil, R. Srianand, H. Chand, C. Ledoux and P. Boisse, "A New Constraint on the Time Dependence of the Proton-to-Electron Mass Ratio. Analysis of the Q 0347-383 and Q 0405-443 spectra", *Astronomy and Astrophysics* **440**(1), 45–52 (2005); e-print arXiv:astro-ph/0507174.
453. S. Weinberg, "The Cosmological Constant Problem", *Reviews in Modern Physics* **61**(1), 1–23 (1989).
454. V. Sahni and A. Starobinsky, "The Case for a Postive Cosmological λ-Term", *International Journal of Modern Physics A* **9**(4), 373–443 (2000).
455. S. M. Carroll, "The Cosmological Constant", *Living Reviews in Relativity* **4**(1) (2001).
456. T. Padmanabhan, "Cosmological Constant — the Weight of the Vacuum", *Physics Reports* **380**(5–6), 235–320 (2003).
457. J. Martin, "Everything You Always Wanted to Know about the Cosmological Constant Problem (but were afraid to ask), *Comptes Rendus Physique* **13**(6–7), 566–665 (2012).
458. J. A. S. Lima, S. Basilakos and J. Solà, "Expansion History with Decaying Vacuum: A Complete Cosmological Scenario", *Montly Notices of the*

Royal Astronomical Society **431**, 923–929 (2013); e-print arXiv:1209.2802 [gr-qc].
459. S. Basilakos, J. A. S. Lima and J. Solà, "From Inflation to Dark Energy Through a Dynamical Lambda: an Attempt at Alleviating Fundamental Cosmic Puzzles", *International Journal of Modern Physics D* **22**(12), 1350038 (2013).
460. E. L. D. Perico, J. A. S. Lima, S. Basilakos and J. Solà, "Complete Cosmic History with a Dynamical $\Lambda = \Lambda(H)$ term", *Physical Review D* **88**, 063531 (2013).
461. J. Solà, "Dark Energy: A Quantum Fossil from the Inflationary Universe?", *Journal of Physics A: Mathematical and Theoretical* **41**(16), 164066 (2008); e-print arXiv:0710.4151 [hep-th].
462. T. Padmanabhan, "The physical principle that determines the value of the cosmological constant", e-print arXiv:1210.4174 [hep-th] (2012).
463. T. Padmanabhan and H. Padmanabhan, "The solution to the cosmological constant problem", e-print arXiv:1302.3226 [astro-ph] (2013).
464. B. Haisch, A. Rueda and H. E. Puthoff, "Inertia as a Zero-Point Field Lorentz Force", *Physical Review A* **49**(2), 678–694 (1994).
465. A. Rueda and B. Haisch, "Contribution to the Inertial Mass by Reaction of the Vacuum to Accelerated Motion, *Foundations of Physics* **28**(7), 1057–1108 (1998).
466. A. Rueda and B. Haisch, "Inertial Mass as Reaction of the Vacuum to Accelerated Motion", *Physics Letters A* **240**(3), 115–126 (1998).
467. B. Haisch, A. Rueda and Y. Dobyns, "Inertial Mass and the Quantum Vacuum Fields", *Annalen der Physik (Leipzig)* **10**(5), 393–414 (2001).
468. Y. Dobyns, A. Rueda and B. Haisch, "The Case for Inertia as a Vacuum Effect: A Reply to Woodward and Mahood", *Foundations of Physics* **30**(1), 59–80 (2000); e-print http://arxiv.org/abs/gr-qc/0002069.
469. H. Sunahata, A. Rueda and B. Haisch, "Quantum vacuum and inertial reaction in non-relativistic QED", e-print arXiv:1306.6036v1 [physics.gen-ph] (2013).
470. H. E. Puthoff, "Gravity as a Zero-Point Fluctuation Force", *Physical Review A* **39**(5), 2333–2342 (1989).
471. S. Carlip, "Comment on Gravity as a Zero-Point Fluctuation Force", *Physical Review A* **47**(4), 3452–3453 (1993).
472. H. E. Puthoff, "Gravity as a Zero-Point Fluctuation Force", *Physical Review A* **47**(4), 3454–3455 (1993).
473. B. Haisch, A. Rueda and H. E. Puthoff, "Physics of the Zero-Point Field: Implications for Inertia, Gravitation and Mass", *Speculations in Science and Technology* **20**, 99–114 (1997).
474. D. C. Cole, A. Rueda and K. Danley, "Stochastic Nonrelativistic Approach to Gravity as Originating from Vacuum Zero-Point Field Van Der Waals Forces", *Physical Review A* **63**(5), 054101-2 (2001).
475. R. H. Dicke, "Gravitation without a principle of equivalence", *Reviews in Modern Physics* **29**, 3, 363–376 (1957); see also R.H. Dicke, "Mach's Principle and Equivalence," in *Proceedings of the International School of*

Physics "Enrico Fermi" Course XX, Evidence for Gravitational Theories, C. Moller ed., Academic Press, New York (1961), pp. 1–49.
476. H. A. Wilson, "An Electromagnetic Theory of Gravitation", Physical Review **17**(1), 54–59 (1921).
477. R. d'E. Atkinson, "General Relativity in Euclidean Terms", Proceedings of Royal Society **272**, 60–78 (1962).
478. Ye Xing-Hao and Lin Qiang, "Inhomogenous Vacuum: An Alternative Interpretation of Curved Space-Time", Chinese Physics Letters **25**(5), 1571–1574 (2008).
479. S. Weinberg, "The Cosmological Constant Problem", Reviews in Modern Physics **61**(1), 1–23 (1989).
480. Sean M. Carroll, "The Cosmological Constant", Living Reviews in Relativity **3**, 1–56 [Online Article] (2001); accessed on Aug 3, 2010, http://www.livingreviews.org/lrr-2001-1 [e-print arXiv:astro-ph/0004075v2].
481. J. A. Frieman, Michael S. Turner and D. Huterer, "Dark Energy and the Accelerating Universe", Annual Review of Astronomy and Astrophysics **46**, 385–432 (2008); e-print arXiv:0803.0982v1 [astro-ph].
482. G. F. R. Ellis, J. Murugan and H. Van Elst, "The gravitational effect of the vacuum", e-print arXiv:1008.1196v1 [gr-qc] (2010).
483. A. Einstein, "Die Feldgleichungen der Gravitation", Sitzungsberichte der Koniglich PreuBischen Akademic der Wissenchaften. Berlin (Mathematical Physics), 844–847 (1915).
484. A. Einstein, "Kosmologische Betrachtungen zur allgemeinen Relativitätstheorie" Sitzungsberichte der Koniglich PreuBischen Akademic der Wissenchaften. Berlin (Mathematical Physics), 142–152 (1917).
485. M. Consoli, A. Pluchino and A. Rapisarda, "Basic Randomness of Nature and Ether-Drift Experiments", Chaos, Solitons and Fractals **44**(12), 1089–1099 (2011); e-print arXiv:1106.1277v2 [physics.gen-ph].
486. P. Rowlands, "What is vacuum?", e-print arXiv:0810.0224v1 [gr-qc] (2008).
487. L. Chiatti, "The transaction as a quantum concept", in Space-Time Geometry and Quantum Events, I. Licata ed., Nova Science Publishers, New York (2014), pp. 11–44; e-print arXiv.org/pdf/1204.6636 (2012).
488. I. Licata, "Transaction and Non-Locality in Quantum Field Theory", European Physical Journal Web of Conferences (2013).
489. I. Licata and L. Chiatti, "The Archaic Universe: Big Bang, Cosmological Term and the Quantum Origin of Time in Projective Cosmology", International Journal of Theoretical Physics **48**(4), 1003–1018 (2009).
490. I. Licata and L. Chiatti, "Archaic Universe and Cosmological Model: 'Big-Bang' as Nucleation by Vacuum", International Journal of Theoretical Physics **49**(10), 2379–2402 (2010); e-print arXiv:genph/1004.1544.
491. G. J. Maclay, "Gedanken Experiments with Casimir Forces, Vacuum Energy and Gravity", Physical Review A **82**, 032106 (2010); e-print arXiv: 1107.0764v1 [physics.gen-ph] (2011).
492. J. Schwinger, "On Gauge Invariance and Vacuum Polarization", Physical Review **82**(5), 664–679 (1951).

493. A. Casher, H. Neuberger and S. Nussinov, "Chromoelectric-Flux-Tube Model of Particle Production", *Physical Review D* **20**(1), 179–188 (1979).
494. H. Neuberger, "Finite Time Corrections to the Chromoelectric-Flux-Tube Model", *Physical Review D* **20**(11), 2936–2946 (1979).
495. R. Brout, S. Massar, R. Parentani, S. Popescu and P. Spindel, "Quantum Source of the back Reaction on a Classical Field", *Physical Review D* **52**(2), 1119–1133 (1994).
496. C. Itzykson and J.-B. Zuber, *Quantum Field Theory*, McGraw-Hill, New York (1980).
497. F. Cooper and E. Mottola, "Quantum back Reaction in Scalar QED as an Initial-Value Proble", *Physical Review D* **40**(2), 456–464 (1989).
498. R. C. Wang and C. Y. Wong, "Finite-size effect in the Schwinger Particle-Production Mechanicsm", *Physical Review D* **38**(1), 348–359 (1988).
499. A. I. Nikishov, "Barrier scattering in Field Theory and Removal of the Klein Paradox", *Nuclear Physics B* **21**, 346–358 (1970).
500. T. D. Cohen and D. A. McGady, "The Schwinger Mechanism Revisited", *Physical Review D* **78**, 036008 (2008); e-print arXiv:0807.1117 [hep-ph].
501. F. R. Klinkhamer and G. E. Volokik, "Dynamics of the Quantum Vacuum: Cosmology as Relaxation to the Equilibrium State", *Journal of Physics: Conference Series* **314**(1), 012004 (2011); e-print arXiv:1102.3152v4 [gr-qc] (2011).
502. F. R. Klinkhamer and G. E. Volovik, "Self-Tuning Vacuum Variable and Cosmological Constant", *Physical Review D* **77**(8), 085015-14 (2008); e-print arXiv:0711.3170.
503. F. R. Klinkhamer and G. E. Volovik, "Dynamic Vacuum Variable and Equilibrium Approach in Cosmology", *Physical Review D* **78**(6), 063528-12 (2008); e-print arXiv:0806.2805.
504. F. R. Klinkhamer and G. E. Volovik, "Towards a Solution of the Cosmological Constant Problem", *JETP Letters* **91**(6), 259–265 (2010); e-print arXiv:0907.4887.
505. F. R. Klinkhamer and G. E. Volovik, "Vacuum Energy Density Kicked by the Electroweak Crossover", *Physical Review D* **80**(8), 083001-7 (2009); e-print arXiv:0905.1919.
506. G. E. Volovik, "On Spectrum of Vacuum Energy", *Journal of Physics: Conference Series* **174**, 012007/1-10 (2009); e-print arXiv:0801.2714.
507. D. Fiscaletti and A. Sorli, "Perspectives about Quantum Mechanics in a Model of a Three-Dimensional Quantum Vacuum Where Time is a Mathematical Dimension", *SOP Transactions on Theoretical Physics* **1**(3), 11–38 (2014).
508. D. Fiscaletti and A. Sorli, "Space-Time Curvature of General Relativity and Energy Density of a Three-Dimensional Quantum Vacuum", *Annales UMCS Sectio AAA: Physica* **LXIX**, 55–81 (2014).
509. D. Fiscaletti and A. Sorli, *The Infinite History of Now. A Timeless Background for Contemporary Physics*, Nova Science Publishers, New York (2014).

510. D. Bohm, *Wholeness and Implicate Order*, Routledge & Kegan Paul, London (1980).
511. A. D. Sacharov, "Vacuum Quantum Fluctuation in Curved Space and the Theory of Gravitation", *General Relativity and Gravitation* **32**, 365–367 (2000).
512. L. M. Caligiuri, "The Emergence of Spacetime and Gravity: Entropic of Geometro-Hydrodynamic Process? A Comparison and Critical Review", *Quantum Matter*, **3**(3), 246–252 (2014).
513. F. Wilczek, "Origins of Mass", http://arxiv.org/pdf/1206.7114.pdf (2012).
514. A. Loinger, "The Gravitational Waves are Fictitious Entities", http://xxx.lanl.gov/abs/astro-ph/ 9810137 (1998).
515. A. Loinger, "The Gravitational Waves are Fictitious Entities-II", http://arxiv.org/vc/astro-ph/papers/9904/9904207v1.pdf (2004).
516. I. Ciufolini and V. Gorini, "Gravitational Waves, Theory and Experiment (An Overview)", http://bookmarkphysics.iop.org/fullbooks/0750307412/ciufoliniover.pdf (2004).
517. H.-J. Schorn, "New Effect for Detecting Gravitational Waves by Amplification with Electromagnetic Radiation", *International Journal of Theoretical Physics* **40**(8), 1427–1452 (2001).
518. Y. Melamed and M. Lin, "Principle of Sufficient Reason", in *The Stanford Encyclopedia of Philosophy*, E. N. Zalta ed., Summer 2011 Edition, http://plato.stanford.edu/archives/sum2011/entries/su_cient-reason/.
519. J. Schaffer, "Monism: The priority of the Whole", *Philosophical Review* **119**, 31–76 (2010).
520. B. Hiley and R. Callaghan, "The Clifford algebra approach to quantum mechanics B: The Dirac particle and its relation to the Bohm approach", e-print arXiv:1011.4033v1 [math-ph] (2010).
521. M. Kaku, *Quantum Field Theory*, Oxford University Press, Oxford (1993).
522. C. Kiefer, "Time in Quantum Gravity", in *The Oxford Handbook of Philosophy of Time*, C. Callender ed., pp. 663–678, Oxford University Press, Oxford (2011).
523. B. Hiley and R. Callaghan, "Clifford Algebras and the Dirac-Bohm Quantum Hamilton–Jacobi Equation", *Foundations of Physics* **42**(1), 192–208 (2012).
524. I. Licata and D. Fiscaletti, *Quantum Potential. Physics, Geometry and Algebra*, Springer, Heidelberg (2013).
525. W. M. Stuckey, M. Silberstein and M. Cifone, "Reconciling Spacetime and the Quantum: Relational Blockworld and the Quantum Liar Paradox", *Foundations of Physics* **38**(4), 348–383 (2008); e-print arXiv:quant-ph/0510090.
526. M. Silberstein, W. M. Stuckey and M. Cifone, "Why Quantum Mechanics Favours Adynamical and Acausal Interpretations Such as Relational Blockworld Over Backwardly Causal and Time-Symmetric Rivals", *Studies in History and Philosophy of Modern Physics* **39**(4), 736–751 (2008).
527. M. Silberstein, W. M. Stuckey and T. McDewitt, "Being, becoming and the undivided universe: a dialogue between relational blockworld and

the implicate order concerning the unification of relativity and quantum theory", e-print arXiv:1108.2261v3 [quant-ph] (2012).
528. J. G. Cramer, "Generalized Absorber Theory and the Einstein–Podolsky–Rosen Paradox", *Physical Review D* **22**(2), 362–376 (1980).
529. J. G. Cramer, "The Arrow of Electromagnetic Time and the Generalized Absorber Theory", *Foundations of Physics* **13**(9), 887–902 (1983).
530. J. G. Cramer, "The Transactional Interpretation of Quantum Mechanics", *Reviews of Modern Physics* **58**, 647–688 (1986).
531. J. G. Cramer, "An Overview of the Transactional Interpretation", *International Journal of Theoretical Physics* **27**(2), 227–236 (1988).
532. R. E. Kastner, "de Broglie Waves as the 'Bridge of Becoming' Between Quantum Theory and Relativity", *Foundations of Science* (2011).
533. R. E. Kastner, "On Delayed Choice and Contingent Absorber Experiments", *ISRN Mathematical Physics* **2012**, 617291 (2012).
534. R. E. Kastner, "The Broken Symmetry of Time", *AIP Conference Proceedings* **1408**, 7–21, (2011).
535. R. Kastner, *The New Transactional Interpretation of Quantum Theory: The Reality of Possibility*, Cambridge University Press, Cambridge (2012).

Subject Index

3D quantum vacuum, 16

aberration of light, 114, 143–145
absolute time, 3, 6, 8–10, 17, 28, 47, 53, 109–110, 156–157
Aharonov-Bohm effect, 209, 213–223
Anandan's geometry for quantum theory, 203–205
anomalous electromagnetic interaction energy, 291–294
a-temporal quantum gravity space theory, 16, 253, 269–303, 394, 408
atomic clocks, 92, 94, 116

Barbour-Foster-O' Murchadha (BFO) approach of geometrodynamics, 34, 36, 38, 100, 105
Bell inequalities, 187, 191
quantum Einstein equations, 239–241, 245, 313, 374
Bohm's theory, 196
Busch's distinction of three roles of time in quantum theory, 159–163

Cahill's model of special relativity, 122–127, 129
canonical quantum gravity, 25, 27, 103, 108, 405
Caticha's approach of time, 65, 83–89, 95–98, 405
changes of the three-dimensional quantum vacuum energy density, 370, 372–373, 377, 408–409

Chiatti's and Licata's approach of transactions, 351–353, 359–360, 365
Chiou's timeless path integral approach, 64–71
clock rate, 6, 119, 120, 121, 263, 339
collapse of the wavefunction, 154–156, 163–164, 167–168, 223–224, 228–229
confined-time-of-arrival operators, 165–167
conformal equivalence principle, 285–287
conformal factor of the a-temporal space metric, 279, 281, 283, 285
Consoli's model of physical vacuum, 306, 345–347, 366, 372
Copenhagen interpretation of quantum mechanics, 151–153
correspondence principle, 184–185, 197, 202
cosmic space, 269–278, 280, 282, 284–289, 291–292, 294–301, 407–408
cosmological constant, 21, 23, 33, 232, 242–244, 282, 285, 309–310, 313, 315–317, 319, 321–322, 324–325, 347–350, 356, 372, 374
curvature, 33–34, 37, 40, 111–112, 231, 235, 236, 241, 278, 282, 286–287, 294–297, 305, 310, 314–315, 338–339, 347–348, 366, 372–373, 375–377, 379, 382–384, 394–395

dark energy, 265–266, 309–310,
  314–315, 322–323, 372–373,
  375–376, 379, 385
de Broglie-Bohm interpretation,
  154–156, 175, 177, 196, 236, 242
density of cosmic space, 269–272,
  274–278, 280, 282, 284–289,
  291–292, 294–301, 407–408
direct information medium, 151,
  168–169, 198, 208, 227–229
discreteness of the spatial geometry,
  13, 253–264, 268–269, 300
Doppler effect, 119, 143, 145–148
double-slit experiment, 170, 188, 190,
  209–213
duration, 2–3, 15, 17, 26–28, 30, 32,
  34, 46–47, 83–84, 86–87, 95,
  99–100, 102, 104–105, 107,
  132–137, 139, 143, 148, 160, 269,
  336, 361, 405–406, 409
duration of material change, 132–134,
  136–137, 139, 143, 148, 406
duration of the material motion, 17,
  133, 136–137, 269
dynamical time, 6–7, 10, 125, 160

electric density of space, 287–289,
  291–292
electromagnetic interaction, 287–294,
  296–297
electromagnetic quantum vacuum,
  305–308, 325–327, 331–333, 359
Ellis', Murugan's and Van Elst's
  model on vacuum energy density
  and cosmological constant, 348–350
Elze's approach of time, 64, 73–78,
  95–96, 98, 405
Elze's reparametrization invariant
  incident number, 76, 96, 98–99, 405
entropic bohmian mechanics, 182–192
entropic time, 65, 83–86, 88–89, 95,
  97–98, 405
ephemeris time, 30, 39, 42, 46,
  99–101, 104–106, 109

EPR-type experiments, 168, 171–174,
  181, 187, 191–193, 197–200, 202,
  402–406
Everett's many-worlds interpretation,
  154, 156–159
Feynman's path integral method, 8,
  11, 65, 69, 397, 399
Fisher information, 177–180
Friedmann equations of general
  relativity, 265, 308–309, 314, 350,
  375
Friedmann–Robertson–Walker
  metric, 311, 314, 370

Galapon's model on the appearance
  of a particle, 164–168, 228
Galilean transformations, 130, 141,
  143, 145
gauge invariance for the definition of
  clock time, 93, 97–98
general relativity, 1, 4–8, 10–13,
  18–20, 22, 24–27, 33–37, 39–41, 47,
  50, 82, 100, 105, 109–112, 114, 117,
  127, 171, 236, 244, 248, 251,
  254–255, 270, 278, 286, 305, 310,
  314–315, 326, 331, 338–344,
  346–348, 369–372, 376, 381–384,
  401, 409, 411
generalized Dirac equation for the
  density of cosmic space, 272,
  274–275, 287–291, 293–297
generalized Klein-Gordon equation
  for the density of cosmic space,
  274–275, 295–298
Girelli's, Liberati's and Sindoni's toy
  model of a nondynamical timeless
  space, 64–65, 79–83, 95–96, 98, 405
gravitational frequency shift of the
  photon, 341–342, 382
gravitational space, 14–15, 246, 249,
  253, 269–271, 275, 277–279, 282,
  284, 287, 298, 303, 377, 379, 385,
  407–409
gravity, 1, 5–6, 10–16, 19–25, 27,
  34–35, 37, 40, 47–49, 53–54, 58, 62,
  71–73, 81–82, 101, 103, 105, 108,

230–232, 234–236, 238, 241,
244–251, 253–257, 259–263,
265–269, 271–272, 274–279,
281–282, 284–295, 298–303,
305–306, 309–310, 325, 337–339,
341–342, 346–348, 353–354,
366–367, 371–372, 377–378,
381–383, 394–395, 397, 400,
404–405, 407–410
Gryb's approach of
Jacobi-Barbour-Bertotti path
integrals, 46, 101–103, 106–108, 405

Haisch's and Rueda's model on
inertial mass and gravitational
mass as effects of the
electromagnetic quantum vacuum,
305–306, 325–337
Heisenberg's uncertainty relation,
176, 186, 207–208, 307, 327,
357–358
Higgs mechanism, 305, 316, 320–321,
324, 326, 366
Hiley's view of implicate order,
397–401
holographic space-time foam,
262–265, 302

immediate information medium, 16,
153, 171–172, 174, 181, 190–193,
197, 199, 201, 202, 206, 208, 222,
228, 230, 245, 249–250, 298, 393,
406–407
immediate symmetric interpretation
of quantum non-locality, 203

Jacobi-Barbour-Bertotti action, 29,
34, 36, 100, 105
Jacobi-Barbour-Bertotti clock, 45–46
Jacobi-Barbour-Bertotti path
integrals, 46, 102–103, 107–108
Jacobi-Barbour-Bertotti theory, 26,
29–31, 34, 41–45, 99–101, 103–106,
108, 405
Jacobi-Barbour-Bertotti time, 28,
45–47

Jupiter's satellites occultation, 114,
119, 148–149

Kaluza-Klein formalism, 11, 268, 285,
290–291, 294
Kastner's possibilist transactional
interpretation of quantum
mechanics, 402–404

Leibniz's view of the space-time
background, 3–4
length contraction, 116, 122, 125, 137,
141–143
light clocks, 114, 139–142
loop quantum gravity, 13, 47, 71, 232,
253–256, 259–260, 262–263, 269,
300–303
Lorentz transformations, 115–116,
119, 121, 127, 132, 136, 138, 148,
406

Mach's view of time, 4, 18–19, 22–23,
26, 28, 30, 32, 64, 99–100, 105,
286–287
metric of the a-temporal space,
283–285, 296–297, 300
Minkowski's space-time, 4–5, 9, 13,
58, 74–75, 80, 82, 110, 113–114,
117–118, 122, 128, 130, 138, 141,
259, 268, 289, 292, 308, 312, 314,
356, 374, 399

Newton's view of the space-time
background, 2–6, 8–9, 11
Ng's model of quantized space-time
foam, 263–266, 300, 302, 303
non-Euclidean structure of the
three-dimensional space, 174–190,
206–212, 221–223, 407
nonlocal geometry, 171, 192–193, 198,
200, 202, 207, 215, 253, 260,
265–269, 287, 298–300, 302–303,
360, 364, 393, 406–407
Novello's, Salim's and Falciano's
approach based on Weyl integrable
space, 174–177, 180–181

numerical order of material changes, 15–17, 26, 93–95, 98–112, 114, 130–131, 133, 139, 142–143, 172, 174, 195, 199, 228, 269, 364, 405–406, 409, 412

Palmer's view of invariant set postulate, 64, 71–73, 95, 98, 405
parameter time, 45, 65, 89–92, 94–95, 102, 106
photon motion, 113, 139
physical time, 5, 7, 132–133, 195, 365, 405
Planck energy density, 309, 359, 366–367, 372
Planck-Fiscaletti granules, 271, 275, 299–300, 408
Planck scale, 11–13, 47, 80, 261–262, 264–265, 269, 275, 300, 303, 323
polarization of the vacuum, 338–343, 366, 379–380
Prati's model of physical clock time, 65, 89–99, 104–105, 107–108, 405
problem of time, 10, 15, 17–25, 37, 47, 58, 70, 89, 242
proper time, 6, 28, 34, 41, 62–63, 73, 75–76, 78, 328, 333–336
Puthoff's polarizable vacuum model of gravitation, 306, 338–343, 345, 380, 382

quantum chromodynamics (QCD), 309, 315–316, 318, 324, 354
quantum cosmology, 24, 71, 230–232, 235–236, 239, 241–242, 267, 404, 407
quantum electrodynamics (QED), 294, 308, 316, 318, 333, 335, 337
quantum entanglement, 172–173, 227, 401
quantum-entropic length, 181, 185–188, 190–192, 199, 202, 207–208, 211–212, 220–223, 227, 407
quantum entropy, 16, 181–185, 188, 190–192, 199–202, 206, 208–209, 211–212, 215–218, 221–223, 227–228, 246, 249–252, 298–303, 406–407
quantum entropy associated with the wavefunction of quantum-gravity space, 298–303
quantum entropy for the gravitational field, 246, 249
quantum field theory, 13, 49, 54, 58–61, 71, 89–91, 255, 272, 304, 307, 315–317, 320–321, 325, 347–351, 354, 399
quantum geometrodynamics, 2, 8, 243, 245
quantum geometry, 12, 14, 201, 203–208, 211–212, 220–223, 227
quantum graphity models, 267–268
quantum gravity, 1, 10–16, 19–25, 27, 40, 47–49, 53–54, 58, 71, 103, 108, 230–232, 234–236, 241, 245–250, 253–256, 259–263, 265–267, 269, 271–279, 281–282, 284–294, 298–303, 305, 347–348, 394, 397, 400, 404–405, 407–408
quantum-gravity space, 16, 253, 269, 271–272, 274–279, 281–282, 284–285, 287–294, 298–300, 302–303, 394, 407–408
quantum Hamilton-Jacobi equation, 175, 180, 194, 216
quantum information, 11, 16, 188, 190, 221, 223
quantum length associated with the metric of the a-temporal space, 300–301, 303
quantum measurement, 152–153, 214, 225
quantum mechanics, 7–13, 15–16, 17, 47–51, 60, 64–65, 70, 73, 83, 109, 114, 151–153, 163–164, 167, 170, 174–177, 179–182, 184–185, 188, 191, 197, 228, 230–231, 235, 238, 245, 247, 249, 253, 255, 270, 275, 306, 309, 351, 357, 368, 372, 386, 391–392, 397–399, 401–403, 409–411

Index 447

quantum physics, 1, 8, 11, 48, 72, 94, 117, 155, 157, 159, 350, 388, 401
quantum potential, 16, 168–172, 174–175, 177–188, 191–200, 202–203, 206–209, 211–212, 215–219, 221, 223, 227, 235–236, 406–407
quantum potential associated with the wavefunction of quantum-gravity space, 276–277, 281–282, 284–287, 299
quantum potential for the gravitational field, 237–246, 248
quantum potential of the three-dimensional quantum vacuum, 360–361, 363–365, 392–395
quantum-time-of-arrival-problem, 163–164, 167–168, 228
quantum vacuum, 1, 12, 15–16, 304–308, 310, 312, 314, 317, 320, 325–327, 330–333, 337, 339, 343, 347, 350, 353–357, 359–361, 363–382, 384–388, 390–395, 404, 407–410
quantum vacuum energy density, 306, 353, 355, 357, 359, 366–368, 370, 372–382, 384–386, 408–410
quantum vacuum fluctuations, 304–307, 310–317, 320–321, 327, 330, 346, 350, 354, 368, 371–376, 379

radar ranging of the planets, 114, 119, 143, 149–150
rate of clocks, 130, 138–139, 144, 148
renormalization group, 317–320, 325
rotation-orientation of space, 270, 291, 407–408
Rovelli's approach of time, 47–51, 54–64, 67, 69, 95, 99, 102–104, 107–108, 405
Rovelli's thermal time hypothesis, 49, 54–63, 102–104, 107–109, 405
RS processes of creation/annihilation of quantum particles, 16, 306, 353,
359, 361, 365–368, 384, 386, 388, 391–395, 407–410
run of clocks, 137

Sacharov's model of gravitation, 305–306, 337–338, 368
Santos' model of quantized general relativity, 310–315, 372–373, 376, 378
scaling physical function of the numerical order, 132–136, 406
Schrödinger equation, 8, 17, 48, 88, 152, 154, 163, 166, 168, 234–235
Schwarzschild metric, 260, 262, 313, 341–342, 345, 348, 374, 382
Selleri's approach to special relativity, 118–122, 129–131
Selleri's formalism on the transformations of the speed of clocks, 131, 142
shadowing of the gravitational space, 340, 379, 381
Shyam's and Ramachandra's model of presympletic dynamics, 37–39, 101, 103, 105–106, 108, 405
Silberstein's, Stuckey's and McDevitt's Relational Blockworld approach, 400–402
space, 1–5, 7–16, 17, 19, 20, 23–24, 26–38, 40–43, 45, 48–51, 55, 58–63, 65, 69–75, 77–83, 85, 87, 89, 91–96, 98–101, 104–112, 113–119, 121–133, 137–139, 141–144, 151, 153, 155, 157–161, 164–169, 171–178, 180–185, 187–188, 190–193, 195, 197–209, 212, 215, 218–223, 226–229, 230, 235–236, 239, 241–249, 253–255, 257–260, 264, 266, 268–303, 304–308, 310–311, 314–315, 327, 335, 337, 339–340, 342–343, 353, 357–359, 366–368, 377–379, 381–382, 384–385, 388, 393–394, 398, 400–402, 405–409
space-time, 2, 4, 5, 8, 10, 12–14, 20, 23–24, 26–28, 34, 36, 61–62, 71, 79–83, 110, 113–114, 117–118, 122,

125–129, 131, 133, 138, 142, 155, 159, 204–205, 208, 215, 219, 222, 242–245, 253, 255, 260, 263–269, 285–287, 289, 301–303, 304–305, 310, 314–318, 320–321, 325, 331, 335, 338, 343–346, 348, 352, 356, 359, 361, 368, 371–373, 375–377, 379, 383, 393–395, 397–405, 409
special relativity, 4–5, 10–11, 15, 109–110, 113–114, 116–118, 121–122, 127, 129–133, 136, 139, 141–143, 145, 156, 174, 254, 259, 296, 404–406, 411
standard Dirac equation, 274–275, 288–291, 295
standard Klein-Gordon equation, 274–275, 295
Standard Model, 308–309, 316, 320, 324, 351, 356
stochastic electrodynamics, 307, 325, 327–328, 332, 338, 350
superposition of Boltzmann entropies, 177–181
symmetrized extension, 193, 194, 195, 196, 199
symmetrized extension of bohmian version of Wheeler-deWitt equation, 248
symmetrized extension of Wheeler-deWitt equation, 230
symmetrized quantum-entropic length, 202, 211–212, 227, 407
symmetrized quantum entropy for gravity, 250–251
symmetrized quantum Hamilton-Jacobi equation, 175, 180, 194
symmetrized quantum potential, 16, 192–200, 202–203, 206–209, 211–212, 215–216, 219, 221, 223, 227, 245, 247–250, 252, 406–407
symmetrized quantum potential approach, 16, 192–223, 227–229, 245–252, 360–366, 406–407

symmetrized quantum potential for gravity, 245, 248–250
symmetrized quantum-entropic length, 202, 211, 212, 227, 407

three-dimensional Euclid space, 15, 113, 129–131, 139, 141–144, 406
three-dimensional non-Euclid space, 153, 168, 227–228
three-dimensional quantum vacuum, 359–361, 363–366, 368–370, 372–373, 377–380, 384–388, 390–395, 408–409
three-dimensional quantum vacuum energy density model, 16, 305–306, 353–395, 407–410
three-dimensional space, 15–16, 114, 121–124, 131, 137, 171–175, 181, 184–185, 188, 190–193, 197–199, 201–202, 208, 222, 227, 235, 241, 405–406, 407
three-dimensional timeless quantum vacuum, 304, 305, 361, 366–368, 388, 392, 395, 407–410
time, 1–16, 17–42, 45–51, 53–65, 70–112, 113–120, 122–133, 135–140, 142–144, 148–149, 153–168, 170, 172–174, 179, 181–182, 192–205, 207–210, 212, 215–219, 221–229, 234, 236, 238, 241–242, 246–249, 251, 263, 269, 277, 305, 310, 312, 323–324, 328–330, 333–337, 339, 343, 349–350, 352–353, 355–356, 361–362, 364–365, 369–370, 378, 385–386, 389, 391–392, 396–397, 399–402, 404–412
time dilation, 116, 125, 137
time in general relativity, 5–7, 10–11
time in quantum theory, 8–11, 159–168, 224–229
timeless approach, 22, 98–99, 411–412
timeless background, 15–16, 83, 95–96, 98, 174, 230, 360, 404–405, 407, 410–411

Index

timeless description of physical events, 17–18, 21–112, 114, 129–150, 168–174, 181, 184–188, 190–212, 215–229, 245–252, 253, 269–303, 352–353, 357–395, 398–412

timeless path integral, 50, 64, 65, 70–71

timeless space, 1, 65, 71, 93, 95, 98, 110, 114, 172–174, 181, 184, 202, 206, 230, 269, 271, 298, 405–406

time-symmetric extension of Klein-Gordon equation, 362–363, 368, 386, 394

time-symmetric extension of Wheeler-deWitt equation, 247–252

trace-free Einstein's equations, 348–349

transformation of the speed of clocks, 119, 131, 136, 142, 406

two point correlation function, 310, 312, 376, 378

unification of generalized Klein-Gordon and Dirac equations, 295–298

unifying re-reading of the Jacobi-Barbour-Bertotti theory and of Rovelli's approach, 15, 98–109, 405

wavefunction of quantum-gravity space, 269, 271–272, 274, 276–279, 289–290, 292, 294

Weyl-like gauge potential, 177, 179

Wheeler-deWitt equation, 19–21, 23, 25, 48, 52, 230, 232, 236–237, 240–242, 244–249, 251, 407

Wise's approach of holographic special relativity, 127–129

zero-point energy, 307, 317–320, 325, 327, 331

Printed in the United States
By Bookmasters